The Application of Mathematics to the Sciences of Nature

The Application of Mathematics to the Sciences of Nature

Critical Moments and Aspects

Edited by

Paola Cerrai

University of Pisa
Pisa, Italy

Paolo Freguglia

University of L'Aquila
L'Aquila, Italy
and Domus Galilaeana
Pisa, Italy

and

Claudio Pellegrini

University of California, Los Angeles
Los Angeles, California

Kluwer Academic / Plenum Publishers
New York, Boston, Dordrecht, London, Moscow

Library of Congress Cataloging-in-Publication Data

The application of mathematics to the sciences of nature: critical moments and aspects/ edited by Paola Cerrai ... [et al.].
p. cm.
Includes bibliographical references and index.
ISBN 0-306-46694-5
1. Science—Mathematics. 2. Mathematical models. I. Cerrai, Paola.

Q172 .A66 2001
501′.51—dc21

2001053914

Proceedings of the conference The Application of Mathematics to the Sciences of Nature: Critical Moments and Aspects, held June 28 to July 1, 1999, in Arcidosso, Tuscany, Italy

ISBN 0-306-46694-5

233 Spring Street, New York, New York 10013

http://www.wkap.nl

10 9 8 7 6 5 4 3 2 1

A C.I.P. record for this book is available from the Library of Congress

Printed in the United States of America

CONTRIBUTORS

L. Accardi, Centro Vito Volterra, Università degli Studi Roma "Tor Vergata", Roma, Italia

Luciano Andreozzi, Università degli Studi di Bari, Bari, Italia

Rosalba Banchetti, Dipartimento di Etologia, Ecologia ed Evoluzione, Università di Pisa, Pisa, Italia

A. Bazzani, Departimento di Physics, Università di Bologna, and INFN Sezione di Bologna, Bologna, Italia

Vieri Benci, Centro Interdisciplinare per lo Studio dei Sistemi Complessi and Dipartimento di Matematica Applicata "U. Dini", Università di Pisa, Pisa, Italia

Jean Dhombres, EHESS, Paris, France

Fabrizio Erra, Dipartimento di Etologia, Ecologia ed Evoluzione, Università di Pisa, Pisa, Italia

Marco Forti, Dipartimento Matematica Applicata "U. Dini", Università di Pisa, Pisa, Italia

Paolo Freguglia, Department of Pure and Applied Mathematics, Università dell'Aquila and Domus Galilaeana, Pisa, Italia

Leone Fronzoni, Dipartimento di Fisica dell' Università di Pisa, CISSC and INFN, Pisa, Italia

Stefano Galatolo, Dipartimento di Matematica dell' Università di Pisa, Pisa, Italia

Lodovico Galleni, Centro Interdisciplinari per lo Studio dei Sistemi Complessi, Università di Pisa, Pisa, Italia

Niccolò Guicciardini, Dipartimento di Filosofia, Università di Bologna, Bologna, Italia

Giorgio Israel, Dipartimento di Matematica, Università di Roma "La Sapienza", Roma, Italia

Gabriele A. Losa, Laboratory of Cellular Pathology, Institute for Interdisciplinary Scientific Studies (ISSI), Center for Research in Mathematics and Physics, Locarno, and Faculty of Natural Sciences, University of Lausanne, Lausanne, Switzerland

Danilo Merlini, Cerfim and Istituto di Studi Scientifici Interdisciplinari, Locarno, Switzerland

Sergio Monteiro, Department of Physics, Moorpark College, Moorpark, California 93021

Marco Panza, University of Nantes, Nantes, France

Theodore M. Porter, Department of History, University of California, Los Angeles, Los Angeles, California

M. Regoli, Centro Vito Volterra, Università degli Studi Roma "Tor Vergata", Roma, Italia

Nicola Ricci, Dipartimento di Etologia, Ecologia ed Evoluzione, Università di Pisa, Pisa, Italia

Luca Rusconi, Cerfim and Istituto di Studi Scientifici Interdisciplinari, Locarno, Switzerland

Giovanni Santangelo, Centro Interdisciplinare per lo Studio dei Sistemi Complessi, Department of Ethology, Ecology and Evolution, Pisa University, Pisa, Italia

Walter Scandale, CERN, Geneva, Switzerland

Tito M. Tonietti, Dipartimento di Matematica, Università di Pisa, Pisa, Italia

Giorgio Turchetti, Dipartimento di Fisica, Università di Bologna, Bologna, Italia

PREFACE

The historical and epistemological reflection on the applications of mathematical techniques to the Sciences of Nature – physics, biology, chemistry, and geology – today generates attention and interest because of the increasing use of mathematical models in all sciences and their high level of sophistication. The goal of the meeting and the papers collected in this proceedings volume is to give physicists, biologists, mathematicians, and historians of science the opportunity to share information on their work and reflect on the way mathematics and mathematical models are used in the natural sciences today and in the past. The program of the workshop combines the experience of those working on current scientific research in many different fields with the historical analysis of previous results. We hope that some novel interdisciplinary, philosophical, and epistemological considerations will follow from the two aspects of the workshop, the historical and the scientific.

This proceedings includes papers presented at the meeting and some of the results of the discussions that took place during the workshop.

We wish to express our gratitude to Sergio Monteiro for all his work, which has been essential for the successful publication of these proceedings. We also want to thank the editors of Kluwer Academic/Plenum Publishers for their patience and constant help, and in particular Beth Kuhne and Roberta Klarreich.

Our thanks to the following institutions:

-Amministrazione Comunale di Arcidosso
-Comunità Montana del Monte Amiata
-Center for the History of Physics, UCLA
-Centre F. Viète, Nantes
-Centro Interdisciplinare per lo Studio dei Sistemi Complessi (CISSC), University of Pisa
-Domus Galilaeana, Pisa
-Istituto e Museo di Storia della Scienza, Firenze
-Banca Monte dei Paschi di Siena

And with "**Alto patronato del Presidente della Republica Italiana**" and with the "**Patrocinio della Regione Toscana e della Provincia di Grosseto.**"

CONTENTS

PHYSICS

BIOLOGY

HISTORY OF SCIENCE

PHYSICS

QUANTUM PROBABILITY AND THE INTERPRETATION OF QUANTUM MECHANICS: A CRUCIAL EXPERIMENT

L. Accardi and M. Regoli

Centro Vito Volterra
Università degli Studi Roma "Tor Vergata"
Via di Tor Vergata, snc
00133 Roma
accardi@volterra.mat.uniroma2.it

ABSTRACT

In the past few years the quantum probabilistic point of view on the foundations of quantum theory has been subject to critiques from several sides. Some of these[1] are based on arguments which, although wrong[2], still remain within the usual framework of scientific debate; some others, altough more pertinent to the domain of rethoric or of sociology than to science, might nevertheless have some interest as examples of that "scientific irrationalism" which was object of impotent irony in[3] (examples are[4] and even more the arguments, used by the director of "Le Scienze"[(1)], Prof. Enrico Bellone, to deny to the supporters of quantum probability the rights of expression he gave to its opponents: those interested in this and other examples of "sociology of academic repression" may consult the Volterra Center's WEB page: *http://volterra.mat.uniroma2.it.*)

One way to close a scientific debate is to produce a crucial experiment in which two contradictory theses are confronted. In our case the two theses are: the quantum probability one, that by the "chameleon effect" one can violate Bell's inequality with classical macroscopic systems, and the opposite thesis maintaining that nonlocality is intrinsic to quantum mechanics. In the present paper a simple experiment is described in which two experimenters by performing independent, local, binary choices on a common source of (pseudo)–randomness, generated by a personal computer, and computing the empirical correlations among their results, arrive to a violation of Bell's inequalities. The situation we describe here can be realized on any personal

(1) Italian translation of "Scientific American".

The Application of Mathematics to the Sciences of Nature, edited by Cerrai *et al.*
Kluwer Academic/Plenum Publishers, New York, 2002.

computer powerful enough to produce pseudo–random sequences of points in the unit (2–dimensional) disk (an order of 50.000 points are already sufficient to have an idea of the results) and those who want see how the experiment concretely works may consult the above mentioned Volterra Center's WEB page from where the program can be downloaded.

INTRODUCTION

The debate on the interpretation of quantum mechanics can fruitfully illustrate the main topic of the present workshop, although in this case, I would speak of "mutual interaction" of mathematics with physics rather than of "applications" of one to the other.

Starting from the late 70's a radically new point of view begun to emerge in this debate as a consequence of a mathematical analysis of the paradoxes of quantum theory which led to the individuation of a common probabilistic root in all of them. This marked the birth of "quantum probability", as opposed to "non–commutative probability" which has an older history and which indicates a purely mathematical (non–commutative) generalization of the classical probabilistic formalism having its roots in the pioneer papers of Heisenberg, Schrödinger, Born, Dirac, von Neumann, and many others .... With quantum probability it begins to emerge the consciousness that this new mathematical formalism is the expression of the existence of a multiplicity of models for the laws of chance, comparable to the multiplicity of models for the laws of space which was discovered almost 150 years before.

The novelty of the probabilistic formalism used in quantum theory was immediately appreciated[5] and the importance of the subsequent question: "is this new mathematical formalism the necessary expression of some new conceptual assumptions and of the experimental data?" motivated a huge literature on the mathematical foundations of quantum mechanics starting with the papers of Birkhoff and von Neumann[6] and Mackey's monograph[7]. However only in the early 80's, and in the context of quantum probability, this question received a satisfactory answer first with the emergence of the "statistical invariants" (probabilisitc counterpart of the "geometrical invariants" which allow to decide of the applicability of different mathematical models to a given set of experimental data such as angles or transition probabilities)[2] and then with the mathematical deduction of the quantum formalism from axioms with a direct physical interpretation[8,9,10].

These ones, together with the clarification of the common probabilistic root of the quantum paradoxes, were interesting mathematical achievements, solving some problems which had been open for more than 50 years and to which a huge literature had been devoted. However a further step was needed in order to pass from a negative statement (the non applicability of classical probability calculus proved using the statistical invariants), to a positive one: why in certain physical situations it is natural to expect that classical probability should not be applied?

This step was done in 1993 with the introduction of the *chameleon effect* and the developement of a theory of measurement capable to include this effect[11,12].

A synthesis of this line of thought, mainly concentrated on the problem of interpretation of quantum theory and written in a language accessible to everybody, is contained in the book[3]. Only in the last chapter of this book (appendix) the basic ideas of the "chameleon effect" were explained in somewhat more detail. From the mathematical point of view this effect simply means that the interaction Hamiltonian, of a system with an apparatus, depends on the observable one wants to measure. This principle, which can be the starting point of a *dynamical theory of adaptive systems*, together with the previous probabilistic analyses, explains why violation of Bell's inequality (or similar ones) can be perfectly compatible not only with quantum but with classical dynamical systems as well.

A counterexample, illustrating this statement in the framework of quantum theory, is contained in[13] and a more detailed description is in[14]. In the present paper we discuss a counterexample, in the framework of classical dynamical systems. In fact the situation we describe here can be realized on a personal computer and those who want to have an idea of how the experiment concretely works may consult the Volterra Center's WEB page: *http://volterra.mat.uniroma2.it.*

THE OBSERVABLES

Let us consider the unit disk in the plane

$$D := \{p = (x, y) \in \mathbf{R}^2 : x^2 + y^2 \leq 1\} \tag{1}$$

and the unit circle

$$S^{(1)} = \{(x, y) \in \mathbf{R}^2 : x^2 + y^2 = 1\} \tag{2}$$

We fix an oriented system of orthogonal axes in the plane and each space direction is characterized by the angle it forms with the x–axis, as shown in Figure 1.

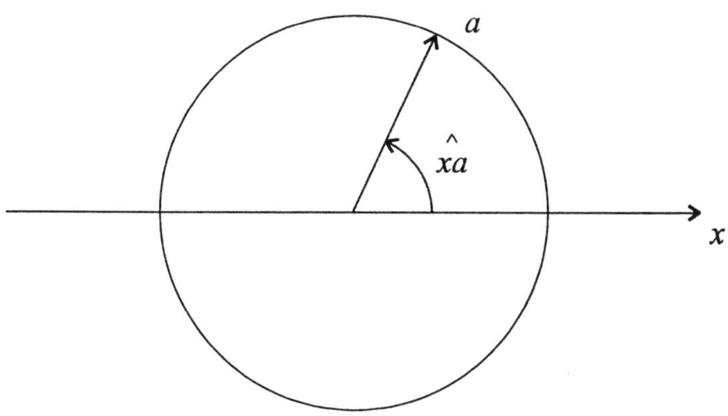

Figure 1.

The angles are measured counterclockwise starting from the positive orientation of the x–axis. Therefore the choice of a direction $a \in S^{(1)}$, corresponding to the angle θ, divides the circle into two halves: an upper (+) and a lower (−) one. The observable $S_a^{(1)}$ takes values +1 in the upper semicircle with respect to a and −1 in the lower semicircle. The observable $S_a^{(2)}$ is determined by the rule shown in Figure 2.

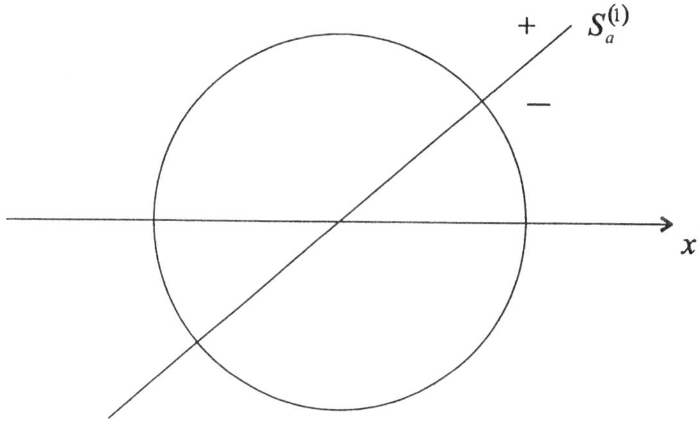

Figure 2.

$S_a^{(2)} = -S_a^{(1)}$ (1)

Let us denote:

$D_+ =$ the upper semi–disk referred to the x–axis

$D_- =$ the lower semi–disk referred to the x–axis

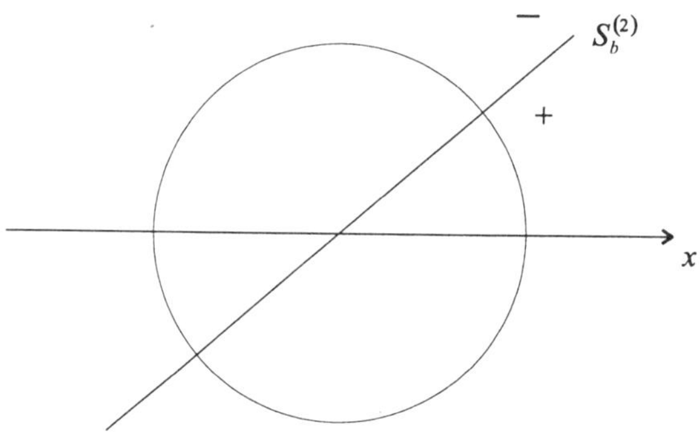

Figure 3.

The calculation of the value of $S_a^{(1)}$ in the point p of the circle is done as follows:

– one rotates the point p, counterclockwise, of an angle θ

$$R_{-\theta}p \quad ; \quad R_{-\theta} = \begin{pmatrix} \cos\theta & \sin\theta \\ -\sin\theta & \cos\theta \end{pmatrix} \tag{2}$$

– one chooses the value

$$S_a^{(1)}(p) = +1 \quad \text{if} \quad R_{-\theta}p \in D_+ \tag{3}$$

$$S_a^{(1)}(p) = -1 \quad \text{if} \quad R_{-\theta}p \in D_- \tag{4}$$

– The value of $S_a^{(2)}(p)$ is determined by the prescription (1).

GENERATION OF RANDOM POINTS IN D AND CORRESPONDING CORRELATIONS

The starting point is an algorithm which produces random points in the unit circle in $\mathbf{R}^2$.

The generating algorithm has been taken from[15] and is implemented by the generation routine: "Genera punti"

It generates up to 1 Million points but already with 30,000 points the results are very clear.

Each observer (Obs 1, Obs 2 in the program) has the following choices from the menu

	a	b	c
1	θ		
2			
3			

where the column indicates the space direction and the row index labels the experiment.

The program sends each point to the two computers (observers).

After this, each observer chooses which observable $S_x^{(j)}$ to measure or, equivalently, each observer chooses arbitrarily an angle between 0 and 2π for each experiment

- the index $j = 1, 2$ labels the observer or, equivalently, the particle in the pair;
- x is an arbitrary direction in $\mathbf{R}^2$, represented by a unit vector in $\mathbf{R}^2$.

Moreover x is identified with $-x$.

To fix the ideas let us suppose that observer 1 chooses S_a^1 and observer 2 chooses S_b^2.

The central unit produces a sequence of N random points in $S^{(2)}$

$$p_1, \ldots, p_N \tag{1}$$

and the observer 1 (resp. 2) evaluates the function $S_a^1(p_j)$ (resp. $S_b^2(p_j)$), $j = 1, \ldots, N$.

Notice that the evaluation must be done on the same point, corresponding to the experimental fact that, in the EPR type experiments each measurement must be done on the same singlet pair.

After that the central unit is informed of the results.

This corresponds to the fact that, in order to compute the empirical correlations

$$\langle S_a^1 S_b^2 \rangle = \frac{1}{N} \sum S_a^{(1)}(p_j) S_b^{(2)}(p_j) \tag{2}$$

the two experimentalists must know the results of both experiments. Notice that the sequence (p_j) depends on the experiment, so for 3 experiments we will have 3 sequences $(p_j^{(1)})$, $(p_j^{(2)})$, $(p_j^{(3)})$.

In the Bell inequality, 3 correlations are involved.

$$\langle S_a^1 S_b^2 \rangle \,, \quad \langle S_b^2 S_c^1 \rangle \,, \quad \langle S_a^1 S_c^2 \rangle \tag{3}$$

Thus in the three experiments observer 1 has to measure S_a^1, S_c^1 and observer 2 has to measure S_b^2, S_c^2. Therefore the directions a and b can be chosen arbitrarily by the two observers and it is not necessary that observer 1 is informed of the choice of observer 2 or conversely.

However the direction c has to be chosen by both observers and therefore at least on this direction there should be a preliminary agreement among the two observers.

In our experiment this preliminary information is not necessary and it can be replaced it by the following procedure:

i) each observer chooses at will the two directions along which the measurement will be done. Thus observer 1 will choose, say:
   - S^1_a in the first experiment
   - S^1_x in the second experiment
   - S^1_a in the third experiment

   while observer 2 will choose, say:
   - S^2_b in the first experiment
   - S^2_b in the second experiment
   - S^2_c in the third experiment

ii) after the experiments have been done, the central unit is informed of the results and she only considers those choices for which it happens (by chance) that

$$x = c \tag{4}$$

If condition (4) is not satisfied it is not possible to check the violation of the Bell inequality.

Whichever procedure has been chosen the central unit receives the results of the experiments which are summarized in the table

$1-st$ experiment	$2-d$ experiment	$3-d$ experiment	$1-st$ experiment	$2-d$ experiment	$3-d$ experiment
$S_a^{(1)}$	$S_c^{(1)}$	$S_a^{(1)}$	$S_b^{(2)}$	$S_b^{(2)}$	$S_c^{(2)}$
$S_a^{(1)}(p_1^{(1)})$	$S_c^{(1)}(p_1^{(2)})$	$S_a^{(1)}(p_1^{(3)})$	$S_b^{(2)}(p_1^{(1)})$	$S_b^{(2)}(p_1^{(2)})$	$S_c^{(2)}(p_1^{(3)})$
$\vdots$	$\vdots$	$\vdots$	$\vdots$	$\vdots$	$\vdots$
$S_a^{(1)}(p_N^{(1)})$	$S_c^{(1)}(p_N^{(2)})$	$S_a^{(1)}(p_N^{(3)})$	$S_b^{(2)}(p_N^{(1)})$	$S_b^{(2)}(p_N^{(2)})$	$S_c^{(2)}(p_N^{(3)})$

The fact that the same N is chosen for the 3 experiments is irrelevant and one could choose different N, say N_1, N_2, N_3 for the 3 experiments, with the only constraint that N_1, N_2, N_3 are large enough to obtain a meaningful statistics.

At this point the central unit computes the 3 empirical correlations

$$\langle S_a^{(1)} S_b^{(2)} \rangle = \frac{1}{N} \sum_{j=1}^N S_a^{(1)}(p_j^{(1)}) S_b^{(2)}(p_j^{(1)}) \tag{5}$$

$$\langle S_b^{(2)} S_c^{(1)} \rangle = \frac{1}{N} \sum_{j=1}^N S_c^{(1)}(p_j^{(2)}) S_b^{(2)}(p_j^{(2)}) \tag{6}$$

$$\langle S_a^{(1)} S_c^{(2)} \rangle = \frac{1}{N} \sum_{j=1}^N S_a^{(1)}(p_j^{(3)}) S_c^{(2)}(p_j^{(3)}) \tag{7}$$

As explained in section (1) of[3], cf. also in the comments below, in order to apply Bell's argument to the empirical data, one has to "clean" these data as follows:

all the points $p_j^{(1)}$, $p_j^{(2)}$ *such that*

$$S_b^{(2)}(p_j^{(1)}) \neq S_b^{(2)}(p_j^{(2)}) \tag{8}$$

have to be eliminated from the calculation of the empirical correlations (5), (6), (7).

If the experiments are mutually independent, this elimination should not alter significantly the empirical correlations.

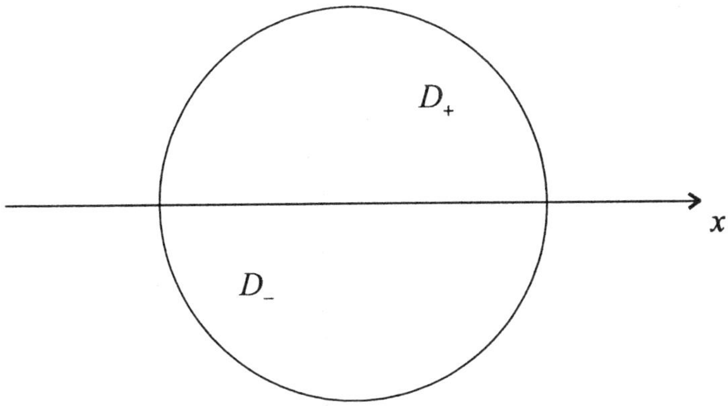

Figure 4.

This is an elementary fact of classical probability whose proof is given, for completeness, in the Appendix of the present paper.

After the "cleaning" one obtains two new sequences of points:

$$p^{(1)}_{j_1}, \ldots, p^{(1)}_{j_{N_1}} \tag{9}$$

$$p^{(2)}_{j_1}, \ldots, p^{(2)}_{j_{N_1}} \tag{10}$$

with the property that

$$S^{(2)}_b(p^{(1)}_{j_\alpha}) = S^{(2)}_b(p^{(2)}_{j_\alpha}) \ ; \quad \alpha = 1, \ldots, N_1 \tag{11}$$

Obviously $N_1 \leq N$. In fact one expects theoretically (and this expectation is confirmed by the simulation experiments) that N_1 is approximatively $N/2$.

The central unit computes the new correlations with the "cleaned" sequences:

$$\langle S^{(1)}_a S^{(2)}_b \rangle_{Cl} = \tfrac{1}{N_1} \textstyle\sum_{\alpha=1}^{N_1} S^{(1)}_a(p^{(1)}_{j_\alpha}) S^{(2)}_b(p^{(1)}_{j_\alpha}) \tag{12}$$

$$\langle S^{(2)}_b S^{(1)}_c \rangle_{Cl} = \tfrac{1}{N_1} \textstyle\sum_{\alpha=1}^{N_1} S^{(1)}_c(p^{(2)}_{j_\alpha}) S^{(2)}_b(p^{(2)}_{j_\alpha}) \tag{13}$$

and compares them with the original ones (5), (6).

We have checked that the difference between the original correlations and the "cleaned" ones is of orders of magnitude smaller than the violation of the inequality. Therefore this violation cannot be attributed to the "cleaning" operation which, on the other hand is not an assumption related to our experiment, but only represents the explicitation of an assumption hidden in any comparison of the Bell inequality with the experimental data.

In fact it is somewhat surprising that, in the vast literature on the Bell inequality, the necessity of this *cleaning operation* has not been sufficiently emphasized.

According to the Bell inequality one should have

$$|\langle S^{(1)}_a S^{(2)}_b \rangle - \langle S^{(1)}_c S^{(2)}_b \rangle| \leq 1 + \langle S^{(1)}_a S^{(2)}_c \rangle \tag{14}$$

The program compares the difference

$$|\langle S^{(1)}_a S^{(2)}_b \rangle - \langle S^{(1)}_c S^{(2)}_b \rangle| - 1 - \langle S^{(1)}_a S^{(2)}_c \rangle \tag{15}$$

with zero.

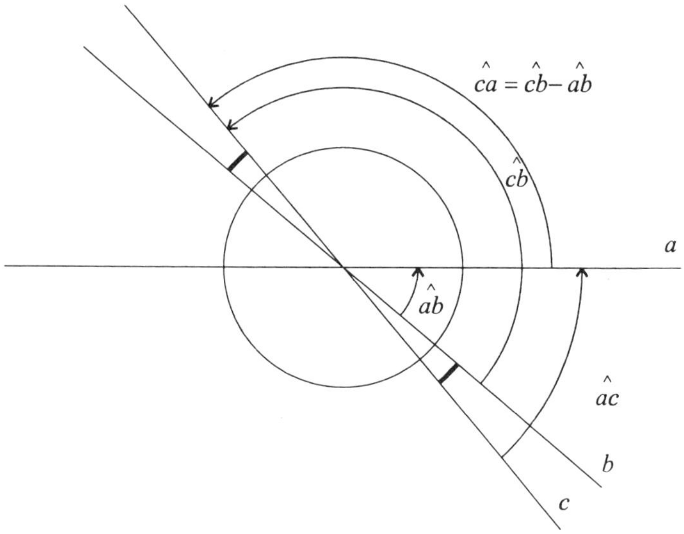

Figure 5.

Violation of Bell's inequality corresponds to a value of the difference (15) strictly greater than zero. The simulation experiments show that, if the directions a, b, c are chosen as in Figure (5) below, then the difference (15) is strictly positive, hence the Bell inequality is violated. An analytical proof of this statement end of section (7). The correlations (5), (6), (7), for this choice of a, b, c are illustrated in Figures (6), (7), (8) below

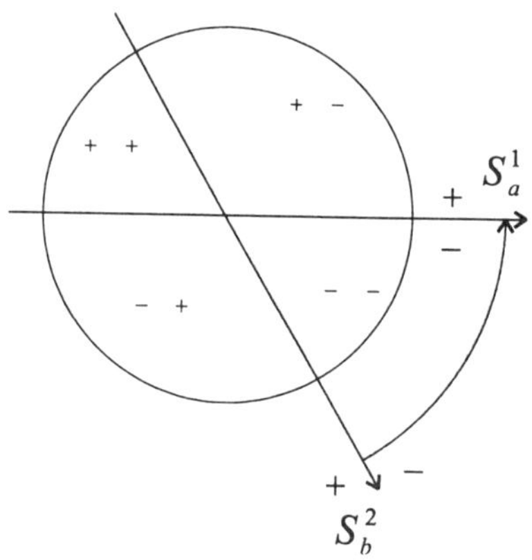

Figure 6.

The reason of the violation is the following: let us denote $\widehat{ab}$ the sector of the $(-,-)$ concordance in the correlation $\langle S_a^{(1)} S_b^{(2)} \rangle$ (cf. Figure 6). Similarly we denote $\widehat{cb}$ the sector of the $(-,-)$ concordance in the correlation $\langle S_c^{(1)} S_b^{(2)} \rangle$ (cf. Figure 7) and $\widehat{ac}$ the sector of the $(-,-)$ concordance in the correlation $\langle S_a^{(1)} S_c^{(2)} \rangle$ (cf. Figure 8).

In the following we shall use the same symbol to denote a sector and its area. With these notations the correlations (5), (6), (7) are given by:

$$\langle S_a^{(1)} S_b^{(2)} \rangle = -1 + 4\widehat{ab}$$

$$\langle S_c^{(1)} S_b^{(2)} \rangle = -1 + 4\widehat{cb}$$

$$\langle S_a^{(1)} S_c^{(2)} \rangle = -1 + 4\widehat{ac}$$

So the two sides of the Bell inequality are respectively:

$$|\langle S_a^{(1)} S_b^{(2)} \rangle - \langle S_c^{(1)} S_b^{(2)} \rangle| = 4|\widehat{ab} - \widehat{cb}| = 4(\widehat{cb} - \widehat{ab})$$

$$1 + \langle S_a^{(1)} S_c^{(2)} \rangle = 4\widehat{ac}$$

and we are reduced to compare

$$\widehat{cb} - \widehat{ab} = \widehat{ca} \qquad \text{and} \qquad \widehat{ac}$$

But, because of our choice of the axes (cf. Figure 5), one has

$$\widehat{ca} > \widehat{ac}$$

Thus we conclude that

$$|\langle S_a^{(1)} S_b^{(2)} \rangle - \langle S_c^{(1)} S_b^{(2)} \rangle| > 1 + \langle S_a^{(1)} S_c^{(2)} \rangle$$

which violates the Bell inequality (14).

In order to establish a connection with the chameleon effect, we interpret, according to the general scheme proposed in[9], the rotations, performed by the two experimenters, as dynamical evolutions and we denote T^1, T^2 the dynamics of the first and the second particle respectively.

In these notations we have, in the first experiment (to measure $\langle S_a^{(1)} S_b^{(2)} \rangle$):

$T_I^1 = id$ (the identity rotation)

$$T_I^2 = R_{-\widehat{ab}}$$

In the second experiment (to measure $\langle S_c^{(1)} S_b^{(2)} \rangle$) we have

$$T_{II}^1 = R_{-\widehat{ab}}$$

$$T_{II}^2 = R_{-ca}$$

In the third experiment, to measure $\langle S_a^{(1)} S_c^{(2)} \rangle$, the orientation of c is reversed and therefore the dynamics are

$$T_{III}^1 = id$$

$$T_{III}^2 = R_{-(2\pi - ca)} = R_{ca}$$

In particular we see that, although in each experiment one has

$$S_x^{(1)} = -S_x^{(2)}$$

for any x in the unit circle, it is not true that

$$S_c^{(1)}(II) = -S_c^{(2)}(III)$$

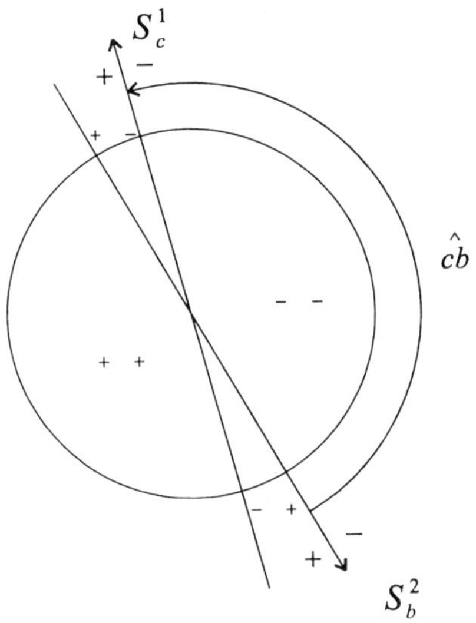

Figure 7.

COMMENTS

From the definition of $S_a^{(1)}$ we see that, if $\chi_{D_\pm}(p) = 1$ for $p \in D_\pm$ and equal to zero otherwise, then

$$S_a^{(1)} = \chi_{D_+} \circ R_{-\theta} - \chi_{D_-} \circ R_{-\theta} \tag{1}$$

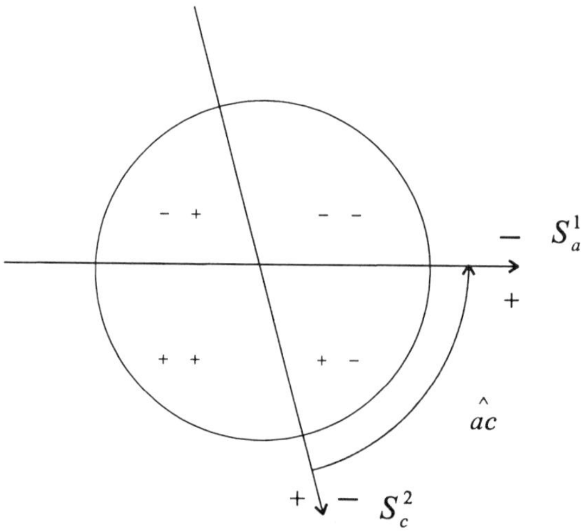

Figure 8.

But for any set D one has

$$\chi_D(R_{-\theta}p) = 1 \Leftrightarrow R_{-\theta}p \in D \Leftrightarrow p \in R_\theta D$$

therefore

$$\chi_D \circ R_{-\theta} = \chi_{R_\theta D}$$

Thus we have

$$S_a^{(1)} = \chi_{R_\theta D_+} - \chi_{R_\theta D_-} \tag{2}$$

and we define

$$S_a^{(2)} = -S_a^{(1)} = S_a^{(1)} \circ R_\pi \tag{3}$$

Therefore the random variables

$$S_x^{(1)}, S_y^{(2)} : (D, \mu) \to \{\pm 1\}$$

where μ is the uniform measure on D, are defined on the same probability space and therefore they satisfy the Bell inequality. However this does not contradict the results of the experiment because the fact that we realize the random variables $S_x^{(1)}$, $S_y^{(2)}$ *on the same probability space* means that the Bell inequality must be satisfied whenever the three correlations

$$\langle S_a^{(1)} S_b^{(2)} \rangle, \langle S_c^{(1)} S_b^{(2)} \rangle, \langle S_a^{(1)} S_c^{(2)} \rangle \tag{4}$$

are compared with experimental results drawn from *the same sample*. More precisely if one generates a random (or pseudo–random) set of points

$$p_1, \ldots, p_N$$

in D, then we are entitled to expect that the empirical correlations

$$\tfrac{1}{N} \textstyle\sum_{j=1}^N S_a^{(1)}(p_j) S_b^{(2)}(p_j) \tag{5a}$$

$$\tfrac{1}{N} \textstyle\sum_{j=1}^N S_c^{(1)}(p_j) S_c^{(2)}(p_j) \tag{5b}$$

$$\tfrac{1}{N} \textstyle\sum_{j=1}^N S_a^{(1)}(p_j) S_c^{(2)}(p_j) \tag{5c}$$

are good approximations of the theoretical correlations (4). However, if the empirical correlation (5a, b, c) are evaluated on different samples, this leads to:

$$\tfrac{1}{N} \textstyle\sum_{j=1}^N S_a^{(1)}(p_j^{(1)}) S_b^{(2)}(p_j^{(1)}) \tag{6a}$$

$$\tfrac{1}{N} \textstyle\sum_{j=1}^N S_c^{(1)}(p_j^{(2)}) S_b^{(2)}(p_j^{(2)}) \tag{6b}$$

$$\tfrac{1}{N} \textstyle\sum_{j=1}^N S_a^{(1)}(p_j^{(3)}) S_c^{(2)}(p_j^{(3)}) \tag{6c}$$

and there is no reason to postulate that these correlations satisfy the Bell inequalities.

In fact, if we try to mimick the same argument, leading to the Bell inequalities, for the empirical correlations (6a, b, c), we are led to consider the difference

$$\left| \tfrac{1}{N} \textstyle\sum_{j=1}^N S_a^{(1)}(p_j^{(1)}) S_b^{(2)}(p_j^{(1)}) - \tfrac{1}{N} \textstyle\sum_{j=1}^N S_b^{(2)}(p_j^{(2)}) S_c^{(1)}(p_j^{(2)}) \right| \tag{7}$$

After the *cleaning* we can suppose that

$$S_b^{(2)}(p_j^{(1)}) = S_b^{(2)}(p_j^{(2)}) \tag{8}$$

So that the expression (7) is equal to

$$\left|\frac{1}{N}\sum_{j=1}^{N}(S_a^{(1)}(p_j^{(1)})S_b^{(2)}(p_j^{(1)}) - S_b^{(2)}(p_j^{(2)})S_c^{(1)}(p_j^{(2)}))\right|$$

and, using (8) this becomes equal to

$$\left|\frac{1}{N}\sum_{j=1}^{N}(S_a^{(1)}(p_j^{(1)})S_b^{(2)}(p_j^{(1)}))(1 - S_a^{(1)}(p_j^{(1)})S_c^{(1)}(p_j^{(2)}))\right|$$

$$\leq \frac{1}{N}\sum_{j=1}^{N}|1 - S_a^{(1)}(p_j^{(1)})S_c^{(1)}(p_j^{(2)})| = 1 - \frac{1}{N}\sum_{j=1}^{N} S_a^{(1)}(p_j^{(1)})S_c^{(1)}(p_j^{(2)}) \tag{9}$$

Now we use the identity (3) and find

$$1 + \frac{1}{N}\sum_{j=1}^{N} S_a^{(1)}(p_j^{(1)})S_c^{(2)}(p_j^{(2)}) \tag{10}$$

But, since the samples $(p_j^{(1)})$, $(p_j^{(2)})$ are different, there is no reason to believe that

$$\frac{1}{N}\sum_{j=1}^{N} S_a^{(1)}(p_j^{(1)})S_c^{(2)}(p_j^{(2)}) \tag{11}$$

is any form of approximation of the correlation

$$\langle S_a^{(1)}S_b^{(2)}\rangle \tag{12}$$

In fact the law of large numbers only claims that the correlation (12) is well approximated by a sum of the form

$$\frac{1}{N}\sum_{j=1}^{N} S_a^{(1)}(p_j)S_c^{(2)}(p_j)$$

relative to *the same sample.*

MATHEMATICAL MODEL OF THE EXPERIMENT

Mathematically the situation can be described as follows.

The three experiments to measure the correlations (4.4) are mutually incompatible and independent. This means that the 3 sequences of points

$$(p_j^{(1)}) \ , \qquad (p_j^{(2)}) \ , \qquad (p_j^{(3)}) \tag{1}$$

are samples from 3 independent identically distributed random variables

$$\xi_1, \xi_2, \xi_3 : (\Omega, \mathcal{F}, P) \to D \tag{2}$$

which we can always think to be realized on a single probability space and taking values in the unit disk D with uniform probability. Therefore both in the case of the real and of the simulation experiments, one has to do with the following random variables:

$$S_a^{(1)}(\xi_1) \ , \quad S_b^{(2)}(\xi_1) \tag{3a}$$

$$S_c^{(1)}(\xi_2) \ , \quad S_b^{(2)}(\xi_2) \tag{3b}$$

$$S_a^{(1)}(\xi_3) \ , \quad S_c^{(2)}(\xi_3) \tag{3c}$$

In this context, the *cleaning operation*, described in section (3), corresponds to replacing the correlations

$$\langle S_a^{(1)}(\xi_1)S_b^{(2)}(\xi_1)\rangle \qquad , \qquad \langle S_c^{(1)}(\xi_2)S_b^{(2)}(\xi_2)\rangle \tag{4}$$

by the *conditioned correlations*

$$E\left(S_a^{(1)}(\xi_1)S_b^{(2)}(\xi_1)|S_b^{(2)}(\xi_1) = S_b^{(2)}(\xi_2)\right) \tag{5a}$$

$$E\left(S_c^{(1)}(\xi_2)S_b^{(2)}(\xi_2)|S_b^{(2)}(\xi_1) = S_b^{(2)}(\xi_2)\right) \tag{5b}$$

where E denotes the P–expectation.

Therefore, if we try to apply the Bell argument to the correlations (5a), (5b) we find .

$$
\begin{aligned}
&|E\left(S_a^{(1)}(\xi_1)S_b^{(2)}(\xi_1) - S_c^{(1)}(\xi_2)S_b^{(2)}(\xi_2))|S_b^{(2)}(\xi_1) = S_b^{(2)}(\xi_2)\right)| \\
&= |E\left(S_a^{(1)}(\xi_1)S_b^{(2)}(\xi_1)\{1 - S_a^{(1)}(\xi_1)S_c^{(1)}(\xi_2)\}|S_b^{(2)}(\xi_1) = S_b^{(2)}(\xi_2)\right)| \\
&\leq E\left(|1 - S_a^{(1)}(\xi_1)S_c^{(1)}(\xi_2)|\ |S_b^{(2)}(\xi_1) = S_b^{(2)}(\xi_2)\right) \\
&= 1 - E\left(S_a^{(1)}(\xi_1)S_c^{(1)}(\xi_2)|S_b^{(2)}(\xi_1) = S_b^{(2)}(\xi_2)\right)
\end{aligned}
$$

Now, using the relation $S_c^{(1)}(\xi_2) = -S_c^{(2)}(\xi_2)$, we arrive to the inequality

$$
\begin{aligned}
&|E\left(S_a^{(1)}(\xi_1)S_b^{(2)}(\xi_1) - S_c^{(1)}(\xi_2)S_b^{(2)}(\xi_2)|S_b^{(2)}(\xi_1) = S_b^{(2)}(\xi_2)\right)| \\
&\leq 1 + E\left(S_a^{(1)}(\xi_1)S_c^{(2)}(\xi_2))|S_b^{(2)}(\xi_1) = S_b^{(2)}(\xi_2)\right)|
\end{aligned}
\tag{6}
$$

If we compare this with the Bell inequality

$$|E(S_a^{(1)}S_b^{(2)} - S_c^{(1)}S_b^{(2)})| \leq 1 + E(S_a^{(1)}S_c^{(2)}) \tag{7}$$

we see that we can suppose with good approximation that

$$E\left(S_a^{(1)}(\xi_1)S_b^{(2)}(\xi_1)|S_b^{(2)}(\xi_1) = S_b^{(2)}(\xi_2)\right) = E(S_a^{(1)}(\xi_1)S_b^{(2)}(\xi_1)) \tag{8a}$$

$$E\left(S_c^{(1)}(\xi_2)S_b^{(2)}(\xi_2)|S_b^{(2)}(\xi_1) = S_b^{(2)}(\xi_2)\right) = E(S_c^{(1)}(\xi_2)S_b^{(2)}(\xi_2)) \tag{8b}$$

because these two approximations are well confirmed by the experiments as well as theoretically justified (cf. Appendix). However there is no reason to suppose that the correlation

$$E\left(S_a^{(1)}(\xi_1)S_c^{(2)}(\xi_2)|S_b^{(2)}(\xi_1) = S_b^{(2)}(\xi_2)\right) \tag{9}$$

is a good approximation of

$$E\left(S_a^{(1)}(\xi_3)S_c^{(2)}(\xi_3)\right) = E(S_a^{(1)}S_c^{(2)}) \tag{10}$$

because (9) refers to a situation in which $S_a^{(1)}$ and $S_c^{(2)}$ are evaluated *in different experiments*, while (10) refers to a situation in which they are evaluated *in the same experiment*.

That the two correlations (9) and (10) are in general different, is obvious. For example, in the extreme case in which $a = c$ and b is orthogonal to a, as in Figure (9).

Then we have

$$\langle S_a^{(1)}S_c^{(2)}\rangle = \langle S_a^{(1)}S_a^{(2)}\rangle = -1 \tag{11}$$

On the other hand the condition

$S_b^{(2)}(\xi_1) = S_b^{(2)}(\xi_2)$ only says that the points ξ_1, ξ_2 are either both in the half–disk $D_+(b)$ or both in the half disk $D_-(b)$ (cf. Figure 9).

But this gives no information on the values of $S_a^{(1)}$ because with equal probability the two points can be either in $D_+(a)$ or in $D_-(a)$.

So, as far as the values of $S_a^{(1)}$ are concerned the conditional probability should give the same result as the probability without condition, i.e.

$$E(S_a^{(1)}(\xi_1)S_a^{(2)}(\xi_2)|S_b^{(1)}(\xi_1)=S_b^{(2)}(\xi_2))=E(S_a^{(1)}(\xi_1)S_a^{(2)}(\xi_2))=$$
$$=E(S_a^{(1)}(\xi_1))E(S_a^{(2)}(\xi_2))=0$$

which is quite different from (11).

THE CHAMELEON EFFECT

In this and the following sections we describe how the preceeding construction can be considered as a particular case of the general theory of quantum measurement proposed in[9,3,13]. We shall keep the notations of sections (2), (3) and for any subset $I\subseteq \mathbf{R}^2$ we denote

$$\chi_I(p)=\begin{cases}0\,, & \text{if} \quad p\notin I\\ 1\,, & \text{if} \quad p\in I\end{cases} \tag{1}$$

the characteristic function of I. Define the function

$$\varphi:=\chi_{D_+}-\chi_{D_-}:D\to\{\pm1\} \tag{2}$$

Given $\alpha\in[0,2\pi]$ we denote R_α the counterclockwise rotation in the (x,y)–plane of an angle α.

Notice that R_π is the reflection with respect to the origin, i.e. for any $p\in D$

$$R_\pi p=R_{-\pi}p \tag{3}$$

If we add angles modulo 2π, the rotations R_α are an abelian group. For each $a\in S^{(1)}$ define

$$\varphi_a:=\varphi\circ R_{-x\hat{a}}=\chi_{R_{\widehat{xa}}D_+}-\chi_{R_{\widehat{xa}}D_-} \tag{4}$$

Notice that, for any $a\in S^{(1)}$

$$\varphi_a\circ R_{-\pi}=-\varphi_a \tag{5}$$

In fact

$$\varphi_a\circ R_{-\pi}=\chi_{R_\pi R_{\widehat{xa}}D_+}-\chi_{R_\pi R_{\widehat{xa}}D_-}=\chi_{R_{\widehat{xa}}D_-}-\chi_{R_{x\hat{a}}D_+}=-\varphi_a$$

The correlations between two functions φ_a, φ_b with respect to the uniform measure on the unit disk D are computed as follows.

In the unit circle the area of a sector bound by an angle $\gamma>0$ is

$$\int_0^1\int_0^\gamma \rho d\rho d\theta=\tfrac{\gamma}{2} \tag{6}$$

Moreover

$$\int_D \tfrac{dp}{\pi}\,\varphi(p)\varphi(R_{-\gamma}p)=\tfrac{1}{\pi}\,\left[2\tfrac{\pi-\gamma}{2}-2\tfrac{\gamma}{2}\right]=1-\tfrac{2\gamma}{\pi} \tag{7}$$

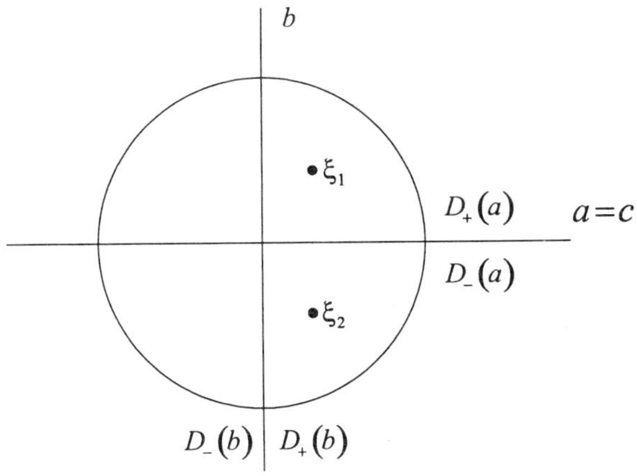

Figure 9.

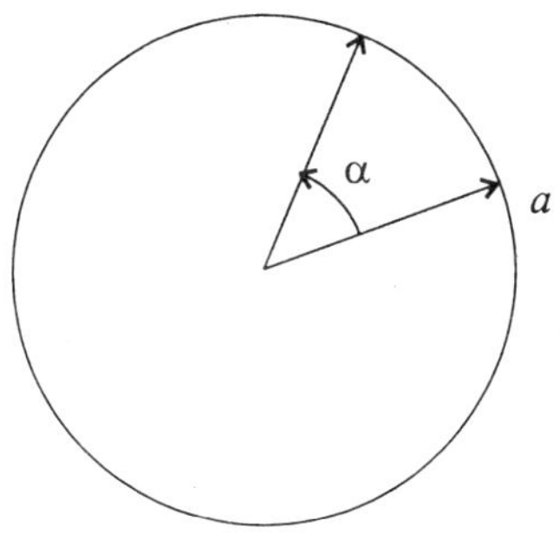

Figure 10.

So, by rotation invariance

$$\int_D \frac{dp}{\pi}\,\varphi(R_{-\alpha}p)\varphi(R_{-\beta}p) = \int_D \frac{dp}{\pi}\,\varphi(p)\varphi(R_{-(\beta-\alpha)}p) = 1 - \frac{2(\beta-\alpha)}{\pi} \tag{8}$$

Notice that this formula holds for $\beta - \alpha \in (0, \pi)$.

STATE SPACE AND SAMPLE SPACE

In the notations of sections (6), $D \times D$ will be interpreted as the state space of a system composed of 2 particles. The configuration space is given by the functions from $S^{(1)} \times S^{(1)}$ to $D \times D$

$$\Omega := \prod_{(x,y)\in S^{(1)}\times S^{(1)}} D \times D = \mathcal{F}(S^{(1)} \times S^{(1)}, D \times D) =$$

$$\mathcal{F}(S^{(1)}, D) \times \mathcal{F}(S^{(1)}, D) \tag{1}$$

The elements ω of Ω will be denoted

$$\omega = (p^{(1)}, p^{(2)}) \in \Omega \tag{2}$$

where

$$p^{(1)}, p^{(2)} : S^{(1)} \to D \tag{3}$$

are arbitrary functions and we use the notations

$$p^{(j)}(x) =: p_x^{(j)} \in D \; ; \quad x \in S^{(1)}$$

The coordinate functions are defined as usual, for any $a \in S^{(1)}$ and $j = 1, 2$:

$$X_a^{(j)} : (p^{(1)}, p^{(2)}) \in \Omega \to X_a^{(j)}(p^{(1)}, p^{(2)}) := p_a^{(j)} \in D \tag{4}$$

Moreover D, hence $D \times D$ and Ω, are compact Hausdorff spaces. Therefore

$$\mathcal{C}(\Omega) = \mathcal{C}\left(\prod_{(x,y)\in S^{(1)}\times S^{(1)}} D \times D\right) \equiv \tag{5}$$

$$\otimes_{(x,y)\in S^{(1)}\times S^{(1)}} \mathcal{C}(D \times D) \equiv [\otimes_{x\in S^{(1)}} \mathcal{C}(D)] \otimes \left[\otimes_{y\in S^{(1)}} \mathcal{C}(D)\right]$$

LEMMA (1). There exists a unique linear, positive, surjective identity preserving map

$$E : \mathcal{C}(\Omega) \to \mathcal{C}(D \times D) \equiv \mathcal{C}(D) \otimes \mathcal{C}(D) \tag{6}$$

with the property that, for any family of continuously functions

$$f_x, g_y : D \to \mathbf{C} \; ; \quad x, y \in S^{(1)} \tag{7}$$

such that almost all (i.e. all but a finite number) of f_x, g_y are equal to 1, one has, for any p, $p' \in D$:

$$E(\otimes_{x\in S^{(1)}} f_x \otimes \otimes \; \otimes_{y\in S^{(1)}} g_y)(p, p') = \prod_{x\in S^{(1)}} f_x(p) \cdot \prod_{x\in S^{(1)}} g_y(p') \tag{8}$$

PROOF. Clear.

COROLLARY (2). For any probability measure μ on $D \times D$ there exists a unique probability measure $\tilde{\mu}$ on Ω such that, for f_x, g_y as in Lemma (1), one has

$$\tilde{\mu}(\otimes_{x\in S^{(1)}} f_x \otimes \; \otimes_{y\in S^{(1)}} g_y) = \mu\left(\prod_{x\in S^{(1)}} f_x \otimes \pi_{y\in S^{(1)}} g_y\right)$$

More explicitly

$$\begin{aligned} &\int_\Omega (\otimes_{x\in S^{(1)}} f_x \otimes \; \otimes_{y\in S^{(1)}} g_y)(\omega) \tilde{\mu}(d\omega) = \\ &= \int\int_{D\times D} \mu(dp, dp') \prod_{x\in S^{(1)}} f_x(p) \prod_{y\in S^{(1)}} g_y(p') \end{aligned} \tag{9}$$

PROOF. Clear.

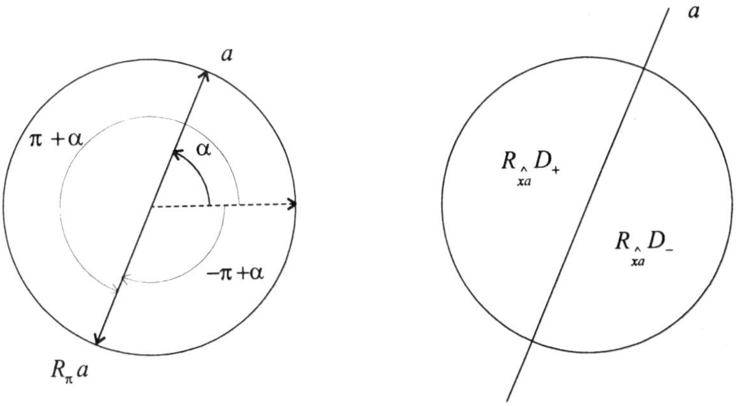

Figure 11.

DYNAMICS ON D AND DYNAMICS ON Ω

For each $a \in S^{(1)}$ we define two maps, of the disk D into itself, by:

$$T_a^1 := R_{-\alpha(a)} \tag{1}$$

$$T_a^2 := R_{-\pi} R_{-\alpha(a)} = R_{-\alpha(a)} R_{-\pi} = R_{-\pi+\alpha(a)} = R_{-\pi} T_a^1 \tag{2}$$

where

$$\alpha : [0, \pi) \to [0, \pi)$$

is an arbitrary function. In the experiment we shall choose

$$\alpha(a) = \widehat{xa} \tag{3}$$

but the arbitrariness of α can be used to improve the results.

In the notations (1), (2) and with the identification (3) we define the map

$$T : \Omega \to \Omega$$

by

$$T(p^{(1)}, p^{(2)})(x, y) := (T_x^1 p_x^{(1)}, T_y^2 p_y^{(2)}) \ ; \quad (p^{(1)}, p^{(2)}) \in \Omega \ ; \quad x, y \in S^{(1)} \tag{4}$$

and, for each a, $b \in S^{(1)}$ we define the map

$$T_{a,b} : \Omega \to \Omega$$

by

$$T_{a,b}(p^{(1)}, p^{(2)})(x, y) := (T_a^1 p_x^{(1)}, T_b^2 p_y^{(2)}) \tag{5}$$

The maps T and $T_{a,b}$ induce automorphisms on $\mathcal{C}(\Omega)$, denoted respectively u and $u_{a,b}$, characterized by

$$u(\otimes_{x \in S^{(1)}} f_x \otimes \ \otimes_{y \in S^{(1)}} g_y) := [\otimes_{x \in S^{(1)}} f_x \circ T_x^1] \otimes [\otimes_{y \in S^{(1)}} g_y \circ T_y^2] \tag{6}$$

$$u_{a,b}(\otimes_{x \in S^{(1)}} f_x \otimes \ \otimes_{y \in S^{(1)}} g_y) := [\otimes_{x \in S^{(1)}} f_x \circ T_a^1] \otimes [\otimes_{y \in S^{(1)}} g_y \circ T_b^2] \tag{7}$$

PROPOSITION (1). Let μ be a probability measure on $D \times D$ and let $\tilde{\mu}$ the probability measure on Ω, associated to μ according to Corollary (8.2). For $\tilde{\mu}$ as in Corollary (7.2), define the probability measures

$$P := \tilde{\mu} \circ u \tag{8}$$

$$P_{a,b} := \tilde{\mu} \circ u_{a,b} \; ; \quad a, b \in S^{(1)} \tag{9}$$

Then, for each a, $b \in S^{(1)}$, the joint probabilities of the random variables $X_a^{(1)}$, $X_b^{(2)}$ for the measure P and for the measure $P_{a,b}$ coincide.

PROOF. For any f, $g \in \mathcal{C}(D)$ one has

$$P(f(X_a^{(1)})g(X_b^{(2)})) = \tilde{\mu}(u(f(X_a^{(1)}))u(g(X_b^{(2)}))) = \tilde{\mu}(f(T_a^1 X_a^{(1)})g(T_b^2 X_b^{(2)})) =$$

$$= \int\int_{D\times D} \mu(dp_a, dp_b) f(T_a^{(1)} p_a) g(T_b^2 p_b)$$

On the other hand

$$P_{a,b}(f(X_a^{(1)})g(X_b^{(2)})) = \tilde{\mu}(u_{a,b}(f(X_a^{(1)}))u_{a,b}(g(X_b^{(2)}))) =$$

$$= \tilde{\mu}(f(T_a^1 X_a^{(1)})g(T_b^2 X_b^{(2)})) = \int\int_{D\times D} \mu(dp_a, dp_b) f(T_a^1 p_a) g(T_b^2 p_b)$$

Therefore the two coincide.

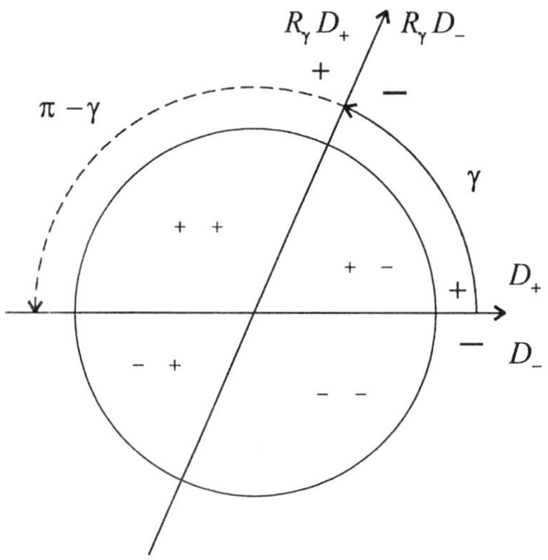

Figure 12.

COROLLARY (2). Let a, $b, c \in S^{(1)}$ and let f_a, f_c, g_b, $g_c : D \to \mathbf{C}$ be arbitrary measurable functions and define

$$P_{a,b}(f_a(X_a^{(1)})g_b(X_b^{(2)})) =: \kappa_{a,b}$$

$$P_{c,b}(f_c(X_c^{(1)})g_b(X_b^{(2)})) =: \kappa_{c,b}$$

$$P_{a,c}(f_a(X_a^{(1)})g_c(X_c^{(2)})) =: \kappa_{a,c}$$

Then, if P is defined by (8) one has

$$P(f_a(X_a^{(1)})g_b(X_b^{(2)})) = \kappa_{a,b}$$
$$P(f_c(X_c^{(1)})g_b(X_b^{(2)})) = \kappa_{c,b}$$
$$P(f_a(X_a^{(1)})g_c(X_c^{(2)})) = \kappa_{a,c}$$

PROOF. Clear from Proposition (1).

ENTANGLED MEASURES

Now we fix on $D \times D$ the measure

$$\mu(dp, dp') = \tfrac{1}{\pi}\,\delta(p - p')dpdp'$$

that is

$$\mu(F) = \int_D F(p,p)\tfrac{dp}{\pi}\,; \quad F \in \mathcal{C}(D \times D) \tag{1}$$

LEMMA (1). For $\tilde{\mu}$ as in (1) let μ denote the probability measure on Ω defined by Corollary (7.2). Then, for any $a \in S^{(1)}$:

$$X_a^{(1)} = X_a^{(2)}\ ; \quad \tilde{\mu} - \text{a.e.} \tag{2}$$

Proof. Using (1) we find

$$P(X_a^{(1)} \neq X_a^{(2)}) = \int \int_{D\times D} \chi_{D\times D\setminus\{(p',p'')\in D\times D: p'=p''\}}(p,p)\tfrac{dp}{\pi} = 0$$

LEMMA (2). Let $\tilde{\mu}$ be as in Lemma (1). Then for any a, $b \in S^{(1)}$ and for any $\varphi \in \mathcal{C}(D)$ one has:

$$u(\varphi(X_a^{(2)})) = -u(\varphi(X_a^{(1)})) \tag{3}$$

$$u_{a,b}(\varphi(X_a^{(2)})) \neq -u_{a,b}(\varphi(X_a^{(1)})) \quad \text{if} \quad a \neq b \tag{4}$$

where both identities are meant $\tilde{\mu}$–a.e. ((4) should be understood in the sense that: *it is not true that the two sides of (4) are equal $\tilde{\mu}$–a.e.)*

Proof. For a, b, φ as in the statement we know that

$$u(\varphi(X_a^{(1)})) = \varphi(T_a^1 X_a^{(1)})) \tag{5}$$

and, using the identities (8.2) and (6.5), we find

$$u(\varphi(X_a^{(2)})) = \varphi(T_a^2 X_a^{(2)}) = \varphi(R_\pi T_a^1 X_a^{(2)}) = -\varphi(T_a^1 X_a^{(2)})$$

But by Lemma (1) we know that $X_a^{(1)} = X_a^{(2)}$ $\tilde{\mu}$–a.e., therefore $\tilde{\mu}$–a.e. one has

$$u(\varphi(X_a^{(2)})) = -\varphi(T_a^1 X_a^{(2)}) = -u(\varphi(X_a^{(1)}))$$

which is (3). On the other hand

$$u_{a,b}(\varphi(X_a^{(2)})) = \varphi(T_b^2 X_a) = \varphi(R_\pi T_b^1 X_a^{(2)}) = -\varphi(T_b^1 X_a^{(2)})$$

and this is equal, $\tilde{\mu}$–a.e., to $-\varphi(T_b^1 X_a^{(1)})$. However, if $b \neq a$,

$$-\varphi(T_b^1 X_a^{(1)}) \neq -\varphi(T_a^1 X_a^{(1)}) = u_{a,b}(\varphi(X_a^{(1)}))$$

and this proves (4).

CONNECTION WITH THE EXPERIMENT

Let a, b, $c \in S^{(1)}$ be mutually different vectors and, in the notations of section (10) consider the 3 dynamics

$$u_{a,b}\ ,\quad u_{b,c}\ ,\quad u_{a,c}$$

given by (9.6) and (9.7). The correlations

$$\langle S_a^{(1)} S_b^{(2)} \rangle\ ,\quad \langle S_c^{(1)} S_b^{(2)} \rangle \tag{1}$$

have to be interpreted in the (Heisenberg) sense for the dynamics $u_{a,b}$, $u_{c,b}$ respectively.

In other words

$$\langle S_a^{(1)} S_b^{(2)} \rangle = \tilde{\mu}(u_{a,b}(\varphi(X_x^{(1)})\varphi(X_x^{(2)})) \tag{2}$$

$$\langle S_c^{(1)} S_b^{(2)} \rangle = \tilde{\mu}(u_{c,b}(\varphi(X_x^{(1)})\varphi(X_x^{(2)})) \tag{3}$$

where x denotes the x–axis which in the experiment is fixed once for all. This means that, in the experiment in which a, b are measured we are dealing with the random variables

$$u_{a,b}(\varphi(X_x^{(1)}))\ ,\quad u_{a,b}(\varphi(X_x^{(2)})) \tag{4}$$

where φ is given by (6.2); analogously if c, b are measured we deal with

$$u_{c,b}(\varphi(X_x^{(1)}))\ ,\quad u_{c,b}(\varphi(X_x^{(2)})) \tag{5}$$

and, if a, c is measured, with

$$u_{a,c}(\varphi(X_x^{(1)}))\ ,\quad u_{a,c}(\varphi(X_x^{(2)})) \tag{6}$$

If $\tilde{\mu}$ is the probability measure on Ω defined by Corollary (7.2), from Lemma (9.2) we know that the following identities take place $\tilde{\mu}$–a.e.:

$$S_a^{(1)} := u_{a,b}(\varphi(X_x^{(1)})) = \varphi(T_a^1 X_x^{(1)}) = u_{a,c}(\varphi(X_x^{(1)})) \tag{7}$$

$$S_b^{(2)} := u_{c,b}(\varphi(X_x^{(1)})) = \varphi(T_b^2 X_x^{(2)}) = u_{c,b}(\varphi(X_x^{(2)})) \tag{8}$$

$$S_c^{(1)} := u_{c,b}(\varphi(X_x^{(2)})) = \varphi(T_c^1 X_x^{(1)}) \tag{9}$$

Using the identities (7), (8) we can, if we want, realize the 3 random variables $S_a^{(1)}$, $S_b^{(2)}$, $S_c^{(1)}$ on the same probability space without altering the correlations (1). Notice that this corresponds to the "cleaning procedure" of the experiment. However, due to Lemma (9.2), since $c \neq b$, we have

$$u_{c,b}(\varphi(X_x^{(2)})) \neq -u_{c,b}(\varphi(X_x^{(1)})) \tag{10}$$

and therefore in this realization we cannot impose the additional condition

$$S_c^{(1)} = -S_c^{(2)}$$

which, as explained in [Ac99] is essential in the deduction of the Bell inequality.

REFERENCES

1. Selleri F., *Gli impossibili camaleonti*, Preprint, Bari (1998).
2. Accardi L., Fedullo A., On the statistical meaning of complex numbers in quantum theory, *Lettere al Nuovo Cimento* 34, 161-172 (1982).

3. Accardi L., *Urne e camaleonti Dialogo sulla realtà, le leggi del caso e la teoria quantistica*, Il Saggiatore (1997).
4. Costantini D., *Urne e camaleonti: Dialogo sulla realtà, le leggi del caso e l'interpretazione della teoria quantistica*, di Luigi Accardi, il Saggiatore, Milano, 1997; in: Le Scienze, N. 359, 107 (1998).
5. Hilbert H., Nordheim L. and Von Neumann, Über die Grundlagen der Quantenmechanik in: *J. von Neumann, Coll. Works, Pergamon Press*, vol. I, 104–133 (1926).
6. Birkhoff G., Neumann J.V., *The Logic of Quantum Mechanics.*
7. Mackey G.W., *Mathematical foundations of quantum mechanics*, Addison–Wesley (1976).
8. Accardi L., *Some trends and problems in quantum probability*, In: Quantum probability and applications to the quantum theory of irreversible processes, ed. L.Accardi et al., Springer LNM n. 1055 1–19 (1984).
9. Accardi L., *Can mathematics help solving the interpretational problems of quantum theory?*, Il Nuovo Cimento 110 B (1995) 685–721, cf. also:
10. Di Castro C., Guerra F., Jona–Lasinio G., *Mesoscopic Physics and fundamental problems in quantum mechanics*, (1995).
11. Accardi L.: *Einstein-Bohr: one all. in: The interpretation of quantum theory: where do we stand?*, Acta Enciclopedica, Istituto dell'Enciclopedia Italiana (1994), Volterra preprint N. 174 (1993).
12. Accardi L., The quantum probabilistic approach to the foundations of quantum theory: urns and chamaleons. In: *Philosophy of Science in Florence, 1995.* M. Dalla Chiara, R. Giuntini, F. Laudisa eds., Kluwer (1997), proceedings of the 10–th International Congress of Logic, methodology and Philosophy of Science.
13. Accardi L.: *Su una nota di Franco Selleri*, Volterra Preprint (1998) N. 350.
14. Accardi L., *On the EPR paradox and the Bell inequality* Volterra Preprint (1998) N. 350; Rejected by: Phys. Rev. Lett.
15. William H., *Numerical recepees in C*, The art of scientific computing Press H. William, Teukolsky Saul A., Cambridge University Press (1993).

AN OPTICAL GEOMETRIC MODEL OF THE BETATRONIC MOTION

A. Bazzani[†] and P. Freguglia[⋆]

† Dipartimento di Fisica, Università di Bologna
and INFN Sezione di Bologna
Via Irnerio 46
40126 Bologna (ITALIA)

⋆ Dip. Matematica Pura e Applicata, Università dell'Aquila
Via Vetoio (Coppito)
67100 L'Aquila (ITALIA)
and Domus Galilaeana, Pisa (ITALIA)

ABSTRACT

We discuss the analogy between the betatronic motion in a particle accelerator and the geometric optics by using a Lagrangian formalism. We show that each element in a magnetic lattice is equivalent to a an optical lens in the thin lens approximation. We use this analogy to introduce a geometric approach to the description of the betatronic motion and we propose an optical experiment to study the nonlinear effects of the beam Physics.

INTRODUCTION

An epistemological topic in the Theoretical Physics, which derives from the classical "*analogia entis*" is the study of the analogy among different physical phenomena by using the corresponding mathematical formulations[1]. The fact that two physical theories can be expressed by an identical mathematical formalism can be not very remarkable. Indeed from a syntactical point of view (i.e. from the point of view of symbolic expressions as an algebraic or a differential equation or a geometrical synthetic representation) a mathematical law can describe different phenomena. However this fact is a necessary condition for a more appropriate mastering about the analogy between two physical phenomena. More precisely we can assume two different points

The Application of Mathematics to the Sciences of Nature, edited by Cerrai *et al.*
Kluwer Academic/Plenum Publishers, New York, 2002.

of view in front of an analogy of different physical systems. At first we can consider the experiment A as an analogic model of a more complex experiment B and use the experimental data obtained from A to improve the analysis of the experiment B. At second we can transfer the theoretical techniques that have been developed to analyze the phenomenon A to the study of the phenomenon B and get a deeper comprehension of the physics of B.

In this paper we consider the analogy between the betatronic motion of a particle magnetically confined in a accelerator[2] and an optical ray which is focused by lenses according of the Snell's law. The Physics of betatronic motion has been recently considered since the particle accelerators started to be used in the experiment of high energy Physics[3], whereas the geometric Optics has been studied by using geometry and analysis since the eighteenth century[4].

The experimental onset of the betatronic motion is extremely complex and expensive since it needs high performance magnets, vacuum techniques and particle injection and accelerating systems; an optical ray experiment is relatively simple and not expensive. In this paper we show that the Lagrangian formalism allows to introduce an analogy between these two different phenomena[5]: more precisely both the betatronic dynamics of a charge particle in a magnetic element and the dynamics of an ray through optical lenses can be described by a second order difference equation. We propose to use this analogy in the project of a optical experiment which points out the effect of nonlinear fields on the beam stability. Indeed we can simulate the nonlinear magnets of a particle accelerator by using lenses with a suitable curved surfaces. This experiment could have great advantages on the analogous experiment in a particle accelerator since it allows a better measure of some dynamical quantities and it is much less expensive. The characterization of different dynamical regimes (regular, resonant and chaotic) may have a direct implication in the studies of the transverse beam stability in a particle accelerator.

Moreover the geometric Optics can be studied by means of a synthetic geometrical description which is used to characterize a optical path and corresponds to a first order difference equation in a 4 dimensional space; we propose that the same description is extended to the analysis of the path of a particle in a accelerator in order to get new geometric quantities that can be used in a numerical simulation or in a real experiment of accelerator Physics.

The paper is organized as follows:

1) in section 2 we introduce the Hamiltonian and Lagrangian description of the betatronic motion in a particle accelerator;
2) in section 3 we develop the analogy between the geometric Optics and the betatronic motion, by using the Snell's law and a Lagrangian formalism;
3) in sections 4 and 5 we discuss a geometrical approach to the description of an optical path in the plane and in the space by introducing a formalism based on complex and quaternionic numbers;
4) in section 6 we discuss a possible experiment of geometric optics which allows to study the effect of the nonlinearities in the transverse stability of beam dynamics.

HAMILTONIAN AND LAGRANGIAN FORMULATION OF THE BETATRONIC MOTION

The dynamics of a charge particle confined by the magnetic lattice of a circular accelerator is described by using a cylindrical coordinate system according to the figure 1.

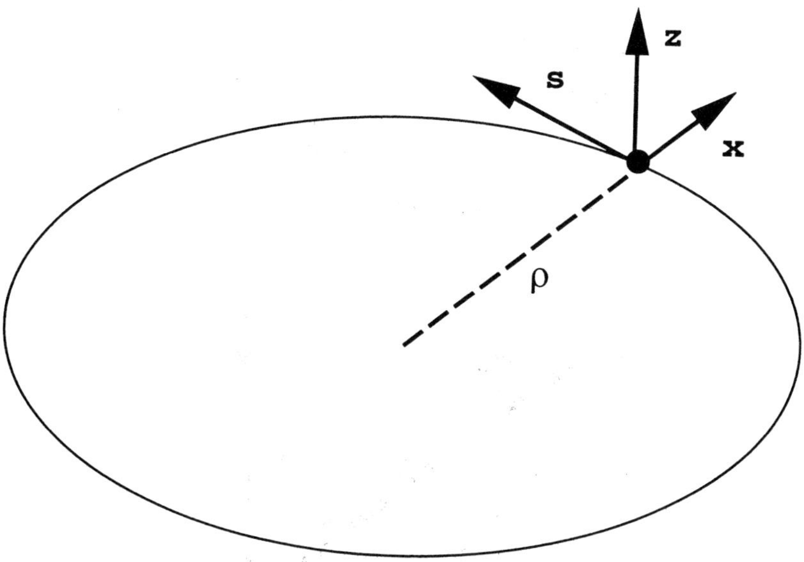

Figure 1: curvilinear coordinate for the betatronic motion in a circular accelerator; s is the arc-length and $\rho(s)$ is the local curvature radius of the reference orbit.

The curvature radius is related to the dipole magnetic field B_0 according to $\rho = cp_0/eB_0$ where p_0 is the total momentum; ρ is ∞ along the straight line and the other magnetic elements. The relativistic formulation of the dynamics is derived from the Hamilton function[2]

$$H = c\sqrt{m^2c^2 + \left(p_s - \frac{eA_s}{c}\right)^2 \left(1 + \frac{x}{\rho}\right)^{-2} + p_x^2 + p_z^2} \tag{1}$$

where (p_x, p_s, p_z) are the canonical momenta and $A_s(x, z; s)$ is the s-component of the vector potential; m and e denote the mass and the charge of a particle and c is the light speed. The s-dependence of the vector potential A_s is discontinuous at the borders of the magnetic elements whereas the dependence on the transverse coordinates (x, z) is analytic inside each magnetic element. For the previous reason it is convenient to study the dynamics of the betatronic motion by using the transfer maps† of the magnets instead of a global Hamiltonian[3].
To reduce the dimensionality of the system, one uses the s coordinate as independent parameter and scales the momenta according to $\hat{p}_x = p_x/p_0$ and $\hat{p}_x = p_x/p_0$. The new Hamiltonian reads

$$K = -\frac{eA_s}{cp_0} - \left(1 + \frac{x}{\rho}\right)\sqrt{1 - \hat{p}_x^2 - \hat{p}_z^2} \simeq -\frac{eA_s}{cp_0} + \frac{\hat{p}_x^2 + \hat{p}_z^2}{2} \tag{2}$$

† We define the transfer map of the magnet the correspondence between the dynamical variables at the beginning of the magnet and the dynamical variables at the end of the magnet.

since in the large accelerators we can consider the limits $\rho \to \infty$ and $p_{x,z} \ll 1$. The corresponding Lagrangian function is calculated by a Legendre transformation of the Hamiltonian (2) and reads

$$L = \frac{(dx/ds)^2 + (dz/ds)^2}{2} + \frac{eA_s}{cp_0} \tag{3}$$

The geometrical meaning of the velocities dx/ds and dz/ds is

$$dx/ds = \text{tg}\theta_x \qquad dz/ds = \text{tg}\theta_z \tag{4}$$

where the angle $\theta_{x,z}$ are defined according to the figure 2

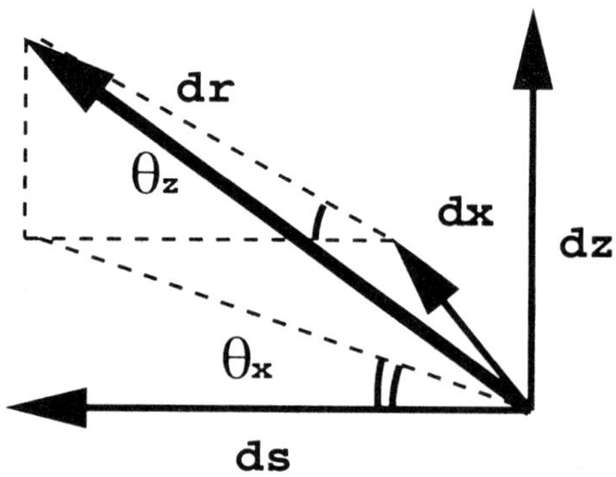

Figure 2: projection of the arc-element dz of a trajectory on the coordinates planes (x,s) and (s,z)

The different elements along the magnetic lattice are characterized by the dependence of the vector potential A_s on the transverse coordinates (x, z). Due to the Maxwell equations the potential A_s is an harmonic function and it can be chosen according to

$$\frac{eA_s(x,z)}{cp_0} = \text{Re}\sum_{n=2}^{\infty} k_n(x+iz)^n \tag{5}$$

The order n terms of the expansion (5) define the normal $2n$-pole potential of a magnetic element: for example the quadratic terms define the quadrupole potential, the cubic terms define the sextupole potential and so on. We remark that the effect of the dipole magnets is not included in the expansion (5), since it is automatically taken into account by the choice of the reference orbit.

In the large circular accelerators, the lengths of a magnets (except for the dipoles) is negligeble with respect the length of the reference orbit and we can use the so called thin-lens approximation to compute the transfer maps. Let $V(x,z) = eA_s(x,z)/cp_0$ the vector potential of a generic magnet of length l, the corresponding transfer map reads

$$\begin{pmatrix} x' \\ dx'/ds \\ z' \\ dz'/ds \end{pmatrix} = \begin{pmatrix} x \\ dx/ds - l\frac{\partial V}{\partial x} \\ z \\ dz/ds - l\frac{\partial V}{\partial z} \end{pmatrix} \tag{6}$$

where the $'$ accent denotes the values at the end of the magnet. The transfer map of an empty section of length L is

$$\begin{pmatrix} x' \\ dx'/ds \\ z' \\ dz'/ds \end{pmatrix} = \begin{pmatrix} x + L\,dx/ds \\ dx/ds \\ z + L\,dz/ds \\ dz/ds \end{pmatrix} \tag{7}$$

The trajectory of a particle can be computed by composing the transfer maps of the magnetic elements and of the empty sections. Due to the periodic structure of the magnetic lattice it is possible to introduce a Poincarè map for the system choosing a section in the phase space and computing the transfer map for one period of the lattice. The Poincarè map is used to study the phase space structure of the betatronic motion.

The basic structure of a magnetic lattice is the FODO cell which is constructed by a focusing and a defocusing quadrupole separated by two dipoles [2].

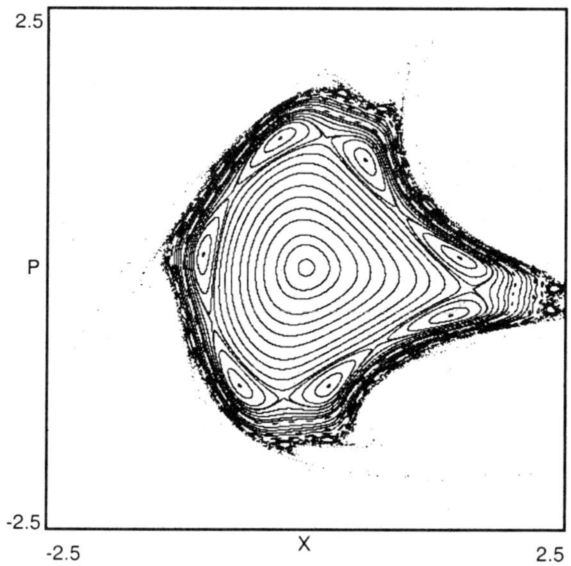

Figure 3: phase space for a FODO lattice in the flat beam approximation in arbitrary unities; the linear normalized coordinates (X,P) are used and the section is chosen at the beginning of the focusing quadrupole.

In the figure 3 we show a section of the phase space for a FODO cell with a thin lens sextupole in the flat beam approximation (1 degree of freedom); the linear normalized coordinates are used

$$\begin{pmatrix} X \\ P \end{pmatrix} = \begin{pmatrix} 1/\sqrt{\beta} & 0 \\ \gamma & \sqrt{\beta} \end{pmatrix} \begin{pmatrix} x \\ dx/ds \end{pmatrix} \tag{8}$$

where the quantities β and γ defines the linear optics of the FODO cell. The nonlinear term causes the presence of nonlinear resonances (island structures) and the chaotic regions. In the next section we shall introduce an optical description for the betatronic motion in a particle accelerator.

OPTICAL MODEL OF THE BETATRONIC MOTION

An optical ray which passes through the border of two different material changes its direction according to the Snell's law

$$\vec{\nu} \times \hat{\tau}_i = k\vec{\nu} \times \hat{\tau}_o \tag{9}$$

where k is the ratio of the refraction coefficients and $\hat{\tau}_i$ (resp. $\hat{\tau}_o$) is the tangent versor to the incoming ray (resp. out coming ray) and $\vec{\nu}$ is the normal at the border surfaces†. We apply the law (9) to compute the path of a optical ray $\vec{r}$ passing through a thin lens in the limit of small deflection angles with respect to a reference trajectory (see fig. 4).

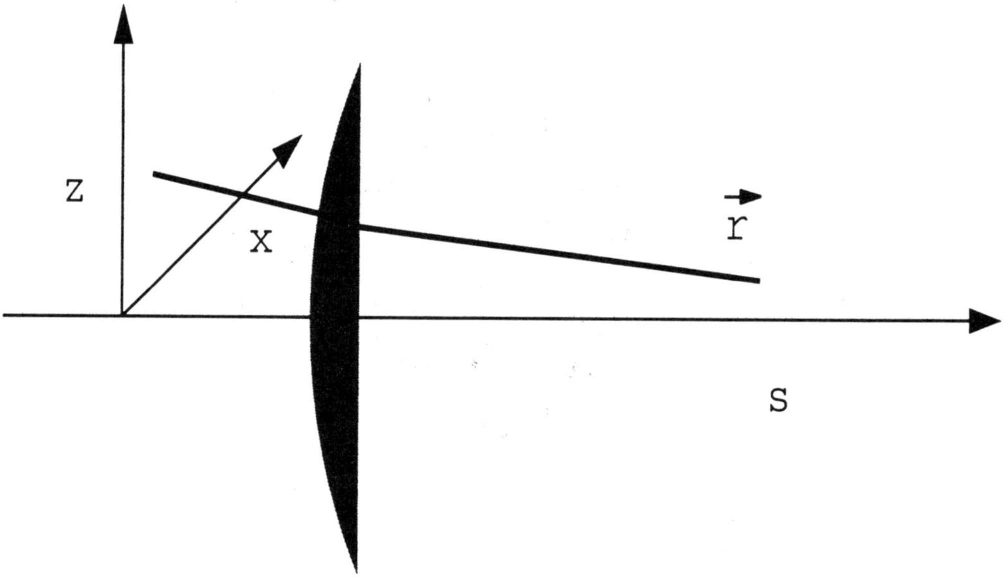

Figure 4: path of an optical ray through a thin lens.

Again we introduce a coordinate system (x, s, z) which is related to a closed reference trajectory according to fig.1. We assume that the discontinuity at the left border is a surface of equation $V(x, z) - s = 0$ and that the coordinate s can be used to parameterized the optical path then we have

$$\vec{\nu} = \left(\frac{\partial V}{\partial x}, -1, \frac{\partial V}{\partial z}\right) \qquad \hat{\tau} = \frac{(dx/ds, 1, dz/ds)}{\sqrt{1 + (dx/ds)^2 + (dx/ds)^2}} \tag{10}$$

† We neglect for the moment the reflection at the border.

In the limit of small deflection angles, we assume dx/ds,dz/ds $\ll 1$ and $\partial V/\partial x, \partial V/\partial z \ll 1$. We denote by an accent $'$ the dynamical variables after the thin lens; then by applying the Snell's law (9) we get the map

$$\begin{pmatrix} x' \\ dx'/ds \\ z' \\ dz'/ds \end{pmatrix} = \begin{pmatrix} x \\ dx/ds + (k-1)\frac{\partial V}{\partial x} \\ z \\ dz/ds + (k-1)\frac{\partial V}{\partial z} \end{pmatrix} \tag{11}$$

which has the same form as the transfer map (6) of a magnetic element in the thin lens approximation. Therefore the path of an optical ray in a suitable device can simulates the betatronic motion of a charge particle in a circular accelerator.

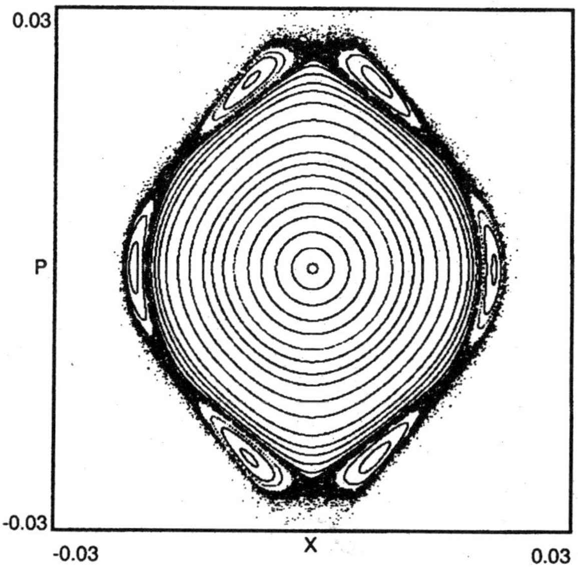

Figure 5: phase space for a thin focusing lens between two empty spaces in arbitrary unities; the linear normalized coordinates $(X,P)$are used.

As an example we consider an optical device constituted by a thin cylindrical lens between two empty spaces; the transfer map of the lens reads

$$\begin{pmatrix} x' \\ \frac{dx'}{ds} \end{pmatrix} = \begin{pmatrix} x \\ \frac{dx}{ds} + (k-1)\frac{x}{\sqrt{R^2-x^2}} \end{pmatrix} \tag{12}$$

whereas the transfer maps of the empty spaces are computed according to eq. (7). We have chosen the ratio $k-1/R$ in order to obtain the same linear motion as in the FODO cell considered in the previous section. The nonlinear terms are introduced by the map (11) which is singular at $x = R$. In fig. 5 we plot the phase space of the Poincarè map of the system by using the linear normalized coordinates (see eq. (8)). The appearance of resonant and chaotic regions in the phase space is clear; the difference between the figures 3 and 5 is due to the different nonlinear terms considered. In a FODO cell only a quadratic nonlinearity is present due to the sextupole effect; in the case of the cylindrical lens we have an algebraic nonlinearity due to the term $x/\sqrt{R^2 - x^2}$.

A GEOMETRIC APPROACH TO THE BETATRONIC MOTION: 2D CASE

A geometrical path in the plane is defined by a sequence of vectors $\{a_t\}$ $(t \in \mathbb{N})$; then we can give a purely geometrical description of an optical ray if we introduce the difference equation[6]

$$w_{t+1} - w_t = a_t \tag{13}$$

where $\{w_t\}$ is another sequence of vectors which are applied to the same point O (we call Λ-determinable a sequence $\{w_t\}$ which has this property). The sequence $\{a_t\}$ univocally defines the Λ-determinable sequence $\{w_t\}$ if we fixed the origin O and the first vector w_1.

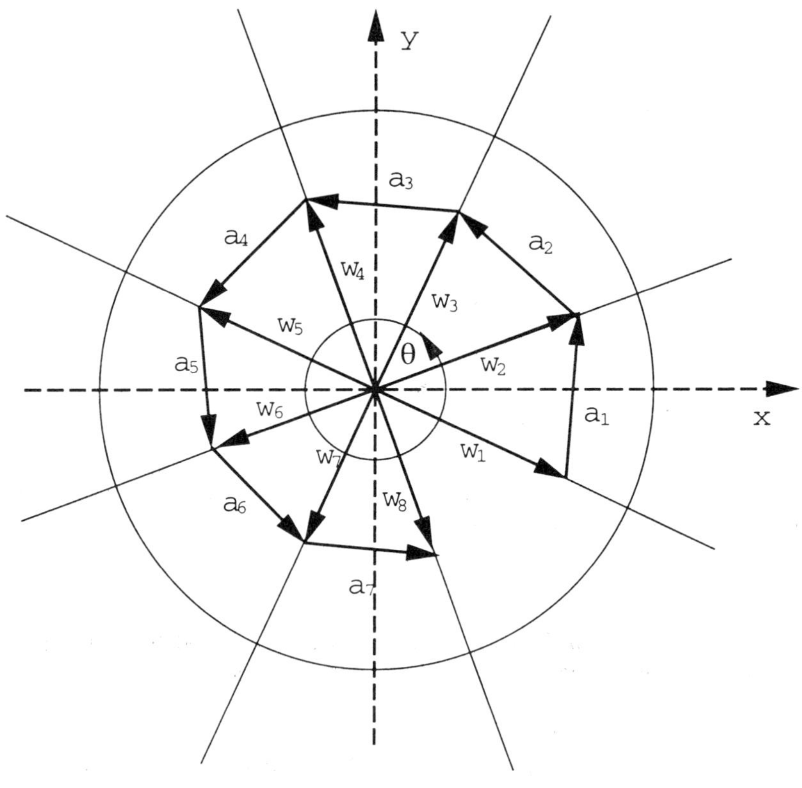

Figure 6: geometrical description of an optical path.

In figure 6 we give an explicit example of the meaning of eq. (13), where the straight lines from the origin give the position of the optical lenses. Assuming that the angle between two nearby lenses is constant, we can represent a periodical optical path by using the solutions $z_j = b^{1/n} \exp 2\pi i j/n$ $j = 1,..,n-1$ of the algebraic complex equation in the Argand's plane

$$z^n + b = 0 \tag{14}$$

where b is a given real number; in such a case the corresponding sequence $\{w_t\}$ has the property $w_{t+n} = w_t$ and we call such a sequence a repeated solution of eq. (14). In a general case we consider a sequence of binomial algebraic equations

$$z^{n_1} + b_1 = 0, \ \ldots\ldots\ldots\ldots, z^{n_t} + b_t = 0, \ \ldots \tag{15}$$

where $b_t \in \mathbf{R}$; then we consider the sequence of complex numbers z_t defined by the h_t-th solution of $z^{n_t} + b_t = 0$ where $0 \le h_t \le n_t$. It follows that to each sequence (15) of binomial algebraic equation it is associated a Λ-determinable sequence of vectors. Let m be the number of lenses in our device (see fig. 6), then any optical path that corresponds to a complete turn can be represented by the matrix

$$\Phi = \begin{pmatrix} n_1 & n_2 & \ldots & n_m \\ b_1 & b_2 & \ldots & b_m \\ h_1 & h_2 & \ldots & h_m \end{pmatrix} \tag{16}$$

where to each vector (n_j, b_j, h_j) we associate the h_j-th solution of the equation $z^{n_j} + b_j = 0$[7]. We observe that the integer numbers n_j and h_j are strictly related to the angular position of the lenses so that if the geometry of our device is fixed, we can consider the matrix Φ as a n-variable function $\Phi(b_1, b_2, ..., b_m)$ into the space of the Λ-determinable sequences.

By using the previous geometrical approach we describe the dynamics of an optical ray in a periodic device where thin lenses are distributed (see figure 7). We assume that there exists a periodic closed orbit which corresponds to the matrix

$$\Phi = \begin{pmatrix} 8 & 8 & \ldots & 8 \\ b & b & \ldots & b \\ 1 & 2 & \ldots & 8 \end{pmatrix} \tag{17}$$

The fixed optical path is a vectorial oriented closed regular polygon. Then we consider a generic path in the device defined by the relations

$$OT_{t+1} - OT_t = T_t T_{t+1} \tag{18}$$

If the ray is confined by the optical lenses, the vectors $T_t T_{t+1}$ lie inside a circular crown and they are generically represented by a vectorial oriented open polygon. As discussed in the previous section the confinement in a circular accelerator is guaranteed by the FODO structure and the corresponding orbit in the phase space is quasi-periodic. Each optical path is defined by the vertexes of a polygon and therefore it can be associated to a Λ-determinable succession of vectors.
It is not difficult to prove the:
Proposition 1: *In the optical geometric model shown in fig. 7, the following relations exist*

$$|OT_{t+1}| = h_t |OT_t| \qquad \textit{where} \quad h_t = \left| \frac{\cos(\delta_t + \alpha_t)}{\sin(\delta_t - \alpha_t)} \right|; \tag{19}$$

the various angles are defined according to the figure 7 and $\theta = \frac{2\pi}{n}$.

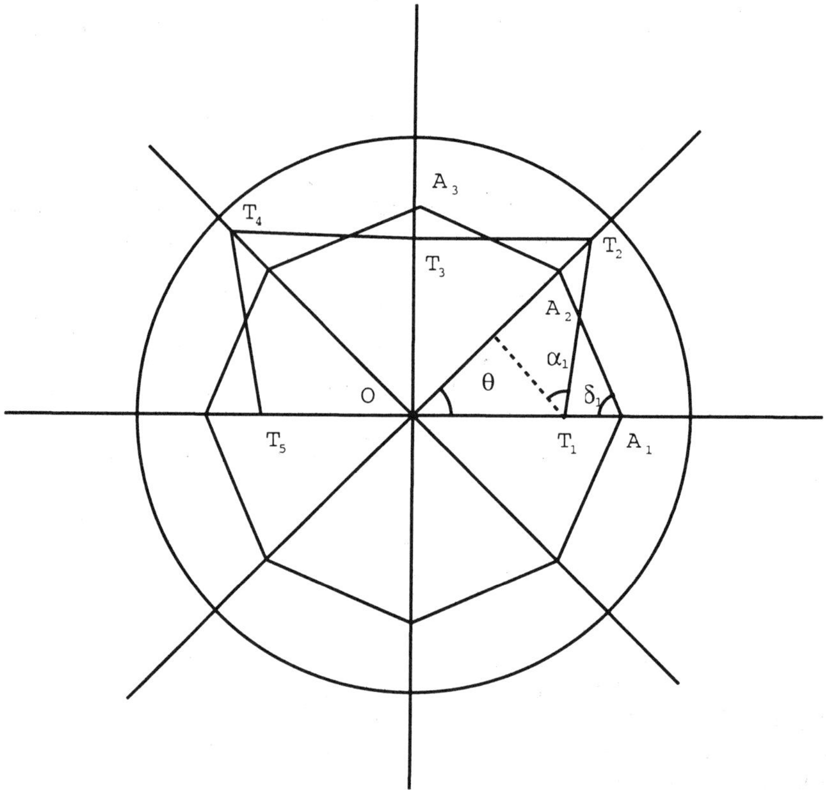

Figure 7: An optical path in a plane circular device.

The angles α_t measure the deviation of the actual orbit with respect to the reference orbit and are related to the incoming and out coming incident angles that change crossing the lens according to the Snell's law. Therefore we recover the complete dynamics of the optical ray if we join eq. (19) with the relation between the angles α_t and α_{t+1} which depends on the refraction index of the lens. By a suitable choice of the compound or of the geometry of the lenses we can get a nonlinear dynamics as discussed in the previous section.
The dynamical point of view is equivalent to a second order difference equation in the vectors OT_t, whereas the geometrical description of the orbit involves a first order difference equation (i.e. eq. (18)).

A GEOMETRIC APPROACH TO THE BETATRONIC MOTION: 3D CASE

When we examine the three dimensional (3D) case we consider an optical path which moves inside a torus in a 3D space: the previous 2D case corresponds to a horizontal section of the torus. Moreover we divide the torus in different sections according to the position of the lenses. A generic optical path $\mathcal{Q}$ can be represented by a succession of points $\{Q_t\}$ corresponding to the crossing points at the lenses. We assume that inside the torus there exists an reference optical path $\mathcal{P}$ defined by the succession of points $\{P_t\}$ (see fig. 8).

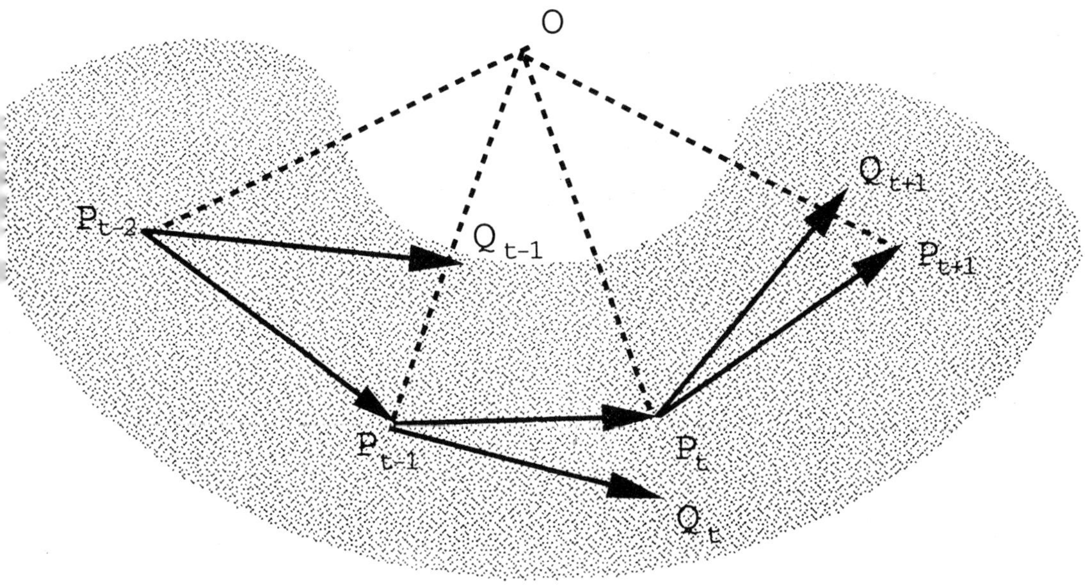

Figure 8: An optical path in a 3D circular device.

If we introduce the vectors $v_t^{(0)} = P_t P_{t+1}$, the generic optical $\mathcal{Q}$ path can be represented by a succession of vectors $v_1, v_2, ..., v_n, ..$ where $v_t = P_t Q_{t+1}$. We associate to the couple of vectors $v_t, v_t^{(0)}$ a quaternion number[8] q_t according to

$$q_t = v_t \cdot v_t^{(0)} \oplus v_t \times v_t^{(0)} \tag{20}$$

where the scalar product is the scalar part of the quaternion and the vector product is the vectorial part of the quaternion. The succession $\{q_t\}$ contains the essential geometrical properties of the path $\mathcal{Q}$ which should be investigated in a 4D space. $\{q_t\}$ can be also defined as the solution of a difference quaternionic equation

$$q_{t+1} - q_t = \alpha_t \tag{21}$$

but we must remember that the algebra of quaternions is different from the complex algebra[9]. The Lagrangian dynamics which determines the evolution of an optical ray should correspond to a second order difference equation in the space of quaternions. One can easily prove that the periodic character of the path $\mathcal{Q}$ corresponds to the periodicity of the quaternions succession $\{q_t\}$.

We consider the set of planes $\{\pi_t\}$ passing through the point O defined by the transverse sections of the torus at the optical lenses. On each plane we introduce a Cartesian system with center at O and then we determine the position of the points P_t by using a sequence of complex numbers. The orbit can be represented by means of a procedure of Λ-determinability and a correspondence between the different planes π_t. Hence we can consider a periodic succession $\{w_t\}$ of complex numbers to characterize the position of the points P_t in the planes π_t and a sequence of quaternions $\{q_t\}$ with origin at P_t which defines the optical path $\mathcal{Q}$; we have the following:

Proposition 2: *Any optical path in a 3D space inside a toroidal surface can be represented by one complex sequence $\{w_t\}$ and one quaternion sequence $\{q_t\}$, by using the procedure of Λ-determinability and the correspondence between a quaternion an a couple of vectors applied at the same point.*

OUTLINE OF A POSSIBLE EXPERIMENT

The analogy between the betatronic motion in a particle accelerator and the geometric optics can be used to project an experiment which studies the effect of nonlinear resonances and chaos in a physical system. The main difficulties to overcome are to avoid or to take the effect of the reflection of a light ray at a discontinuity surface and to limit the diffraction of a light beam which moves along an optical device.

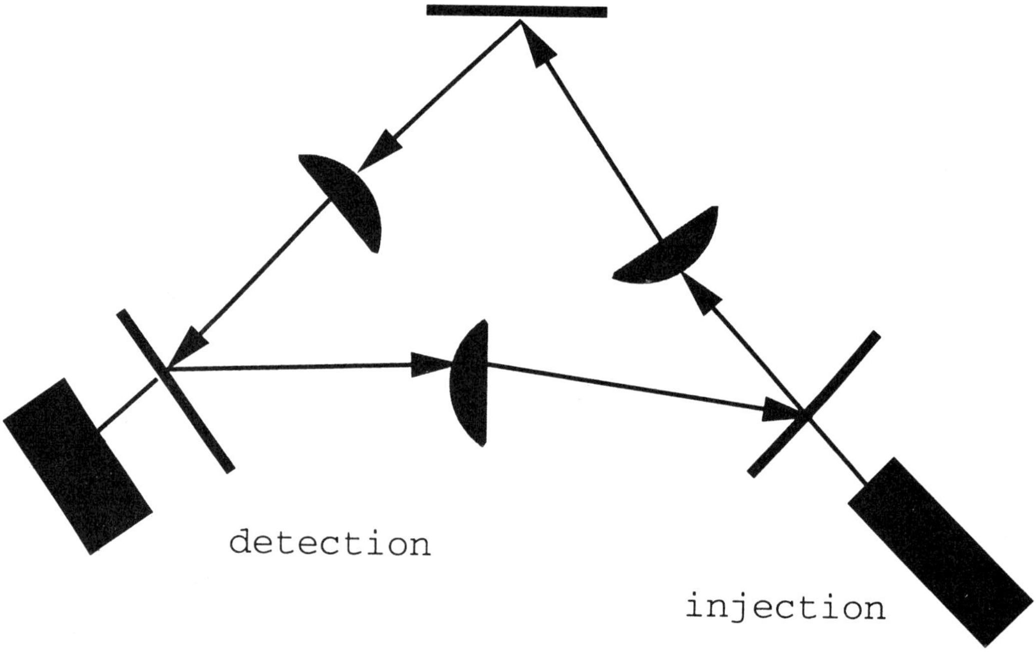

Figure 9: scheme of an optical device for a possible experiment.

We do not discuss such problems, but we outline the scheme of a possible optical device which could be used for an experiment and we consider the possibility of detecting the onset of chaos. In fig. 9 we show the scheme of a possible experiment to simulate the betatronic motion by using the geometrical optics. A closed path is achieved by using three glasses; between each glass we have introduced a cylindrical lens to get the transverse stability of a ray. The injection and the detection of the light are obtained by means of semi-reflecting glasses. The Poincarè map of the system is the transfer map for a single side of the triangle and fig. 5 is an example of phase space. Since it is not possible to follow a single ray and to avoid the dissipative effects we inject and extract continuously the light in the device until a stationary distribution is achieved. In such a case one can assume that the light distributes on an invariant orbit of the dynamics with an intensity proportional to the associated invariant measure.

Therefore one obtains informations on the nature of the orbit (regular, resonant, chaotic) in the phase space by measuring the light intensity at a given section. In the figure 10 we show the density profile $\rho(X)$ of the projection of three orbits plotted in fig. 5 where it is clear the difference between the different type of dynamics.

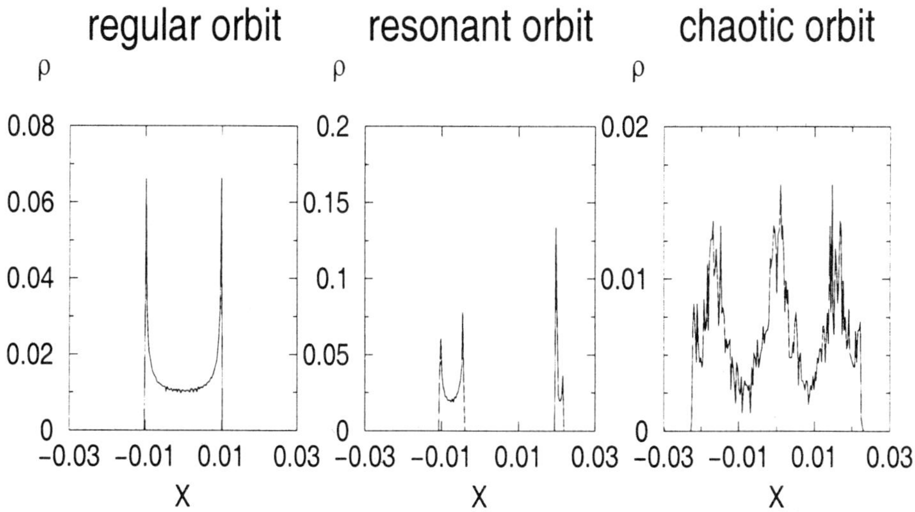

Figure 10: density profile ρ for a regular, resonant and chaotic orbit versus the linear normalized coordinate X; the map used is the same as in fig. 5.

ACKNOWLEDGMENTS

The authors thank Giorgio Turchetti and Leo Fronzoni for useful discussions.

REFERENCES

1. P. Freguglia ,*A synthetic geometrical approach to betatronic motion: some remarks*, IOP Conference Proceedings, S.Chattopadhyay, M.Cornacchia, C. Pellegrini, Eds. vol. **468**, (1999).
2. S.Tuner Ed., *CAS Fifth Advanced Accelerator Physics Course*, Cern report, **CERN 95-06**, (1995).
3. A.Bazzani,E.Todesco,G.Turchetti, *A normal form approach of nonlinear betatronic motion*,**CERN 94-02**, (1994).
4. M.Born,E.Wolf, *Principles of Optic*,Cap. III e IV,Pergamon Press, London, (1959).
5. G.Turchetti, Contribution to this volume.

6. Ronald E.Mickens, (ed.)*Difference Equations Theory and Applications*, Van Nostrand Reinhold, New York, (1990).
7. Bruno Finzi, *Meccanica Razionale*, Vol. I cap. 2, ed. Zanichelli, Bologna (1964).
8. P.G.Tait, *Elementary Treatise on Quaternions*, Univ. Press, Cambridge (1867).
9. S.Mac Lane, G.Birkhoff, *Algebra*,Cap. VII par. 7, Mac Millan, New York, (1965).

SOME REMARKS ON THE THEORY OF RELATIVITY AND THE NAÏVE REALISM

Vieri Benci
Centro Interdisciplinare per lo Studio dei Sistemi Complessi (CISSC)
and
Dipartimento di Matematica Applicata "U. Dini"
Università di Pisa, Via Bonanno 25/b
56126 PISA (Italia)
benci@dma.unipi.it

0.1. INTRODUCTION

The physics of the first part of the twentieth century has changed our image of the universe in a very profound way. Probably those changes have been more dramatic than those created by the birth of modern science during the seventeenth century. In fact during the seventeenth century the Aristotelic philosophy was rejected as the basic framework in which it was possible to organize known facts. The new ideas of this century have changed our image of the universe in an even more drastic way: not only has the Newtonian paradigm been rejected but also so have the basic categories of our knowledge:

- space is not Euclidean
- time does not flow with the same speed in any part of the universe: the notion of contemporareity makes no sense and hence any distinction between past and future is meaningless
- space and time are mixed together
- there is an absolute constant ($\hbar$) which fixes a priori the limits of our knowledge
- the material world does not have any reality *per se*, but instead reality exists as consequence of the operation of measurement performed by an observer (at least at atomic level).

The Application of Mathematics to the Sciences of Nature, edited by Cerrai *et al.*
Kluwer Academic/Plenum Publishers, New York, 2002.

- measurements determine not only the present and the future, but also the past (Delayed Choice Experiments).

The first three assertions are consequences of the theory of Relativity while the last three descend from Quantum Mechanics. Relativity and Quantum Mechanics are theories verified beyond any reasonable doubt in the sense that the predictions deduced by their equations are entirely verified by experiments.

However, we believe that

> *the most radical consequences are not implied by the theory and the experiments, but rather they reflect the cultural environment where they were born.*

In this paper we try to prove this statement for the theory of relativity by a simple method: using the basic assumptions of modern physics (**A-1** and **A-2** of section 0.2), we will construct a model of a relativistic universe which can be very easily interpreted in the spirit of the physics of the nineteenth century.

The same method of analysis can be applied to Quantum Mechanics, but the questions involved are more delicate; in this paper we will consider only some aspects of QM which naturally arise from the model considered here.

0.1.1. The special theory of relativity

The Special Theory of Relativity originated from the following fact:

the Maxwell equations are invariant for the Poincaré group (0.1)

Essentially the Michelson-Morley experiment is nothing, but a verification of this fact. To date, the content of the full theory can be summarized with the following statement:

all the fundamental equations of physics are invariant for the Poincaré group (0.2)

This (formally) simple statement implies many facts which contradicts our intuition and which are verified by many experiments. The most dramatic of them are:

- **Fact 1:** the contractions of lengths
- **Fact 2:** the dilation of time
- **Fact 3:** the increasing of mass
- **Fact 4:** equality between mass and energy.

There are other consequences of the theory such as those listed in the previous section, but they are not implied by (0.2) but by the

$$\textit{Einstein convention of syncronization of clocks} \tag{0.3}$$

Clearly (0.2) and (0.3) are two different assertions and (0.3) is not a logical consequence of (0.2); it is only a reasonable consequence of (0.2) and of the *neopositivistic* philosophy which was dominating the cultural environment in which Einstein was operating.

We will show that facts 1,...,4 are consistent not only with (0.3) but also with the following assertion

$$\textit{there is an absolute time and an absolute space} \tag{0.4}$$

We will come back later on these points in section 0.4.5. First we will discuss the basic assumptions of modern physics (section 0.2) and the model which descends from them (section 0.3). In section 0.4, we will discuss some aspects of quantum mechanics shared by this model (it is very surprising that there are some quantum phenomena which occur in a classical model, namely by objects described by partial differential equations defined in the 4-dimensional space-time).

The reader who is not interested in Mathematics can go directly to section 0.4.5; however he must remember that all the "philosophical" remarks are suggested by the strucure and the properties of the equations considered in the previous sections.

0.2. THE BASIC ASSUMPTIONS

Our model is based on the following two assumptions which, today, are shared by every fundamental theory in Physics.

- **A-1.** *The equations describing the fundamental laws of physics are variational.*
- **A-2.** *The equations describing the fundamental laws of physics are invariant for the Poincaré group.*

Next, we will discuss these assumptions in some detail.

0.2.1. Variational principles

Assumption **A-1** states that the fundamental equation of Physics are the Euler-Lagrange equations of a suitable functional. This fact is quite surprising. There is no logical reason for this. It is just an empirical fact: all the fundamental equations which have been discovered until now derive from a variational principle. If the Creator of the Universe were a mathematician, He would work in Variational Calculus!

For example, the equations of motion of k particles whose positions at time t are given by $x_j(t),\ x_j \in \mathbf{R}^3,\ j = 1, ..., k$ are obtained as the Euler Lagrange equations relative to the following functional

$$\mathcal{S} = \int \left(\sum_j \frac{m_j}{2} |\dot{x}_j|^2 - V(t, x_1, ..., x_k) \right) dt \tag{0.5}$$

where m_j is the mass of the j-th particle and V is the potential energy of the system.

More generally, the equations of motion of a finite dimensional system whose generalized coordinates are $q_j(t),\ j = 1, ..., k$ are obtained as the Euler Lagrange equations relative to the following functional

$$\mathcal{S} = \int L(t, q_1, ..., q_k, \dot{q}_1, ..., \dot{q}_k)\, dt$$

The function L is called Lagrangian function of the system and determines its dynamics.

Also the Dynamics of fields can be determined by Variational principles. The basic fields of Physics can be regarded as a modification of an entity which, in nineteenth century was called "ether" and which is now called "vacuum". From a mathematical point of view a field is a function

$$\psi : \mathbf{R}^{3+1} \to \mathbf{R}^k, \qquad \psi = (\psi_1, ..., \psi_k).$$

where $\mathbf{R}^{3+1}$ is the space-time continuum and $\mathbf{R}^k$ is called the internal parameters space. The space and time coordinates will be respectively denoted by $x = (x_1, x_2, x_3)$ and t respectively. The function $\psi(t, x)$ describes the *internal* state of the ether (or vacuum) at the point x and time t.

From a mathematical point of view, assumption **A-1** states that the field equations are obtained by the variation of the action functional defined as follows:

$$\mathcal{S} = \int \int \mathcal{L}\left(t, x, \psi, \nabla\psi, \frac{\partial \psi}{\partial t} \right) dx\, dt.$$

The function $\mathcal{L}$ is called Lagrangian density function but in the following, for simplicity we will call it just Lagrangian function

0.2.2. The Poincaré group

Assumption **A-2**, as we have discussed in the previous section is the real content of relativity. We recall that the Poincaré group is a 10 parameter Lie group generated by the following one-parameter transformations:

- Space translations in the directions x, y, z :

$$\begin{array}{c} x' = x + x_0 \\ y' = y \\ z' = z \\ t' = t \end{array} ; \quad \begin{array}{c} x' = x \\ y' = y + y_0 \\ z' = z \\ t' = t \end{array} ; \quad \begin{array}{c} x' = x \\ y' = y \\ z' = z + z_0 \\ t' = t. \end{array}$$

This invariance guarantees that space is homogeneous, namely that the laws of physics are independent of space: if an experiment is performed here or there, it gives the same results.

- Space rotations:

$$\begin{array}{c} x' = x \\ y' = y\cos\theta_1 - z\sin\theta_1 \\ z' = y\sin\theta_1 + z\cos\theta_1 \\ t' = t \end{array} ; \quad \begin{array}{c} x' = x\cos\theta_2 - z\sin\theta_2 \\ y' = y \\ z' = x\sin\theta_2 + z\cos\theta_2 \\ t' = t \end{array} ; \quad \begin{array}{c} x' = x\cos\theta_3 - y\sin\theta_3 \\ y' = x\sin\theta_3 + y\cos\theta_3 \\ z' = z \\ t' = t. \end{array}$$

This invariance guarantees that space is isotropic, namely that the laws of physics are independent of orientation.

- Time translations:

$$\begin{array}{c} x' = x \\ y' = y \\ z' = z \\ t' = t + t_0. \end{array}$$

This invariance guarantees that time is isotropic; namely that the laws of physics are independent of time: if an experiment is performed earlier or later, it gives the same results.

- Lorentz transformations:

$$\begin{array}{c} x' = \gamma\left(x - v_1 t\right) \\ y' = y \\ z' = z \\ t' = \gamma\left(t - \frac{v_1}{c^2}x\right) \end{array} ; \quad \begin{array}{c} x' = x \\ y' = \gamma\left(y - v_2 t\right) \\ z' = z \\ t' = \gamma\left(t - \frac{v_2}{c^2}y\right) \end{array} ; \quad \begin{array}{c} x' = x \\ y' = y \\ z' = \gamma\left(z - v_3 t\right) \\ t' = \gamma\left(t - \frac{v_3}{c^2}z\right) \end{array} .$$

where

$$\gamma = \frac{1}{\sqrt{1 - \frac{v^2}{c^2}}}. \tag{0.6}$$

with $v = v_i$, $i = 1, 2, 3$. This invariance is an empirical fact and, as it will be shown below, it implies the remarkable facts of the theory of relativity. Here c is to be interpreted as a parameter having the dimension of *velocity*.

The Poincaré group is the 10 parameters Lie group generated by the above transformations (plus the time inversion, $t \to -t$, and the parity inversion $(x, y, z) \to (-x, -y, -z)$).

An equation for a scalar field ψ is said to be invariant for the Poincaré group if the following happens: if $\psi(t, x, y, z)$ is a solution of our equation, then $\psi(t', x', y', z')$ is also a solution of our equation where (t', x', y', z') are obtained from (t, x, y, z) by applying a Poincaré transformation. The simplest equation invariant for the Poincaré group is the D'Alembert equation:

$$\Box\psi = 0 \tag{0.7}$$

where

$$\Box\psi = \frac{\partial^2\psi}{\partial t^2} - c^2\Delta\psi \text{ and } \Delta\psi = \frac{\partial^2\psi}{\partial x^2} + \frac{\partial^2\psi}{\partial y^2} + \frac{\partial^2\psi}{\partial z^2}.$$

It is easy to check that if $\psi(x, y, z, t)$ is a solution of this equation, then

$$\psi(x + x_0, y, z, t), \ldots\ldots, \psi\left(\gamma\left(x - v_1 t\right), y, z, \gamma\left(t - \frac{v_1}{c^2}x\right)\right) \ldots\ldots$$

are solutions of the equation.

The D'Alembert equation is the simplest *variational* field equation invariant for the Poincaré group. In fact it is obtained from the variation of the action

$$\mathcal{S} = \frac{1}{2}\int\int\left[\left(\frac{\partial\psi}{\partial t}\right)^2 - c^2\left|\nabla\psi\right|^2\right] dx\, dt.$$

In this case, the Lagrangian function is given by

$$\mathcal{L} = \frac{1}{2}\left(\frac{\partial\psi}{\partial t}\right)^2 - \frac{1}{2}c^2\left|\nabla\psi\right|^2$$

A more general Lagrangian invariant for the Poincaré group can be written in the following form

$$\mathcal{L} = -\rho\left(\sigma, \psi\right) \tag{0.8}$$

where

$$\psi : \mathbf{R}^{3+1} \to \mathbf{R}^k, \qquad \psi = (\psi_1, \ldots., \psi_k).$$

is a k-uple of scalar field and

$$\begin{aligned} \sigma &= \frac{1}{2}c^2\left|\nabla\psi\right|^2 - \frac{1}{2}\left(\frac{\partial\psi}{\partial t}\right)^2 \\ &= \frac{1}{2}\left(c^2\left|\nabla\psi_1\right|^2 - \left(\frac{\partial\psi_1}{\partial t}\right)^2, \ldots., c^2\left|\nabla\psi_k\right|^2 - \left(\frac{\partial\psi_k}{\partial t}\right)^2\right) \end{aligned}$$

The equations of motion relative to the Lagrangian (0.8) are the following:

$$\frac{\partial}{\partial t}\left(\frac{\partial\rho}{\partial\sigma}\cdot\frac{\partial\psi}{\partial t}\right) - c^2\nabla\cdot\left(\frac{\partial\rho}{\partial\sigma}\nabla\psi\right) + \frac{\partial\rho}{\partial\psi} = 0 \tag{0.9}$$

where "$\nabla\cdot$" denotes the divergence operator.

Clearly, if $\rho(\sigma, \psi) = \sigma$, then (0.8) reduces to the D'Alembert equation. If

$$\rho(\sigma, \psi) = \sigma + W(\psi), \tag{0.10}$$

we get the nonlinear wave equation

$$\Box\psi + W'(\psi) = 0 \tag{0.11}$$

We end this section recalling the notion of 4-vector. A 4-tuple of quantities

$$\mathbf{F}(t, x) = (F_0, (t, x), F_1(t, x), F_2(t, x), F_3(t, x))$$

is called a 4-vector if, under a Lorentz transformation, it changes as do the quantities (ct, x_1, x_2, x_3), namely

$$\mathbf{F}'(t', x') = (F_0', (t', x'), F_1'(t', x'), F_2'(t', x'), F_3'(t', x'))$$

where

$$\begin{aligned} F_0', (t', x') &= \gamma\left(F_0, (t', x') - \tfrac{v}{c}F_1(t', x')\right) \\ F_1'(t', x') &= \gamma\left(F_1, (t', x') - \tfrac{v}{c}F_0(t', x')\right) \\ F_2'(t', x') &= F_2(t', x') \\ F_3'(t', x') &= F_3(t', x') \end{aligned}$$

For example, if φ is scalar field, then $\left(\frac{1}{c} \cdot \frac{\partial\varphi}{\partial t}, \nabla\varphi\right)$ is a 4-vector field. Another example is given by the gauge potential for the electromagnetic field $(\phi, \mathbf{A})$; the Maxwell equations are obtained by the simplest invariant Lagrangian function of the variables $(\phi, \mathbf{A})$:

$$\mathcal{L} = \frac{1}{2}|\nabla\phi|^2 - \frac{1}{2}\left(\frac{\partial\phi}{\partial t}\right)^2 - \frac{1}{2}|\nabla\mathbf{A}|^2 + \frac{1}{2}\left(\frac{\partial\mathbf{A}}{\partial t}\right)^2. \tag{0.12}$$

0.2.3. Fields and particles

Newtonian Physics describes the universe as being composed of particles interacting with each other via an action at distance; the Dynamics is described by a Lagrangian as (0.5) in which the potential energy $V(t, x_1, \ldots\ldots, x_k)$ depends instantaneously on the positions $x_1, \ldots\ldots, x_k$ of the particles.

In last century, thanks to the studies in electromagnetism, people introduced the notion of field and gave up the notion of instantaneous action: in this new paradigm the particles do not act directly on each other, but act on a field, and the modification of this field influences the motion of the other particles.

Moreover, if we assume that the equations are invariant for the action of the Poincaré group the modification of any field cannot propagate faster than the constant c appearing in the Poincaré group.

The first consequence of this fact is that rigid bodies cannot exist; therefore if we want to admit the existence of entities which are different from a field (such as particles) they must be pointwise entities: material points.

The notion of material point not only presents a lot of philosophical difficulties, but it is also inconsistent with assumption 1 and 2. We will try to explain this inconsistency, making a very simple example: a nonrelativistic charged material point which moves in an electric field $\mathbf{E} = -\nabla\phi$. For simplicity we ignore the magnetic field. The action of the particle is given by

$$\mathcal{S}_p = \int \frac{1}{2} m_0 \left|\dot{x}\right|^2 dt - e \int \phi\left(x\right)\, dt$$

where $x(t)$ is the trajectory of the material point, m_0 is its mass and e is its charge. The variation of $x(t)$ of the above functional gives the correct equation:

$$m_0 \ddot{x} = -e\nabla\phi.$$

The action of the field is given by

$$\mathcal{S}_f = \frac{1}{8\pi} \int\int \left|\nabla\phi\right|^2 dx\, dt$$

The variation of the field gives us the equation

$$\frac{1}{4\pi}\Delta\phi = 0$$

which is the correct one.

However, if we want the field to interact with matter we have the action

$$\mathcal{S} = \mathcal{S}_p + \mathcal{S}_f = \int \frac{1}{2} m_0 \left|\dot{x}\right|^2 dt - e \int \phi\left(x\left(t\right)\right)\, dt - \frac{1}{8\pi} \int\int \left|\nabla\phi\right|^2 dx\, dt$$

Using the Dirac δ-measure, the action can be written as follows:

$$\mathcal{S} = \int \frac{1}{2} m_0 \left|\dot{x}\right|^2 dt - e \int\int \phi\left(x\right)\, \delta\left(x\left(t\right)\right)\, dx\, dt - \frac{1}{8\pi} \int\int \left|\nabla\phi\right|^2 dx\, dt \tag{0.13}$$

The variation of the field in the above functional gives the equation

$$\Delta\phi = 4\pi e \delta\left(x\left(t\right)\right) \tag{0.14}$$

which has the solution

$$\phi\left(x\right) = 4\pi e \frac{1}{\left|x - x\left(t\right)\right|}.$$

But this solution is notacceptable since the integral $\frac{1}{8\pi}\int\int \left|\nabla\phi\right|^2 dx\, dt$ diverges, and the integral $\int\int \phi\left(x\right)\, \delta\left(x\left(t\right)\right)\, dx\, dt$ does not even makes sense. If we add the magnetic field and if we use a relativistic Lagrangian instead of $\frac{1}{2} m_0 \left|\dot{x}\right|^2$ things do not get better. The strange thing about this story is that the variational principles give all the fundamental equations of physics in a natural way and are used in most text books (see e.g. [19]); nevertheless the Mathematics used does not make sense if we let fields and pointwise particles interact. In other words, now, at the end of this century, we do not know how to couple the Maxwell equation with matter within a mathematically consistent theory.

If we abandon the idea of using a variational principle, equation (0.14) makes sense, but we are left with the problem of pointwise particles having infinite energy. Given the relativistic equivalence between mass and energy, pointwise particles would have infinite mass and of course this is not a secondary problem in building any consistent theory (see e. g. [19] page 116 for a brief discussion on this point). Even if in this century Quantum Mechanics made people forget this problem, it seems interesting to know if....a theoretically consistent theory called Classical Physics exists.

Concluding this discussion, if we want a mathematically simple theory consistent with Assumptions **A-1**and **A-2**, it is better to give up the obscure notion of material point and to assume the following point of view:

$$\textit{material particles are bumps of the field} \tag{0.15}$$

Following this point of view, material particles, and more in general material bodies are nothing else but regions of space-time where the field energy concentrates and forms a *stable* structure.

In modern literature, mathematical objects of this type are called *solitary waves* and/or *solitons*: a soliton is a solution of a field equation whose energy travels as a localized packet and which preserves its form under perturbations; if the form is not perfectly preserved and if the degree of stability is not very high, usually the word solitary wave is preferred.

Thus, elementary particles can be modelled by solitons; material bodies can be considered as solitary waves.

0.2.4. Conservation laws

Noether's theorem states that any invariance for a one-parameter group of the Lagrangian implies the existence of an integral of motion, namely of a quantity which is preserved with time (see e.g.[17]). Thus Eq. (0.9) has 10 integrals. We describe the most important ones: energy and momentum; the angular momentum is not relevant for our discussion.

- **Energy**. Energy, by definition, is the quantity which is preserved by virtue of the time invariance of the Lagrangian; it has the following form

$$E = \int \left(\frac{\partial \mathcal{L}}{\partial \frac{\partial \psi}{\partial t}} \frac{\partial \psi}{\partial t} - \mathcal{L} \right) dx$$

In particular, if we take the Lagrangian (0.8), we get

$$E=\int\left(\frac{\partial\rho}{\partial\sigma}\cdot\left(\frac{\partial\psi}{\partial t}\right)^2+\rho\right)dx \tag{0.16}$$

and in the case (0.10), we get

$$E=\int\left(\frac{1}{2}\left(\frac{\partial\psi}{\partial t}\right)^2+\frac{1}{2}c^2\left|\nabla\psi\right|^2+W(\psi)\right)dx$$

- **Momentum.** Momentum, by definition, is the quantity which is preserved by virtue of the space invariance of the Lagrangian; the invariance for translations in the x_i direction gives the following invariant

$$P_i=\int\frac{\partial\mathcal{L}}{\partial\psi_t}\frac{\partial\psi}{\partial x_i}\,dx$$

In particular, if we take the Lagrangian (0.8), we get

$$P_i=-\int\frac{\partial\rho}{\partial\sigma}\frac{\partial\psi}{\partial x_i}\frac{\partial\psi}{\partial t}\,dx$$

and since $\mathbf{P}=(P_1,P_2,P_3)$ is a vector, we can write

$$\mathbf{P}=-\int\frac{\partial\rho}{\partial\sigma}\frac{\partial\psi}{\partial t}\nabla\psi\,dx \tag{0.17}$$

In the case (0.10), the momentum takes the following form:

$$\mathbf{P}=-\int\frac{\partial\psi}{\partial t}\nabla\psi\,dx$$

0.3. THE MODEL

In this section we will describe a model of a universe which is consistent with the experimental facts described in section 0.1.1. This model is based on the two assumptions discussed in section 0.2 and on assumption (0.15) which is one of their consequences.

In order to make this model as simple as possible, we consider only scalar fields and a very simple Lagrangian function, namely the Lagrangian (0.8) which is a Lagrangian for scalar fields. Actually it is possible to couple this Lagrangian with a vector valued one such as (0.12) and in fact it has been done in [9]; in the latter case the Mathematics is more complicated but the conclusions are the same. Thus, in this paper, we restrict ourselves to the Lagrangian (0.8).

Notice that, as a particular case, the Lagrangian (0.8) may assume the following form

$$\mathcal{L}=-\rho(\sigma,\psi)=\frac{1}{2}\left(\frac{\partial\psi}{\partial t}\right)^2-\frac{1}{2}c^2\left|\nabla\psi\right|^2-\frac{1}{2}\Omega^2\left|\psi\right|^2-\rho_1(\sigma,\psi);\ \ \Omega>0$$

where $\rho_1(\sigma, \psi)$ is a polynomial in σ and ψ of degree higher than 2. In this case the Eq. (0.9) takes the following form

$$\Box\psi + \Omega^2\psi = \varepsilon F; \tag{0.18}$$

where F is a nonlinear term which contains ψ and, possibly, its derivative up to the second order. This equation is a nonlinear perturbation of the Klein-Gordon equation. The Klein-Gordon equation is the simplest dispersive scalar equation invariant for the Poincaré group. By our assumptions on ρ_1, the linearization of equation (0.18) near $\psi = 0$ is the Klein-Gordon equation

$$\Box\psi + \Omega^2\psi = 0 \tag{0.19}$$

Therefore, the small amplitude solutions of (0.18) solve the Klein-Gordon up to higher order terms.

0.3.1. Existence of static solitons

In this framework, a static soliton is a stable solution of Eq. (0.9) which does not depend on t; we will denote it by $u = u(x_1, x_2, x_3)$. Thus u solves the following equation:

$$-c^2\nabla \cdot \left(\frac{\partial\rho}{\partial\sigma}\nabla u\right) + \frac{\partial\rho}{\partial u} = 0 \tag{0.20}$$

The solutions of equations (0.20) are the critical points of the energy functional

$$E(u) = \int \rho\left(c^2\nabla u, u\right) dx$$

The stability is guaranteed if this solution is a local minimum of the energy. The existence of finite energy solutions of (0.20) depends on the form of the function ρ; if ρ satisfies (0.10), Eq. (0.20) reduces to the following one:

$$-c^2\Delta u + W'(u) = 0 \tag{0.21}$$

In a remarkable paper, Derrick [16] proved that, if $W \geq 0$, the above equation does not have nontrivial finite energy solutions no matter what the form of W and the dimension of the internal parameter space are. However, it is possible to prove that more general Lagrangians admit soliton solutions. For example,in [10], it has been proved that (0.9) has soliton solutions if the Lagrangian has the following form

$$\mathcal{L} = -\frac{1}{2}\alpha(\sigma) - W(\psi), \tag{0.22}$$

where

$$\psi : \mathbf{R}^{3+1} \to \mathbf{R}^4,$$

$$\alpha(\sigma) = \frac{1}{2}\sigma^2 + \frac{\varepsilon}{4}\sigma^4$$

and W has a singular part as e.g.

$$W(\psi) = \frac{1}{2}\Omega^2 |\psi|^2 + \varepsilon \frac{|\psi|^2}{|\psi - \psi_0|^6}.$$

With this Lagrangian, we obtain a particular case of Eq. (0.18) for four scalar fields. In this equation the radius of solitons is of the order of $\varepsilon^{1/4}$. If we want solitons to have the radius of elementary particles, ε must be very small. Nevertheless the density of energy in the soliton is very high. More general Lagrangian functions which admit soliton like solutions have been considered in [4],...,[12].

In all these papers the existence and some degree of stability of the solitons is guaranteed by topological constraints imposed by the singular part of the potential W.

0.3.2. Travelling solitons

From now on, $u = u(x_1, x_2, x_3)$ will denote a solution of Eq. (0.20). By the Lorentz invariance of $\mathcal{L}$, it follows that

$$\psi_v(x_1, x_2, x_3, t) = u\left(\gamma\left(x_1 - vt\right), x_2, x_3\right), \tag{0.23}$$

is a solution of Eq. (0.9) provided that γ satisfies (0.6). This solution represents a solitary wave which travels with velocity $\mathbf{v} = (v, 0, 0)$ in the x_1 direction. If this solitary wave is "small" and does not have a complex internal structure, it can be considered as an " elementary particle"; if it is "large" and has a rich internal structure, it can be considered as a "material body". In any case, by Eq. (0.23), the following theorem follows:

Theorem 0.1. *Any moving material body experiences a contraction in the direction of its movement of a factor* $1/\gamma$ *with*

$$\gamma = \frac{1}{\sqrt{1 - \frac{v^2}{c^2}}}$$

Thus, Fact 1 of section 0.1.1 is a trivial consequence of our general assumptions.

Next we compute the energy and momentum of travelling solitons. Using (0.16) and (0.23), we get the following expression for the energy:

$$E\left(\psi_v\right) = \gamma^2 v^2 \int \left(\frac{\partial \rho}{\partial \sigma} \cdot \left|\frac{\partial u}{\partial x_1}\right|^2\right) dx + \int \rho \, dx$$

In the above expression the function u and its derivatives are computed at the point

$$y = (y_1, y_2, y_3) = \left(\gamma\left(x_1 - vt\right), x_2, x_3\right)$$

Making the change of variable $x \to y$, we get the following expression:

$$E(\psi_v) = \gamma v^2 \int \left(\frac{\partial \rho}{\partial \sigma} \cdot \left| \frac{\partial u}{\partial y_1} \right|^2 \right) dy + \frac{1}{\gamma} \int \rho \, dy$$

or, just changing the name to the y variable, we get

$$E(\psi_v) = \gamma v^2 \int \left(\frac{\partial \rho}{\partial \sigma} \cdot \left| \frac{\partial u}{\partial x_1} \right|^2 \right) dx + \frac{1}{\gamma} \int \rho \, dx \tag{0.24}$$

where, now, the function u is computed at the point $x = (x_1, x_2, x_3)$

Notice that, if $v = 0$, the above expression reduces to

$$E(\psi_0) = E(u) = \int \rho \, dx$$

thus, $\rho \left(c^2 |\nabla u|^2, u \right)$ represents the energy density which is due to the deformation of the field.

Using (0.17) and (0.23), we get the momentum of our soliton:

$$P_i(\psi_v) = v\gamma^2 \int \frac{\partial \rho}{\partial \sigma} \cdot \left| \frac{\partial u}{\partial x_1} \right|^2 dx$$

where u is computed at the point $(\gamma (x_1 - vt), x_2, x_3)$. Making a change of variable as we did for the energy, we get

$$P_i(\psi_v) = v\gamma \int \frac{\partial \rho}{\partial \sigma} \cdot \left| \frac{\partial u}{\partial x_1} \right|^2 dx$$

where u is computed at the point (x_1, x_2, x_3). Finally, using the vector notation, we can write

$$\mathbf{P}(\psi_v) = \mathbf{v}\gamma \int \frac{\partial \rho}{\partial \sigma} \cdot \left| \frac{\partial u}{\partial x_1} \right|^2 dx \tag{0.25}$$

0.3.3. The energy-momentum as 4-vector

Next we need to prove a simple lemma:

Lemma 0.2. *Let u be a solution of the Eq. (0.20); then*

$$\int \rho \, dx = c^2 \int \left(\frac{\partial \rho}{\partial \sigma} \cdot \left| \frac{\partial u}{\partial x_i} \right|^2 \right) dx; \; i = 1, 2, 3 \tag{0.26}$$

and

$$\int \rho \left(\frac{c^2}{2} |\nabla u|^2, u \right) dx = \frac{c^2}{3} \int \frac{\partial \rho}{\partial \sigma} \cdot |\nabla u|^2 \, dx. \tag{0.27}$$

Proof. First we prove (0.26) for $i = 1$. Let $u_\lambda(x_1, x_2, x_3) = u(\lambda^{-1}x_1, x_2, x_3)$; then, setting $y = (y_1, y_2, y_3) = (\lambda^{-1}x_1, x_2, x_3)$

$$\begin{aligned} E(u_\lambda) &= \int \rho\left(\frac{c^2}{2}\left|\nabla_x u_\lambda\right|^2, u_\lambda\right) dx \\ &= \lambda \int \rho\left(\frac{c^2}{2\lambda^2}\left|\frac{\partial u_\lambda}{\partial y_1}\right|^2 + \frac{c^2}{2}\left|\frac{\partial u_\lambda}{\partial y_2}\right|^2 + \frac{c^2}{2}\left|\frac{\partial u_\lambda}{\partial y_3}\right|^2, u_\lambda\right) dy \end{aligned}$$

Since u is a critical point of E,

$$\left.\frac{d}{d\lambda} E(u_\lambda)\right|_{\lambda=1} = 0 \tag{0.28}$$

Let us compute this expression explicitly:

$$\begin{aligned} \frac{d}{d\lambda} E(u_\lambda) &= \int \rho\left(\frac{1}{2\lambda^2}\left|\frac{\partial u_\lambda}{\partial y_1}\right|^2 + \frac{1}{2}\left|\frac{\partial u_\lambda}{\partial y_2}\right|^2 + \frac{1}{2}\left|\frac{\partial u_\lambda}{\partial y_3}\right|^2, u_\lambda\right) dy \\ &\quad - c^2\lambda^{-2} \int \frac{\partial \rho}{\partial \sigma} \cdot \left|\frac{\partial u_\lambda}{\partial y_1}\right|^2 dy. \end{aligned}$$

For $\lambda = 1$, using (0.28), we get

$$\int \rho(\frac{1}{2}\left|\nabla_y u_\lambda\right|^2, u)\, dy = c^2 \int \frac{\partial \rho}{\partial \sigma} \cdot \left|\frac{\partial u_\lambda}{\partial y_1}\right|^2 dy$$

which, renaming the variable, gives (0.26) for $i = 1$. For $i = 2, 3$, we argue in the same way. Eq. (0.27) is obtained adding Eqs.(0.26) for with $i = 1, 2, 3$.
□

This simple lemma has a wealth of consequences. The first consequence is Derrick's theorem. In fact, if ρ has the form (0.10), Eq. (0.27) reduces to

$$\int \left(\frac{c^2}{2}|\nabla u|^2 + W(u)\right) dx = \frac{c^2}{3} \int |\nabla u|^2\, dx$$

from which we get

$$\frac{c^2}{6} \int |\nabla u|^2\, dx = -\int W(u)\, dx;$$

thus if $W(u) \geq 0$, we get $|\nabla u| = 0$, and hence u is constant, and since the energy if finite, it must assume a constant value $u_0 \in \mathbf{R}^k$ such that $W(u_0) = 0$.

Next we examine another consequence of the above lemma. By Eq. (0.26) with $i = 1$, and Eq. (0.24), we get the following expression for the energy:

$$E(\psi_v) = \left(\gamma\frac{v^2}{c^2} + \frac{1}{\gamma}\right) \int \rho\, dx$$

and by (0.6),

$$E(\psi_v) = \gamma \int \rho\, dx \tag{0.29}$$

Using Eqs. (0.25) and (0.26), we get a nice expression for the momentum:

$$\mathbf{P}(\psi_v) = \frac{\gamma \mathbf{v}}{c^2} \int \rho\, dx \tag{0.30}$$

By (0.29) and (0.30), it follows that the 4-uple $(\mathbf{P}(\psi_v), E(\psi_v)/c)$, is a 4-vector, in fact

$$\begin{bmatrix} \mathbf{P}(\psi_v) \\ \frac{E(\psi_v)}{c} \end{bmatrix} = \left(\frac{1}{c} \int \rho\, dx \right) \cdot \begin{bmatrix} \gamma \frac{\mathbf{v}}{c} \\ \gamma \end{bmatrix}$$

and it is well known that the expression

$$\begin{bmatrix} \gamma \frac{\mathbf{v}}{c} \\ \gamma \end{bmatrix}$$

is a 4-vector in Minkowsky space (called 4-velocity).

0.3.4. The definition of mass

In this section, we will define the mass of a soliton within the framework of this theory. First of all notice that the concept of momentum comes before the notion of mass; in fact the momentum is defined as the integral of motion due to the homogeneity of space. Then mass can be defined as the ratio between momentum and velocity:

$$m = \frac{P}{v}$$

Notice that the mass is a scalar and not a tensor since, by Eq. (0.30), $\mathbf{P}$ and $\mathbf{v}$ are parallel.

If we take this definition of mass, by Eq. (0.30), we get that

$$m(\psi_v) = \frac{\gamma}{c^2} \int \rho\, dx \tag{0.31}$$

Here, we get the second remarkable fact of Theory of relativity (cf. Fact 3, of section 0.1.1):

Theorem 0.3. *The mass increases with velocity by the factor*

$$\gamma = \frac{1}{\sqrt{1 - \frac{v^2}{c^2}}}.$$

For $v = 0$, we get the expression for the rest mass of a soliton:

$$m_0(u) = \frac{1}{c^2}\int \rho\, dx. \tag{0.32}$$

Then, comparing (0.31) with (0.29), we get the celebrated Einstein equation (cf. Fact 4, of section 0.1.1):

Theorem 0.4. *For any material body ψ_v, we have*

$$E(\psi_v) = m(\psi_v)\, c^2$$

0.3.5. Contraction of time

Let us take a clock. By clock, we mean a solution of equation (0.9) which is not only bounded in space, but periodic in time, as for example a stationary wave. The simplest form of a stationary wave is the following one:

$$\psi(t,x) = u(x)\, e^{-i\omega_0 t}. \tag{0.33}$$

Of course in this case we have to assume that the target space is $\mathbf{C}^k$ rather than $\mathbf{R}^k$. In the following considerations, it is not necessary to assume that our "clock" has the form (0.33), however, in order to fix our ideas, it is better to assume a concrete and simple case.

First of all we want to remark that solutions of (0.9) having the form (0.33) exist; they can be found as critical points of the functional

$$\int \rho\left(|\nabla u|^2 - \omega_0^2 \cdot |u|^2, u\right) dx$$

If ρ has the form (0.22), their existence has been proved in [7]. Also equations of the particular type (0.11) have stable solutions of the form (0.33). In fact, Derrick's theorem cannot be applied to stable stationary waves and in fact an existence theorem exists under suitable assumptions on W (the proofs can be found in [26]).

Next we let our clock move. Applying a Lorentz transformation to (0.33) we get a new solution of (0.9):

$$\psi_v(t,x) = \psi_v(t,x_1,x_2,x_3) = u(\gamma(x_1 - vt), x_2, x_3)\, e^{-i\omega_0\gamma(t - vc^{-2}x_1)} \tag{0.34}$$

Let us denote by $\mathbf{q}(t)$ the position of our clock at the time t; a mathematical definition of $\mathbf{q}$ is given by

$$\mathbf{q}(t) = \frac{\int x\rho\, dx}{\int \rho\, dx}.$$

The point $\mathbf{q}(t)$ will be called *barycenter* of the soliton. If we assume that at $t = 0$, $\mathbf{q}(0) = (0,0,0)$, the motion of the clock is given by

$$\mathbf{q}(t) = (vt, 0, 0);$$

then our description of the clock can be made replacing x by $\mathbf{q}(t)$ in Eq. (0.34):

$$\begin{aligned}\psi_v(t, \mathbf{q}(t)) &= \psi_v(t, vt, 0, 0) \\ &= u(0,0,0)\, e^{-i\omega_0\gamma(1-v^2c^{-2})t} \\ &= u(0,0,0)\, e^{-i\omega_0\gamma^{-1}t};\end{aligned}$$

this equation shows that our moving clock is vibrating with a frequency

$$\omega = \frac{\omega_0}{\gamma} = \omega_0\sqrt{1 - \frac{v^2}{c^2}}$$

Since the intervals of time measured by a clock are inversely proportional to the frequency of the vibrations, we get that

$$\Delta T = \frac{\Delta T_0}{\sqrt{1 - \frac{v^2}{c^2}}}.$$

This implies Fact 2 of section 0.1.1:

Theorem 0.5. *A moving clock moves slower than a resting clock by a factor* γ^{-1}.

0.4. SOME REMARKS ON QUANTUM MECHANICS

The model described in section 0.3 is based on very general assumptions and it does not have anything to do with Quantum Mechanics. Nevertheless, it presents some phenomena which are considered typical of Quantum Mechanics. This fact contradicts the idea, shared by most people, that no phenomena of Quantum Mechanics can be described in a world where the observer does not play any role. Actually, the model which we have presented so far presents some of these phenomena.

We think that it is very relevant to understand which phenomena can be explained by nonlinear effects in Classical Mechanics and which ones require the new principles of QM.

0.4.1. The De Broglie relations

In our picture, a solution of type (0.33) represents a particle, but it has the structure of a standing wave: it reminds us of the big discussion in the beginning of this century about the dualism between waves and particles, one of the starting points of Quantum Mechanics. In this section, we will spend few words to see what our model can say about this.

If we look at the solution (0.34) with an apparatus at rest with dimension of the order of a wavelength, it is not adequate to consider only what happens at the barycenter $\mathbf{q}(t)$. We need to consider the space vibrations. Our solution will appear as a wave packet having the following form:

$$\psi_v(t,x) = \psi_v(t,x_1,x_2,x_3) = u\left(\gamma\left(x_1 - vt\right), x_2, x_3\right) e^{i(\mathbf{k}\cdot\mathbf{x}-\omega t)} \tag{0.35}$$

with

$$\mathbf{k} = \gamma \frac{\mathbf{v}}{c^2}\omega_0 \text{ and } \omega = \gamma\omega_0 \tag{0.36}$$

Comparing (0.29), (0.30) and (0.36), we get

$$E\left(\psi_v\right) = \hbar\omega \tag{0.37}$$

and

$$\mathbf{P}\left(\psi_v\right) = \hbar\mathbf{k} \tag{0.38}$$

with

$$\hbar = \frac{1}{\omega_0}\int \rho\, dx. \tag{0.39}$$

where $\hbar$ has the dimension of action and is reminescent of the Planck constant. More or less this is the argument which De Broglie used to deduce his celebrated relations (0.37) and (0.38). It is interesting to note that also (0.37) and (0.38) are a general consequence of **A-1** and **A-2**. Here, it is important to emphasize that in this model $\hbar$ is *not* a universal constant since it depends on ρ and ω_0; different solutions of (0.9) of type (0.33) might give different values of $\hbar$. It is possible that very special assumptions on the form of $\mathcal{L}$ might give a universal value for $\hbar$, but at this moment we do not know if this is possible. Thus, if you want to make a model for quantum mechanics, you need to introduce a new principle:

> *nature selects the solutions of Eq. (0.9) which satisfy the relation (0.39) where $\hbar$ is the actual Planck constant.*

Clearly this is an independent axiom: an assumption *ad hoc.* But Quantum Mechanics is based an a lot of independent Axioms and assumptions *ad hoc,*......and no one is scandalized.

0.4.2. The pilot wave theory

In this section we consider a particular case of Eq. (0.9), namely Eq. (0.11):

$$\Box\psi + W'(\psi) = 0 \tag{0.40}$$

The solutions of type (0.33) can be obtained solving the following elliptic equation:

$$-c^2\Delta u + W'(u) = \omega_0^2 u \tag{0.41}$$

We know the following facts:

- **(i)** If there exists a t such that

$$W(t) - \frac{1}{2}\omega_0^2 t^2 < 0 \quad \text{and} \quad W''(0) - \omega_0^2 > 0$$

then Eq. (0.41) has bounded energy solutions and hence Eq. (0.40) has solutions of the form (0.33) [13].

- **(ii)** Under suitable assumptions, these solutions are also orbitally stable [26].

Any solution $\psi(t,x)$ of Eq. (0.40) can be written in the following form

$$\psi(t,x) = u(t,x)e^{iS(t,x)/\hbar}; \quad u, S \in \mathbf{R} \tag{0.42}$$

here $\hbar$ is a contant which has the dimension of an action, but a this point it is not necessary to interpret it as the constant of Planck. Using the Eq. (0.42), Eq. (0.40) is equivalent to the following couple of equations:

$$\Box u - \frac{1}{\hbar^2}\left[\left(\frac{\partial S}{\partial t}\right)^2 - c^2 \left|\nabla S\right|^2\right] u + W'(u) = 0 \tag{0.43}$$

and

$$\frac{\partial}{\partial t}\left(u^2 \frac{\partial S}{\partial t}\right) - c^2 \nabla \cdot \left(u^2 \nabla S\right) = 0 \tag{0.44}$$

Eq. (0.43) can be rewritten as follows

$$\left(\frac{\partial S}{\partial t}\right)^2 = c^2 \left|\nabla S\right|^2 + \hbar^2 \frac{\Box u + W'(u)}{u} \tag{0.45}$$

In particular, if the function ψ is a solution of type (0.33), by Eq. (0.41), we get

$$\left.\frac{\Box u(y) + W_0'(u(y))}{u(y)}\right|_{y=x-vt} = \frac{-c^2 \Delta u(x) + W'(u(x))}{u(x)} = \omega_0^2$$

If ψ is a perturbation of such a solution, we can write

$$\left.\frac{\Box u(y) + W_0'(u(y))}{u(y)}\right|_{y=x-vt} = \left(\omega_0 + \frac{Q}{\hbar}\right)^2$$

where Q is a small compared to $\hbar\omega_0$. Following De Broglie, we will call it *quantum potential.* Then, Eqs. (0.45) take the form:

$$\left(\frac{\partial S}{\partial t}\right)^2 = (\hbar\omega_0 + Q)^2 + c^2 \left|\nabla S\right|^2 \tag{0.46}$$

Eq. (0.46) is nothing else but the relativistic Hamilton-Jacobi equation relative to the Hamiltonian function

$$H(\mathbf{p}) = \sqrt{(\hbar\omega_0 + Q)^2 + c^2\mathbf{p}^2} \tag{0.47}$$

According to the equations of Planck (0.39) and Einstein (0.32), we have

$$m = \frac{\hbar\omega}{c^2} \quad \text{and} \quad m_0 = \frac{\hbar\omega_0}{c^2} \tag{0.48}$$

Actually, we have derived Eq. (0.32) only for static solitons, but with some extra work, it is possible to deduce Eq. (0.32) also for stationary solitons having the form (0.33). Then, replacing (0.48) in (0.47), we obtain

$$H(\mathbf{p}) = \sqrt{(m_0c^2 + Q)^2 + c^2\mathbf{p}^2} \tag{0.49}$$

The equation (0.44) is the continuity equation for the "fluid" $\rho = u^2 \frac{\partial S}{\partial t}$ which moves according to the velocity field

$$\mathbf{v}(t,x) = -c^2 \frac{\nabla S(t,x)}{\frac{\partial S}{\partial t}(t,x)}. \tag{0.50}$$

If this fluid contains a bump of type (0.33) whose position is $\mathbf{q}(t)$, then we get

$$\dot{\mathbf{q}} = -c^2 \frac{\nabla S(t,\mathbf{q})}{\frac{\partial S}{\partial t}(t,\mathbf{q})} \tag{0.51}$$

Comparing (0.33) and (0.42), at the point $\mathbf{q}(t)$, we have that

$$\frac{\partial S}{\partial t} = -\hbar\omega.$$

Thus, (0.51) becomes

$$\frac{\hbar\omega}{c^2}\dot{\mathbf{q}} = \nabla S(t,\mathbf{q})$$

and by (0.48)

$$m\dot{\mathbf{q}} = \nabla S(t,\mathbf{q}) \tag{0.52}$$

namely

$$\mathbf{p} = \nabla S(t,\mathbf{q})$$

Eqs. (0.46), (0.52) are the equations of a relativistic material point formulated in the Hamilton-Jacobi theory. These equations force our soliton to moves as a particle whose Hamiltonian is given by (0.49).

If $c \to \infty$, then

$$\begin{aligned} H(\mathbf{p}) &= \sqrt{(m_0c^2+Q)^2+c^2\mathbf{p}^2} = m_0c^2\sqrt{1+\frac{\mathbf{p}^2}{m_0^2c^2}+\frac{2Q}{m_0c^2}+\frac{Q^2}{m_0^2c^4}} \\ &\cong m_0c^2+\frac{\mathbf{p}^2}{2m_0}+Q. \end{aligned}$$

This Hamiltonian differs from the free particle Hamiltonian by the term Q. If the mass of this material point is large, namely, if

$$m_0 \gg \frac{Q}{v^2}$$

this term is negligible. Ignoring the lower order terms and the constant term m_0c^2, we get the Hamiltonian of a classical free particle $H(\mathbf{p}) = \frac{\mathbf{p}^2}{2m_0}$. However, if $Q \approx m_0v^2$, the term Q is relevant and the soliton follows the dynamics relative to the Hamiltonian $\frac{\mathbf{p}^2}{2m_0} + Q$. This case will be discussed in the next section considering directly equation (0.40) instead of the Hamiltonian (0.49).

Eq. (0.52) has been first obtained by De Broglie in the linear case (see [15]) and he called it *guidance equation.* It is the base of the "pilot wave theory". Since in his work, he considered only the linear case, the position of the particle $\mathbf{q}(t)$ was represented by a singularity of the solution of the Klein-Gordon equation (0.19). His linear theory encountered some difficulties, mainly due to the fact that the singularities of (0.19) carry infinite energy. De Broglie was very aware of this fact:

> Des considérasions sur lesquelles je rievendrai me conduisent aujord'hui à penser que le corpuscule doit être assimilé non pas à un véritable singularité punctuelle de ψ, mais à un très petite région singulière de l'espace où ψ prenderait une très grande valeur et obéirait à une équation non linéaire dont l'équation linéaire de la Mécanique ondulatoire ne serait qu'une forme approximative valable en dehor de la region singulière. L'idée que l'équation de propagation de ψ, contrairement à l'équation classique du Ψ, est en principe non linéaire m'apparaît même maintenant comme tout à fait essentielle. ([15], Ch. IX,1, page 95.).

But 40 years ago the methods of Nonlinear Analysis were not sufficiently developed to construct a mathematical model and the word *soliton* did not even exist. Now, we have proved that the intuitions of De Broglie were right, at least from the mathematical point of view.

0.4.3. The Schrödinger equation and Böhmian Mechanics

Now let us consider the nonrelativistic limit of equation (0.40). In this case it is interesting to consider a potential which depends also on the space variable x;

thus we replace $W(\psi)$ in Eq. (0.40) with

$$W(t,x,\psi)=\frac{1}{2}\left(\omega_0+\frac{V(t,x)}{\hbar}\right)^2\psi^2+W_0(\psi) \tag{0.53}$$

The constant $\hbar$ makes V to have the dimension of an energy. $V(t,x)$ can be thought as a deformation of the medium.

Using (0.53), Eq. (0.40) takes the form

$$\Box\psi+\left(\omega_0+\frac{V(t,x)}{\hbar}\right)^2\psi+W_0'(\psi)=0; \tag{0.54}$$

We look for solutions ψ having the following form

$$\psi(x,t)=\varphi(x,t)e^{S(x,t)/\hbar}=\left[u(x-\mathbf{q}(t))+\varphi_1(x,t)\right]\,e^{S(x,t)/\hbar} \tag{0.55}$$

where $u(x)e^{-i\omega_0 t}$ is a stationary solution of the equation

$$\Box\psi+\omega_0^2\psi+W_0'(\psi)=0,$$

namely u satisfies the following elliptic equation:

$$-c^2\Delta u+W_0'(u)=0 \tag{0.56}$$

We make the following assumptions:

- (i) $\varphi_1(x,t)$ has a very small amplitude compared to the amplitude of u; on the other hand the support of φ_1 is much larger than the support of u (the support of a field F is the space region where F is sensibly different from zero). From a technical point of view this assumption can be stated as follows:

  $$\|\varphi_1\|_{H^{1,\infty}}\ll\|u\|_{H^{1,\infty}}\,;\quad\|\varphi_1\|_{H^{1,1}}\gg\|u\|_{H^{1,1}}$$

  This assumption express the fact that (0.55) can be considered as a combination of a wave, $\varphi_1(x,t)\,e^{S(x,t)/\hbar}$, which is extended in space and a particle of position $\mathbf{q}(t)$, $u(x-\mathbf{q}(t))\,e^{S(x,t)/\hbar}$ which is very concentrated.

- (ii) the phase has the following form

  $$S(x,t)=-\hbar\omega_0 t+s(x,t), \tag{0.57}$$

  where $\frac{\partial s}{\partial t}$ and ∇s change very little in the support of u, namely

  $$\frac{\partial s}{\partial t}u\approx\hbar\omega u;\ \nabla u\approx\hbar\mathbf{k}u \tag{0.58}$$

  where ω and $\mathbf{k}$ do not depend on x, but might depend on time. Also we assume that $V(t,x)$ varies very little in space. Basically this assumption states that the soliton is spatially very small if compared with the other relevant quantities.

- (iii) $\frac{\partial^2 \varphi}{\partial t^2} \ll c^2$, $\frac{\partial s}{\partial t} \ll c$ and $W_0''(\psi) \ll c^2$: these inequalities are assumed since we are interested in the nonrelativistic limit.

Replacing ψ, given by (0.55), in Eqs. (0.54), we get

$$\hbar^2 \Box \varphi + \left[-\left(\frac{\partial S}{\partial t}\right)^2 + c^2 |\nabla S|^2 + (\hbar\omega_0 + V)^2 \right] \varphi + \hbar^2 W_0'(\varphi) = 0 \tag{0.59}$$

$$\frac{d}{dt}\left(\varphi^2 \frac{\partial S}{\partial t}\right) - c^2 \nabla \cdot (\varphi^2 \nabla S) = 0 \tag{0.60}$$

First we analyze Eq. (0.59). By (0.57), we get

$$\hbar^2 \Box \varphi + \left[2\hbar\omega_0 \frac{\partial s}{\partial t}^2 - \left(\frac{\partial s}{\partial t}\right)^2 + c^2 |\nabla s|^2 + 2\hbar\omega_0 V + V^2 \right] \varphi + \hbar^2 W_0'(\varphi) = 0,$$

Using (0.48) and dividing the equation by $2m_0c^2$, we get

$$\frac{\hbar^2}{2m_0c^2}\frac{\partial^2 \varphi}{\partial t^2} - \frac{\hbar^2 \Delta \varphi}{2m_0} + \left[\frac{\partial s}{\partial t} - \frac{\left(\frac{\partial s}{\partial t}\right)^2}{2m_0c^2} + \frac{|\nabla s|^2}{2m_0} + V + \frac{V^2}{2m_0c^2} \right] \varphi + \frac{\hbar^2 W_0'(\varphi)}{2m_0c^2} = 0;$$

by (iii), the terms $\frac{\hbar^2}{2m_0c^2}\frac{\partial^2 \varphi}{\partial t^2}$, $\frac{1}{2m_0c^2}\left(\frac{\partial s}{\partial t}\right)^2$ and $\frac{V^2}{2m_0c^2}$ are small with respect to the others terms and can be neglected. Rearranging the other terms, we get

$$-\frac{\hbar^2}{2m_0}\Delta \varphi + \left[\frac{\partial s}{\partial t} + \frac{|\nabla s|^2}{2m_0} + V \right] \varphi + \frac{\hbar^2 W_0'(\varphi)}{2m_0c^2} = 0$$

Next, we replace φ with $u + \varphi_1$and we use the Taylor expansion negletting the term $\frac{\hbar^2}{2m_0c^2} W_0''(u + \vartheta\varphi_1)\varphi_1$:

$$-\frac{\hbar^2}{2m_0}[\Delta u + \Delta \varphi_1] + \left[\frac{\partial s}{\partial t} + \frac{|\nabla s|^2}{2m_0} + V \right](u + \varphi_1) + \frac{\hbar^2 W_0'(u)}{2m_0c^2} = 0$$

By assumption (i), u and φ_1 do not "see" each other, then we get the following two equations:

$$-\frac{\hbar^2}{2m_0}\Delta \varphi_1 + \left[\frac{\partial s}{\partial t} + \frac{|\nabla s|^2}{2m_0} + V \right] \varphi_1 = 0 \tag{0.61}$$

$$\frac{\hbar^2}{2m_0}\left[-\Delta u + \frac{W_0'(u)}{c^2} \right] + \left[\frac{\partial s}{\partial t} + \frac{|\nabla s|^2}{2m_0} + V \right] u = 0 \tag{0.62}$$

If we compare this equation with Eq. (0.56), we get

$$\left[\frac{\partial s}{\partial t} + \frac{|\nabla s|^2}{2m_0} + V(t,x) \right] u = 0 \tag{0.63}$$

and by (0.61),

$$\frac{\Delta\varphi_1}{\varphi_1}u = 0. \tag{0.64}$$

By assumption (i), $\int \Delta\varphi_1 u\, dx \approx 0$. Then, to our level of approximation, we can assume that equation (0.64) and hence (0.62) is identically satisfied, at least in the region where $\varphi_1 > 0$.

Next we consider Eq. (0.60). By (0.57), and (0.48) it becomes

$$\frac{d}{dt}\left[(u+\varphi_1)^2\left(-m_0c^2 + \frac{\partial s}{\partial t}\right)\right] - c^2\nabla\cdot\left[(u+\varphi_1)^2\,\nabla s\right] = 0$$

and since $\frac{\partial s}{\partial t} \ll m_0c^2$, we can neglect this term; then

$$m_0\frac{d}{dt}(u+\varphi_1)^2 + \nabla\cdot\left[(u+\varphi_1)^2\,\nabla s\right] = 0$$

By assumption (i), $\int u\cdot\varphi_1 dx \approx 0$, and so φ_1 and u do not interact directly with each other; then the above equation splits in the following two:

$$m_0\frac{d}{dt}\varphi_1^2 + \nabla\cdot\left(\varphi_1^2\nabla s\right) = 0 \tag{0.65}$$

and

$$m_0\frac{d}{dt}u^2 + \nabla\cdot\left(u^2\nabla s\right) = 0 \tag{0.66}$$

Since $u = u(x - \mathbf{q}(t))$, (0.66) becomes

$$-m_0\nabla u^2\cdot\dot{\mathbf{q}} + \nabla u^2\cdot\nabla s + u^2\Delta s = 0. \tag{0.67}$$

then, by (0.58), the above equation can be written as follows

$$\nabla u^2\cdot(m_0\dot{\mathbf{q}} - \mathbf{k}) = 0$$

and is satisfied if $m_0\dot{\mathbf{q}} = \mathbf{k}$. Using again (0.58), we get the guidance equation:

$$m_0\dot{\mathbf{q}} = \nabla s(t,\mathbf{q}) \tag{0.68}$$

Now we state our conclusion. To any solution u of Eq. (0.56) we can associate a "particle" of mass $m_0 = \frac{\hbar\omega_0}{c^2}$. The dynamics of this particle and of the surrounding field can be deduced by the knowledge of the functions $\varphi_1(t,x)$, $s(t,x)$ and $\mathbf{q}(t)$. The functions $\varphi_1(t,x)$ and $s(t,x)$ can be determined by the equations (0.61) and (0.65), and the evolution of $\mathbf{q}(t)$ is determined by Eq. (0.68). Now, if we set

$$\psi_1(t,x) = \varphi_1(t,x)e^{is(t,x)/\hbar},$$

by equations (0.61) and (0.65), the function ψ_1 satisfies the Schrödinger equation

$$i\hbar\frac{\partial\psi_1}{\partial t} = -\frac{\hbar^2}{2m}\Delta\psi_1 + V(t,x) \tag{0.69}$$

and equation (0.68) can be written as follows

$$\dot{\mathbf{q}}(t) = \frac{\hbar}{m} \operatorname{Im} \frac{\nabla \psi_1}{\psi_1}; \tag{0.70}$$

The approximate dynamics of a soliton of the form (0.55) is uniquely determined by equations (0.69) and (0.70). The accuracy of this approximation depends on the inequalities in assumptions (i), (ii) and (iii). Essentially this approximation is good if the size of the particle is small with respect to the wavelength of φ_1, if its velocity is small compared to c and if the amplitude of φ_1 is small compared with u.

The Schrödinger equation (0.69) and the guidance equation (0.70) are the equations of Böhmian mechanics ([14], [18]). It is well known that a particle whose motion is governed by Eqs. (0.69) and (0.70) behaves as a quantum particle, at least in many situations.

Thus, as far as one particle is involved, our model presents all the feature of quantum mechanics. In particular it presents the intersection phenomena in the double slit experiment, the energy level of the hydrogen atom, the Heisenberg uncertainly principle etc. We refer to [14], or to [18] for the description and the interpretation of these phenomena with respect to Eqs. (0.69) and (0.70).

However, in our model, Eqs. (0.69) and (0.70) are only nonrelativistic approximations of Eq.(0.54); if we think of the actual solution ψ as composed by a wave and a soliton, there is an exchange of energy and momentum between them even if the time-average of this exchange is esentially null.

0.4.4. Topological invariants and quantum numbers

As we have mentioned in section 0.3.1 the existence of static stable solitons can be guaranteed by topological constraints which are due to the singular part of W.

We denote by Σ the singular set of W and by $\mathcal{M}$ the part of the internal parameter space where W is regular. The topological constrains are related to the topology of $\mathcal{M}$. More exactly we require that

$$\pi_3(\mathcal{M}) \neq 0$$

where π_3 denotes the third homotopy group.

For example if

$$W(\xi) = \frac{1}{2} \cdot \frac{|\xi|^2}{\prod_{j=1}^{r} |\xi - \xi_j|^6}, \quad \xi_j \in \mathbf{R}^4;$$

then

$$\Sigma = \{\xi_1, \ldots, \xi_r\};$$
$$\mathcal{M} = \mathbf{R}^4 \backslash \Sigma$$

and

$$\pi_3(\mathcal{M}) = \mathbf{Z}^r.$$

In this case the configuration space for equation (0.9) is given by

$$\Lambda = \mathcal{C}^2(\mathbf{R}^3, \mathcal{M}) = \{\psi \in \mathcal{C}^2(\mathbf{R}^3, \mathbf{R}^4) \mid \forall x \in \mathbf{R}^3 : \psi(x) \in \mathcal{M}\}.$$

To each function $\psi \in \Lambda$, we can associate a homotopy class $\psi^\# \in \pi_3(\mathcal{M})$. Indeed we consider a homeomorphism $\pi : S^3 \to \mathbf{R}^3 \cup \{\infty\}$ such that $\pi(*) = \infty$, where $*$ is the base point. Then we denote by $\psi^\#$ the homotopy class of $\psi \circ \pi$, that is

$$\psi^\# = [\psi \circ \pi] \in \pi_3(\mathcal{M}).$$

In this way Λ reflects the structure of $\pi_3(\mathcal{M})$. The connected components of Λ, which are called sectors, are identified by the homotopic invariant we have introduced. After fixing a set of generators $\{\alpha_1, ..., \alpha_r\}$ of $\pi_3(\mathcal{M})$, every sector of Λ is characterized by a r-uple of integer numbers. Clearly the same characterization holds for every configuration $\psi \in \Lambda$. We set

$$q(\psi) = (m_1, ..., m_r).$$

if the following relation holds:

$$\psi^\# = m_1\alpha_1 + ... + m_r\alpha_r.$$

If $\psi(t,x)$ is a regular solution of Eq.(0.9), it does not change sector. Thus the numbers $m_1, ..., m_r$ are discrete integrals of motion which need to be added to the continuous integrals discussed in section 0.2.4. For this reason, we will call them *quantum numbers*.

The picture we have is the following. Let

$$w(t,x) = \sum_{j=1}^{N} u_j\,(t,x)$$

where the $u_j\,(t,x)$ are solution of Eq. (0.9) of type (0.23). We assume that for $t = 0$, the barycenters of the u_j's are far from each other. Under this assumptions, it possible to prove that ([5],[4],[11])

$$q(w(0,\cdot)) = \sum_{j=1}^{N} q\left(u_j\,(0,\cdot)\right),$$

namely, the quantum number of our configuration $w(0,\cdot)$ is the sum of the quantum numbers of each soliton. Now, let $\psi(t,x)$ be a solution of (0.9) with the following initial conditions:

$$\begin{aligned} \psi(0,x) &= w(0,x) \\ \frac{\partial\psi}{\partial t}(0,x) &= \frac{\partial w}{\partial t}(0,x) \end{aligned}$$

Since Eq. (0.9) is not linear, $w(t,x)$ is not a solution, and hence $w(t,x) \neq \psi(t,x)$ for $t \neq 0$. However, since the barycenters of the u_j's are far from each other,.for small t, $w(t,x)$ is a good approximation of a solution and we can write

$$\psi(t,x) = w(t,x) + \psi_1(t,x)$$

where $\psi_1(t,x)$ is a small correction. This correction remains small as long as the nonlinear effects are small, namely as long as the baricenters of the u_j's remains far from each other (the u_j's dacay exponentally and far away from the barycenter they are essentially null). Eventually, the solitons will "collide" and the solution $\psi(t,x)$ will be very different from $w(t,x)$. However, if it is continuous, the numbers $q(w(0,t))$ are preserved. So $w(t,x)$ will assume a form of type

$$w(t,x) = \sum_{j=1}^{M} \tilde{u}_j\,(t,x) + \tilde{\psi}_1(t,x)$$

with

$$\sum_{j=1}^{M} q\left(\tilde{u}_j\,(t,x)\right) = \sum_{j=1}^{N} q\left(\varphi_j\,(0,t)\right).$$

The solitons $\tilde{u}_j$'s appear as new particles "created" by the disintegration of the u_j's: the similarity of the behavior of solitons and elementary particles is remarkable!

0.4.5. DISCUSSION OF THE MODEL

In the previous section, we have constructed a field model in an axiomatic way. We have assumed that the laws of physics are variational, and invariant for the Poincaré group. The surprising facts of the theory of relativity described in section 0.1.1 and some quantum phenomena can be deduced from these assumptions: they are theorems within a classical field theory. Now we want to discuss whether the above facts imply that the traditional conception of time as an absolute entity which flows independently of the observer must be abandoned.

From a strictly logical point of view, the relativity of time is not a consequence of assumption **A-1** and **A-2,** but of the Einstein convention of synchronization of clocks. Next we will discuss this convention and then we will show two other alternative points of view. Finally we will compare these points of view.

At this point, it is important to distinguish Theory of Relativity and Relativism as was done in [25]. The Theory of Relativity is a scientific theory which describes some phenomena (such as those mentioned in section 0.1.1) in a mathematically consistent way and which can be deduced by very general Assumption such as **A-1** and **A-2**. Relativism is a philosophical point of view which states the inexistence of absolute time. It is a logical consequence of (0.3) and it is consistent with the Theory of Relativity, but is not a strictly logical consequence of it.

0.4.6. The Einstein convention for synchronizing clocks

The Einstein convention for synchronizing clocks is justified by the following facts which are consequences of **A-2**:

- an observer in a spaceship moving with constant velocity cannot perform local experiments to detect wether he is moving or not.
- no signal can travel faster than c.

We briefly recall this convention. An observer wants to synchronize clocks in any two different points A and B of his reference frame S'. The frame S' is moving with velocity v with respect to an other reference frame S. This observer can use the following method: he puts an electronic device at the point C which is the middle point of the segment AB and makes C to emit an electromagnetic signal (such as a light ray). When this signal reaches the points A and B the clocks are synchronized.

Using this method the time t' measured in S' is given by.

$$t' = \frac{t - \frac{v}{c^2}x}{\sqrt{1 - \frac{v^2}{c^2}}}$$

where t denotes the time measured in the reference frame S. Note that the above equation is just the Lorentz transformation relative to time .

This point of view has a very big advantage: using the time t' and the relative space coordinates, the equations of Physics in the reference frame S' have the same form as in S (and as in any other inertial frame). This fact, is another way of saying that the equations of physics are invariant for the Poincaré group. In particular we have the following consequence:

the speed of light is the same in any inertial reference frame. (0.71)

But this is not an *empirical* fact; it is just a consequence of the method of synchronization of the clocks. It is not possible to define any speed unless you synchronize two clocks far away from each other. Notice that the experiment of Michelson and Morely proves that the *two ways* velocity of light is independent of the direction of the earth's motion. In fact, it is conceptually impossible to prove the constancy of the speed of light without having adopted a method to synchronize clocks far away from each other!

The basic point is the following: the use of the coordinates (x', t') in the reference frame S' can be considered as just a *technical trick* to make computations easier, or as an *objective fact* related to the very deep nature of time and space.

Today, the last point of view is accepted by the majority of physicists. However, this point of view presents both some philosophical and technical difficulties. Some of the unpleasant philosophical consequences have been discussed by Popper (cf. e.g., [22]): the universe is reduced to a "four-dimensional Parmenidean block in which change is a human illusion, or very nearly so". In this universe there is no room for evolution or for free will.

From the technical point of view there are also difficulties: we refer to a beautiful paper of Selleri [24] (see also [25]) in which an ideal experiment is presented in which the above point of view shows one of its very weak points.

0.4.7. A nineteenth century style description of the world

In this section we will present a different interpretations of assumption **A-1** and **A-2**. We try to give this interpretation in the style of the *naive materialism* of the nineteenth century forgetting the lesson of the *neopositivism* at the origin of today's interpretation of (0.3).

We start with two remarks.

The first remark consists in observing that an equation such as the (0.18) could be the equation of a particular elastic medium. Of course in this case c does not represent the velocity of light but the velocity of sound in that medium. The solitary waves propagating in that medium follow the laws of relativity theory once we interpret the mass as $\frac{momentum}{speed}$ as we did in section 0.3.4.

The second remark is tautological: if there are vibration, there is an entity which is vibrating. In the last century this entity was called *ether.*

Using these remarks, the picture which comes out is the following. The universe is not empty but is filled with a "substance" called ether whose vibrations are determined by two general principles: assumptions **A-1** and **A-2** which give Eq. (0.9) (or (0.18)) or similar equations such as the Maxwell equations if we consider vector fields equations.

This universe is filled with linear waves which correspond to the small amplitude solutions of (0.18) (and other vector equations) and they appear as *radiation.* Moreover there are large amplitude stable solutions of (0.18) which form solitary waves and that is what we call *matter.*

The variable t and x are the coordinates of time and space which are the time and the space given by our intuition. Of course, in this picture, it is necessary to define the absolute space and the absolute time relative to which the ether is at rest. But our old style physicist is clever and he knows how to detect absolute space and absolute time as it will be shown in the following section.

0.4.8. The absolute time

In eighteenth century the universe was thought of as a fixed entity. The only aspect of reality which was thought to evolve was human history. The laws of Nature were supposed to have been fixed once for all by the Creator and the aim of Science was to discover them. In particular the laws of physics were thought to take the form of differential equations. The phenomena are modelled by the solutions of these equations.

It is well known that a differential equation does not have a unique solution. In order to get uniqueness you need to add the initial conditions and in some cases the boundary conditions. Then, if the equation is well-posed, it provides a unique solution which describes the observed phenomena.

In this framework the idea is the following:

- the equations have been established by the Creator and the theoretical scientist should discover them;
- the initial conditions are prepared by the experimental scientist in the laboratory.

Clearly the Creator deserves more respect than an experimental physicist: the equations and not the initial conditions are the essence of science.

This was a nice picture, but became inadequate with the discovery of evolution. After the discovery of evolution in human history (during the Renaissance), evolution was also discovered in Science: the arrow of time in Physics (second law of thermodynamics), the evolution of life (nineteenth century), of earth, (modern Geology, nineteenth century) and of the Universe in its globality (Cosmology, twentieth century).

If you consider evolution, the important facts of Nature are not contained in the equations (the basic law of nature) but in the initial conditions and in the subsequent history. In the *history* of the universe, there are *accidents* which happened only once and which are not contained in the differential equations (e.g. the big bang, the formation of earth, of life, the appearing of the human beings, and lots of less dramatic events). These are called *frozen accidents:* they are called "frozen" since their consequences persist in hystory. The universe we live in cannot be "explained" or even only described by equations, but by equations *and* the frozen accidents which today must be considered as important as the basic laws. The occurrence of frozen accidents cannot be deduced by the basic equations. Equations can only tell us that during the evolution of a system there are *bifurcation points,* and that the choice between the alternatives depends on small *casual* fluctuations. Small fluctuations might change all the subsequent history. In this discussion it is not important to know whether the very deep nature of the universe (or/and of the basic equations) is deterministic or not. The important point is that history depends on small fluctuations: whether these fluctuations are *intrinsically* casual or *determined* by infinitesimal differences in the initial condition is irrelevant for our discussion and is a question more philosophical then scientific.

After this long discussion let us go back to our old-minded physicist who is looking for absolute time and absolute space. Since he takes **A-2** for granted, he knows that the basic equations are not sufficient to detect absolute space and time. But since he knows the importance of Cosmic Evolution, he realizes that absolute space and time must be considered as a property of the universe in its globality, including initial conditions and frozen accidents. And he has also an operative way to detect them:

- absolute time is detected by the absolute temperature of the universe;
- in order to check if you are at rest with respect to absolute space, you control the background radiation of the universe: if it appear homogeneous and isotropic then you are at rest.

Even if he would have lived in the past century before the discovery of the background radiation, he would have had a good method to detect absolute space: an observer is at rest with respect to absolute space if he is at rest with respect to the center of gravity of the matter of this galaxy (he did not know too much of other galaxies). And absolute time can be defined as the time measured by a clock at rest in absolute space. Note that these independent methods of defining absolute space and time give the same answer, and that this answer is consistent with the time of the evolution of life, stars, galaxies etc. Moreover, using the notion of absolute space, he discovers that the major part of the matter is almost at rest (comparing its velocity with c) and that the earth is moving around the Sun: he thinks that Galileo was right!

It seems that the Universe in its globality has completely ignored the principle of Relativity (or better, of Relativism). Even its geometry, according to modern cosmology, is not consistent with this principle. In fact the models of space-time consistent with the principle of Relativism are the Minkowsky space-time and the De Sitter space-time, but our universe is a Friedman space-time (see e.g.[20]). Moreover, space and time have a completely different status: the universe, in the large, is very homogeneous and isotropic in space, but not in time since the arrow of time and evolution appear at any scale.

Absolute time seems *to be real* and *has a very different nature than space.* These statements seem to be trivial, but they were trivial.... only before the twentieth century.

Anyhow, our old-minded physicist believes that the existence of absolute time (and space) is a *real fact* and he is not to picky about the fact that he cannot detect it if he is closed in a spaceship without windows which moves by inertia. Also he believes that ether is not moving in the absolute space. There is no reason to waste so much energy to get a moving ether and/or to have all the matter moving in the ether. May be he also tries to find some phenomena which cannot be described with equations satisfying **A-2** since for him **A-2** is only an empirical fact given by experience and not tied to a methaphysical belief. Maybe he would try to explain the results of Aspect (relative to the EPR type experiments) by using same equations which are not Lorentz invariant and which allow velocities higher than c. And of course, he would not have any problem with the De Broglie pilot-wave theory and its variants since it is consistent with his picture: of course this theory does not explain everything, but no theory in the world has ever explained everything. He would not feel any need to assume the strange facts described in the introduction which do not increase the number of phenomena which can be "explained".

Basically he is very happy with his obsolete and naive philosophy which claims that reality exists even if we do not look at it, that time is real, that the past cannot be changed, that nature can be understood, that logic is universal etc. Concluding, he thinks that the naive realism is a sound ground for Science and leaves the Philosophy to philosophers.

0.4.9. The Noneuclidean Geometry

Before continuing our analysis, it is necessary to discuss one of the major scientific discoveries of the nineteenth century which had a fundamental role for the Theory of Relativity and Quantum Mechanics and whose consequences in all Science are very difficult to be overestimated:

the discovery of Noneuclidean Geometries (1829) (0.72)

The first consequence of (0.72) was the rejection of two basic Kantian ideas:

- space and time as fundamental intuition of our mind,
- Euclidean geometry as a theory which is *synthetic* and *a priori*.

The *empiricist* point of view was enforced: every thing which we know about the world comes from experience. The fact that the sum of the angles of a triangle is 180° became an empirical fact; there is not any logical or *a priori* explanation for it. Euclidean Geometry which has been considered as one of the few sources of absolute truth, became a physical theory which might be *falsified*. Even if the concept of falsification is more recent (Popper), we must remember that Gauss himself made some measurement to check the empirical validity of Euclidean Geometry. If Euclidean Geometry might be false, *a fortiori*, Newtonian Mechanics might be false since it contains Euclidean Geometry as part of it. The first obstacle towards a new conception of space and time was removed.

The second big consequence of (0.72) was the new meaning to be given to the Axiomatic Method whose paradigmatic example is just Euclidean Geometry. Before (0.72) the meaning of the axiomatic method was the following: the Axioms are assertions whose truth is beyond any reasonable doubt; if they are true also their logical consequences are true since the validity of our logic was considered absolute. After (0.72) the meaning of the Axioms changed dramatically: they became arbitrary statements and the truth of their consequences became relative to the truth of the axioms. Also the laws of logic deduction, and the meaning of *truth* were questioned, but we will not discuss this point here. The consequences of this new point of view became very relevant. If, by changing the axioms it is possible to get many geometries and more in general many mathematics, then changing the axioms, we can get many physics. Of course, only "one" physics is true, but this is an empirical fact; it is not a question of logic. Then the emphasis was put on axioms (which in Physics usually assume the name of principles) and not on logical deduction.

The attitude of an old-style physicist was to *explain* the word, namely to *reduce* the observed phenomena to the laws of Newton (the Axioms of classical mechanics): this kind of paradigm is well-exemplified by the reduction of Thermodynamics to Statistical Mechanics, and hence to Newtonian Mechanics.

The new generations of physicists put more and more emphasis on principles. Usually Einstein is considered the beginner of the Physics of Principles. The Special Theory of Relativity is a good example for this point. The experiment of Michelson and Morley showed that velocity implies the contraction of lengths and the dilation of time. The attitude of an old-style physicist such as Lorentz was to explain these facts; it was known that the material bodies are kept together by electromagnetic forces which obey the Maxwell equations. Thus the explanation of these phenomena must be deduced by these equations. Today this explanation can be considered as an almost trivial mathematical theorem relative to *all* equations invariant for the Poincaré group. In fact in section 0.3, we have deduced these theorems in few pages. Thus an old-style physicist would have considered the contraction of lengths and the slowing of clocks as *real* facts occurring in the absolute space-time and due to the particular form of the equations which are involved in the phenomena.

The attitude of Einstein was quite different. He assumed (0.71) as a basic Principle (Axiom) and not as a consequence of (0.3) and **A-2**. In this way, he was able to construct a consistent and elegant theory whose consequences (the facts mentioned in section 0.1.1) have been confirmed by experience beyond any reasonable doubt. Using (0.71) as an axiom, the deduction of the basic facts are almost immediate and involve relatively simple mathematics. This new method, applied to several fields of physics, has been very efficient and gave to this science the possibility to develop very fast for the first half of the twentieth century. All the new phenomena appearing in the microscopic word were explained (or described?) assuming new principles without questioning too much their *meaning* and their *internal consistency.* Actually every discussion about the *meaning* of the new principles was considered *methaphysical* (neopositivism). The birth of Quantum Mechanics represents a good example of this new attitude. Lots of new principles have been introduced as very basic laws of nature: the *Bohr Complementary Principle*, the *Heisenberg Indeterminacy Principle*, the *Pauli Exclusion Principle*, the *correpondency principle* etc. Even the textbook presentations of quantum mechanics reflect this new attitude: axioms are given in a very abstract and general form without any explanation of their possible meaning, their consistency and their mutual relations: nature works this way and there is nothing we can do. These very abstract and universal axioms are justified *a posteriori* by the experimental results in limited and particular cases. Basically Quantum

Mechanics is now a theory in which we do not know what we are talking about: nobody really knows what the ψ-function is and if you ask ten physicists what it is, you hear at least nine different and incompatible answers and most of them, using different motivations, will explain to you that this question does not even make sense. QM is a beautiful theory which provides very powerful tools which make the right predictions on atomic phenomena, but it is imprisoned in a very poor philosophical framework. It does not have the internal consistency of Classical Mechanics or of the Theory of Relativity. This quotation from Roger Penrose [21] makes this point quite clear:

> Quantum theory, it may be said, has two things in its favour and only one against it. First, it agrees with all the experiments. Second, it is a theory of astonishing and profound mathematical beauty. The only thing to be said against the theory is that it makes absolutely no sense.

Now, let us go back to the discussion on the cultural impact of (0.72). We have analyzed the new attitudes toward space and time and toward the Axiomatic method. Now we want to analyze a third point not less important than the other two: the concept of model.

0.4.10. The notion of model

Since the birth of Riemannian Geometry (Riemann, 1854), it was clear that the surfaces in Euclidean space represent models for Noneuclidean Geometry. This point of view became clear with the work of Beltrami (1868). The development of Logic and Set Theory also made clear another fact. If A and B are two sufficiently powerful theories, then we can construct a model A' of A in B but also a model B' of B in A. For example, Goedel (1931) has shown that the Arithmetic of natural number is powerful enough to represent in itself any other formalized Mathematical theory.

This general fact can also be applied to physical theories. This point of view appears very clearly if you consider the theory of General Relativity. The trajectories of material bodies and of light rays can be represented as geodesics of a Lorentzian metric in a 4-dimensional Lorentz manifold, namely they are critical pints of the following functional

$$\int g_{ij}(x)\frac{dx^i}{ds}\frac{dx^j}{ds}ds; \quad x = (x_0, ..., x_3) \in \mathbf{R}^4 \tag{0.73}$$

where the $g_{ij}(x)$ are determined by the matter distribution via the Einstein field equations.

In the standard paradigm, the $g_{ij}(x)$ are considered as the component of a metric tensor. This implies that the universe is not Euclidean and not even flat. However, it is also possible to claim that the universe is Euclidean and that the

$g_{ij}(x)$ describe the "deformations" of the ether due to the presence of matter. A presentation of the Theory of General Relativity based on the Lagrangian of type (0.73) can be found in [6] or in [8]. In [8] there are also some epistemological considerations and they will not repeated here.

In general, we can say that Newtonian Mechanics can be "represented" in a relativistic world, but also that the Relativity Theory can be "represented" in a Newtonian world.

Everybody knows that Newtonian Mechanics is contained in the Theory of Special Relativity as limit case for low velocities. In sections 0.3, 0.4.7 and 0.4.8, we have shown that also the converse is possible. A model for the theory of relativity can be constructed from Eq (0.18). This equation can be interpreted in a very mechanistic way as the equation of a lattice of little rigid balls (ether particles) which interact with each other by an elastic force which presents a small nonlinear term.

0.5. CONCLUSIONS

We have seen that the fundamental facts of the Theory of Relativity are consistent with at least two different models of our physical universe which have been described in sections 0.4.6 and 0.4.7:

- *The relativistic principle*

which is based on the Einstein convention of synchronizing clocks and on the spirit of the neopositivism dominating at the beginning of twentieth century.

- *The absolute time principle*

which is based on our intuition of time and space and which seems more adequate to describe a world in which time determines evolution and is more in the spirit of the physics of nineteenth century.

Moreover, after the discussion of section 0.4.10 an other point of view is possible

- *The Model Principle*

which states that there are many possible models which describe reality and it does not make sense to ask which one is true.

Officially, the majority of the community of physicists declares to accept the first point of view, however more or less explicitly also the other two points of view are present in a rather strange mixture. For example, people say that the "big bang" is an event which

$$\textit{occurred about 15 billions years ago.} \tag{0.74}$$

People do not say that the big bang

$$\textit{occurred about 15 bilions years ago if time is measured by a clock at rest with respect to the center of gravity of our galaxy.} \quad (0.75)$$

Analogously, people say that, on a large scale,

$$\textit{space is homogeneous,} \quad (0.76)$$

and not that

$$\textit{space is homogeneous with respect to the reference frame of our galaxy.} \quad (0.77)$$

Maybe (0.74) and (0.76) are nothing else but a short way of saying (0.75) and (0.77). Nevertheless most people seem to believe that (0.74) and (0.76) are properties of the universe and not properties of our reference frame.

The same attitude can be observed with respect to the third point of view. The influence of neopositivism is still strong, and people say that we can only build models of an unknowable reality: only models and their prediction power is the object of Science and all the rest is meaningless. On the other hand the basic principles of Relativity and Quantum Mechanics are accepted in a rather naive way and considered as universal truths which can be applied to every aspect of reality: from elementary particles to Zen philosophy.

A good criterion to distinguish between two empirically equivalent theories is simplicity: the best theory is the one which does not introduce artificial complications (Occam razor) and assumptions *ad hoc*. Most scientists claim to agree with this point of view. Since it is very difficult to give a measure of *simplicity* everybody claims to believe in the *simplest* theory.

For example, consider the theory of Relativity and the two descriptions given in sections 0.4.6 and 0.4.7. It is possible to claim that the Relativistic Principle simplifies the description of the world since all the equations assume the same form in every reference frame. But also you can affirm that this principle leads to a very complicated theory since it denies the basic intuitions of space and time. You can say that the Relativistic Principle simplifies the base of our knowledge since it introduces **A-2** in a natural way, but you can also say that you are introducing an assumption *ad hoc* which "explains" **A-2**, but which is not consistent with many other aspects of reality such as the arrow of time and evolution.

Quantum Mechanics represent an even better example. The orthodox interpretation of QM is based on the principle that reality in itself is meaningless and it not possible to give any description of the world without introducing an observer: the equations of QM describe only the results of laboratory experiments, but say very little about the nature of the object we are talking about.

Nevertheless there are other interpretations of the equations of QM which are consistent with the experiments. The simplest one, probably, is the De Broglie-Böhm pilot wave theory which we have briefly described in section 0.4.2 and 0.4.3.

You can claim that the pilot wave theory is more complicated since it introduces two different classes of objects, namely waves and particles and hence an extra equation, namely Eq. (0.70). On the other hand you can say that it is simpler since it is very complicated to think of an object which at the same time is both a wave and a particle (*Complementary Principle)* and is not located in any region of space-time. You can say that equation (0.70) is an *ad hoc* assumption but you can also say that the "*wave collapse*" is an *ad hoc* assumption.

Of course a discussion on the many interpretations of Quantum Mechanics deserves a much deeper analysis than few words. Nonetheless these remarks are sufficient to get to the point and we are ready to state the conclusions of this paper:

- The big theories of this century (Relativity and QM) are formulated with a very heavy baggage of philosophical beliefs which are not necessary consequences of the form of the equations (which are the main objects tested by experiments). In fact, we have shown that at least the theory of relativity can be put in a completely different framework and interpreted with a very different philosophy.

- A reasonable consequence of the above statement is that the philosophical part of these theories reflect the European cultural environment of the first half of this century.

- Once a theory has been accepted by the leading part of the scientific community, its philosophical extra-baggage becomes a sort of dogma: the theory can be falsified, but the philosophical dogma cannot. If you do not accept the philosophical dogma you are a "heretic" and that is all (see e.g. [2], page 160).

- Most people are not aware of this heavy philosophical baggage. Moreover, there is a naive optimism which makes people believe that the philosophical framework is improving in proportion to the quantity of knowledge and technical applications. The *weltanschauung* of Einstein has improved that of Poincaré and Lorentz, and the *weltanschauung* of Bohr and Heisenberg has improved that of Einstein, De Broglie and Schrödinger. The winners are always right.

- The philosophical baggage influences the directions of the research: only the works within the mainstream are considered relevant. Thus most people try to solve the new problems within the philosophy of these theories. When some logical inconsistency appears, a lot of ingenuity and work is done to solve these contradictions. For example the distinction of Bohr between a microscopic and a macroscopic world is in conflict with the applications of QM to cosmology or to macrosystem such as the atomic laser. In order to solve these inconsistencies people propose awkward theories such as the Multiuniverse Theory (Everett).

To conclude this paper, I will present my point of view. My impression is that the physics of the second half of twentieth century was a prisoner in a cage constructed by the philosophy of the beginning of the century. I think that some effort should be made to open this cage and I hope that the future generations will succeed in this task.

ACKNOWLEDGMENTS

The ideas exposed in this paper have been developed during long discussions which I had with some friends and colleagues. So they deserve special thanks. A particular acknowledgment is due to Gianfausto Dell'Antonio, Leone Fronzoni, Dino Fortunato, Luigi Galgani and Franco Selleri. Of course, the ideas exposed here are fully my responsibility; most of my friends disagree in part or completely with my points of view.

References

[1] A. Abbondandolo, V.Benci, *Solitary waves and Bohmian mechanics*, to appear on Proceedings of the National Academy of Sciences.

[2] J. Bell, *Speakable and unspeakable in quantum mechanics*, Cambridge University Press, Cambridge, (1987).

[3] V. Benci, *Quantum phenomena in a classical model*, Foundations of Physics, Foundations of Physics, 29, 1-29(1999)

[4] V. Benci, D. Fortunato, *Solitons and Particles*, to appear on Proc. "International Conference on Nonlinear Differential Equations and Applications" Tata Inst. of Fundamental research, Bangalore, India (August 1996).

[5] V. Benci, D. Fortunato, *Solitons and relativistic dynamics*. in *Calculus of Variations and Partial differential equations*, .G.Buttazzo, A.Marino and M.K.V. Murty editors, Springer (1999), 285-326.

[6] V. Benci, D. Fortunato, *A new variational principle for the fundamental equations of classical physics*, Foundations of Physics, 28 (1998), 333-352

[7] V. Benci, D. Fortunato, *On the existence of the impossible pilot wave*, in Calculus of variationd and differential equations, Ioffe, Reich and Shafrir ed., Chapman & Hall/CRC, (1998), 1-20.

[8] V. BENCI, D. FORTUNATO, *On the equality between gravitational and inertial mass*, to appear on *Geometry and Physics*, P.Freguglia, G Turchetti ed..

[9] V. BENCI, D. FORTUNATO, A. MASIELLO, L. PISANI, *Solitons and electromagnetic field*, Math. Zeit.,232, (1999), 73-102.

[10] V. BENCI, D. FORTUNATO, L. PISANI, *Soliton like solution of a Lorentz invariant equation in dimension 3*, Reviews in Mathematical Physics, **3** (1998), 315-344.

[11] V. BENCI, D. FORTUNATO, L. PISANI, *Remarks on topological solitons*, Topological methods in nonlinear Analysis, **7** (1996), 349-367.

[12] V. BENCI, F. GIANNONI, P. PICCIONE, *A variational problem for manifold valued functionals*, Adv. In Diff. Eq., 5, (2000), 369-400.

[13] H. BERESTYCKI, P.L. LIONS, *Nonlinear Scalar Field Equations, I - Existence of a Ground State*, Arch. Rat. Mech. Anal., **82** (4) (1983), 313-345.

[14] D. BÖHM, B. HILEY, *The Undivided Universe*, Routledge, London (1994).

[15] L. DE BROGLIE, *Un tentative d'interprétation causale et non linéaire de la Mécanique ondulatoire: la théorie de la double solution,* Gauthier-Villars, Paris, 1958. English traslation: Non-linear wave mechanics, a causal interpretation, Elsevier, Amsterdam, 1960.

[16] C.H. DERRICK, *Comments on Nonlinear Wave Equations as Model for Elementary Particles*, Jour. Math. Phys. 5 (1964), 1252-1254.

[17] I.M. GELFAND, S.V. FOMIN, *Calculus of Variations*, Prentice-Hall, Englewood Cliffs, N.J. 1963.

[18] P. R. HOLLAND, *The Quantm Theory of motion,* Cambridge University-Press, Cambridge, 1993, (last reprint 1997).

[19] LANDAU L.,LIFCHITZ E., *Théorie du Champ*, Editions Mir, Moscow, 1966.

[20] O' NEILL, *Semiriemannian Geometry with application to Relativity*, Academic Press Inc., New York-London, 1983.

[21] PENROSE R., *Review of "The Quanum World" by J.C. Polkinghorne,* The Times Higher Education Supplement, March 23, 1984.

[22] POPPER K. R., *Unended quest, an intellectual biography*, Fontana-Collins, Glasgow, 1976, p.129.

[23] F. SELLERI, Found. Phys. 26, (1996), p. 641

[24] F. SELLERI, *On a Phisical and Mathematical Discontinuity in Relatvity Theory,* in Open questions in Relativistic Physics, F. Selleri ed., Apeiron, Montreal, (1998), 69-80.

[25] F. SELLERI, *La fisica del Novecento. Per un bilancio critico,* Progedit snc, Bari (1999).

[26] J. SHATAH, *Stable Standing waves of Nonlinear Klein-Gordon Equations,* Comm. Math. Phys., 91, (1983, 313-327.

PATTERN INDUCED BY PARAMETER MODULATION IN SPATIOTEMPORAL CHAOS

Leone Fronzoni

Dipartimento di Fisica dell'Università di Pisa
CISSC and INFN
Pisa, Italia
fronzoni@df.unipi.it

ABSTRACT

We consider the effect of a parametric modulation in a system that shows electrohydrodynamic instability with spatiotemporal chaos. This instability appears when an electric field is applied to a sample of Liquid Crystals confined between two conducting plates. The pattern degenerates in to spatiotemporal chaos when the field exceeds a threshold. A suitable modulation of the electric field induces a new ordered pattern in spite of the presence of chaos. A possible interpretation of this phenomena is discussed on the basis of transient travelling waves (TTW) present in the chaotic regime.

INTRODUCTION

In the last twenty years there has been increasing interest within the scientific community in the science of chaos. Although the foundations of this discipline were laid at the end of the last century by Henry Poincare' [1], only recently, with the development of computers, has chaos dynamics attracted the interest of scientists in every field of science. In principle, it is possible to distinguish two lines in the development of this subject: the study of systems with few degree of freedom and those with a large number of degree of freedom, like hydrodynamic systems. Although low dimensional chaos has been widely studied, much is still to be explored about systems that involve large number of degrees of freedom. Of particular interest is the spatiotemporal dynamics that anticipates full turbulence. These processes are defined as "weak turbulence" and are characterized by the presence of order within chaotic dynamics. In many cases, it would be interesting to modify the state of the system by means of external perturbations, in some cases, it might be particularly useful to

The Application of Mathematics to the Sciences of Nature, edited by Cerrai *et al.*
Kluwer Academic/Plenum Publishers, New York, 2002.

suppress the chaotic dynamics. The name "the control of chaos" has been given to this topic and there are many examples in the literature; two significant examples are shown in [2]. We have "control of chaos" when the perturbation disappears precisely when the chaos is suppressed. This means that the perturbations must act as small negative feedback in the system. This feedback could be additive or multiplicative, depending on whether the reaction is obtained directly in the variable of interest [3] or in the control parameter [4]. In the latter case we have "control of chaos by means of parametric perturbations". However we also consider it interesting, to modify the state of a system by removing chaotic instability by means of an external parametric modulation and not necessarily with feedback and small amplitude. In the extended system, the stabilization of chaos makes it possible to obtain new patterns from the chaotic regime. In other words, the stabilization of chaotic orbits by means of external perturbation generates new spatial configurations. Here we present an example in which a parameter modulation in an electrohydrodynamic instability produces a special pattern in a sample of Nematic Liquid Crystal.

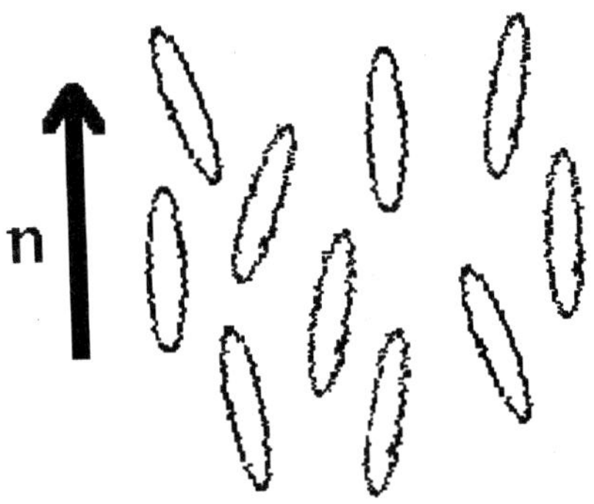

Figure 1. A draft representation of molecules of Nematic Liquid Crystals. The vector " n" means the average orientation.

LIQUID CRYSTAL AND PATTERN FORMATION

Nematic Liquid Crystals are an isotropic liquid constituted of elongated molecules. These molecules are free to move in space, but maintain a mean orientation, as shown in Figure 1. A vector "n" defines the mean orientation. For an overview on Liquid Crystals see[5]. Because of the anisotropy of electric conductivity, the interaction of the nematic with an electric field produces electroconvection. The nematic is placed between two glass plates. A suitable rubbing of the internal wall induces a molecular orientation parallel to the surface of the plates. A metallic substratum on the surfaces plates makes it possible to apply an electric field to the Liquid Crystal. When an electric field is present in the sample, the anisotropy of the conductivity generates spatial charges. These charges interact with the field and move the fluid generating electroconvection. This instability arises when a threshold value of the electric field is reached, named threshold of the Williams' pattern (WD) [6]. The pattern is organized in rod-like configurations with rotating fluid in the form of adjacent ``rolls''. Defects appear as a local loss or gain of rolls and they break the regularity of the structure. An alternating field is usually used to avoid the pollution of the sample by injected ions. There is a particular frequency value (Lifsht point), under which the rolls assume a zig-zag configuration, as shown in Figure 2. This image is the top view of a sample obtained by means of a light polarized microscope. The dark and light lines correspond the two different molecular orientation induced by the fluid velocity. The angle α indicates the angle between the two orientations of the rolls and the circle shows a defect. The number of defects increases as a function of the amplitude of the external electric field and move erratically in the structure [7]. Analysis of the statistics of the motion, in these conditions, indicates the presence of aleatory and regular behaviour depending on the region of the cell. This regime is defined as weak turbulence. At a second threshold the pattern is completely destroyed and the sample appears turbulent. In this case, we can say that the dynamic is fully chaotic. An image of a Liquid Crystal in the presence of full chaos is shown in Figure 8, late B.

TRAVELLING WAVES AND CHAOS

Spectral analysis of the light coming from the sample in the presence of a weak chaos regime shows peaks that indicate the presence of auto-oscillations within the sample. These auto-oscillations are generated by the coupling of the twist with the bend deformation of the nematic orientation [8]. Figure 3 shows a schematic image of these local deformations. The model that describes the dynamics of these two extended mode is not trivial and the oscillations assume the form of travelling waves. Generally speaking, from mathematical point of view, these auto-oscillations are ``limit cycle`` generated by a Hopf bifurcation. For instance, a simple example that generates a limit cycle is given by two differential equation of the form

$$\frac{dz}{dt} = y, \tag{1}$$

$$\frac{dy}{dt} = -z + (1 - z^2)\frac{dz}{dt}, \tag{2}$$

these equations describe the dynamics of an oscillator with nonlinear friction, the Van der Pol oscillator [9]: Of course, in our systems, partial differential equations with Laplacian terms are needs to describe an extended field.

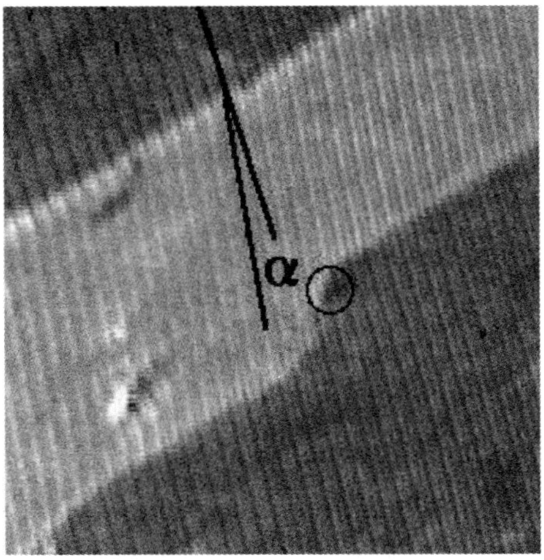

Figure 2. Top view of the cell obtained by means of a polarized microscope. The rolls are highlighted by the black and white straight lines. A defect is present in the sample as indicated by the circle.a denotes the values of the angle between the two " rool " orientations corresponding to the zig and zag states.

COUPLED COMPLEX GINZBURG-LANDAU EQUATIONS (CCGL)

A possible model that reproduces the phenomenology of electroconvection with bend and twist deformation, is based on two coupled partial differential equations of two complex variables A and B. These are the phase and the amplitude of fluid velocity of modulated rolls in the form of "zig-zag" configuration [10] characterized by two modes: the zip-mode, with the axes of the rolls that form an angle β_A with respect to normal axes of the molecular orientation, and the zag-mode with $\beta_B = - \beta_A$.
The field of the velocity U(x,y) in the plane of the cell reads :

$$U = U_0\, e^{iq_0 x}[A e^{iq_0 y} + B e^{iq_0 y}] + cc, \tag{3}$$

where the amplitudes A e B are governed by the following equations:

$$\frac{\partial A}{\partial t} = \varepsilon A + [\xi_x^2\partial_x^2 + \xi_y^2\partial_y^2 + 2\xi_x\xi_y a\partial_x\partial_y]A - (g_1|A|^2 + g_2|B|^2)A, \tag{4}$$

$$\frac{\partial B}{\partial t} = \varepsilon B + [\xi_x^2\partial_x^2 + \xi_y^2\partial_y^2 + 2\xi_x\xi_y a\partial_x\partial_y]B - (g_1|B|^2 + g_2|A|^2)B, \tag{5}$$

the control parameter e is a function of the voltage applied. This model provides a qualitative description near to the threshold of the pattern and is able to interpret the travelling wave and the weak chaos mediated by the nucleation of defects[11]. Unfortunately, it has so far been difficult, to obtain a precise comparison of the experimental data with the theoretical predictions. However, a theory developed by Treiber M. and Kramer L.[11] links the values of the coefficients of Eqs.(3)(4) with physical quantities such a sviscosity, conductivity and dielectricity of the Liquid Crystals.

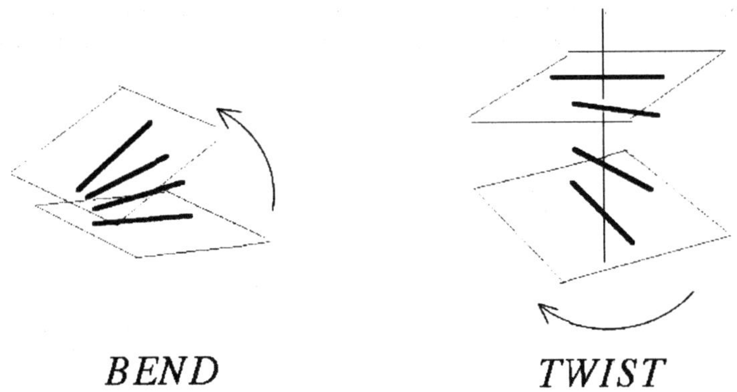

Figure 3. A draft scheme of the ``bend" and ``twist" deformation.

WAVE SYNCHRONIZATION

Let us consider a non-linear system with auto-oscillations driven by an external forcing. In some conditions, one can observe that the oscillator synchronizes with the forcing. The synchronization depends on the amplitude of the forcing and on the difference in internal and external frequencies. A simple model is provided by Strogaz-Morillo[12] which can be summarized as follows: consider two clocks coupled with, respectively, ω_1 and ω_2 , the characteristic frequencies. Their phases are given by

$$\dot{\phi}_1 = \omega_1 - f(\phi_1 - \phi_2), \tag{6}$$

$$\dot{\phi}_2 = \omega_2 + f(\phi_1 - \phi_2), \tag{7}$$

where $f(\phi_1 - \phi_2)$ represents the coupling. This function is a periodic function of the phase difference $\theta = \phi_1 - \phi_2$, which in the first order could be assumed to be of the form $f(\theta) = K \sin \beta$, where K means the intensity of the coupling. Subtracting Eq.(6) from Eq.(7) we get

$$\theta = \Delta\omega - 2K\sin(\theta), \tag{8}$$

where $\Delta\omega$ denotes the difference in frequency $\omega_1 - \omega_2$. A stationary solution of the Eq.(7) is possible if

$$\Delta\omega < 2K. \tag{9}$$

This means that, with a fixed frequency, the synchroniszation is obtained for an intensity of the coupling above a threshold. This threshold increases linearly as a function of Dw. Consider that, in spatial systems, instead of simple oscillation there are travelling and standing waves. In our case, we refer to eqs.(3)(4), that for some parameters, give solutions of the form of travelling waves according to a generalization of the Hopf bifurcation. An external forcing could synchronize these waves by means of the mechanism previously described. In the chaotic regimes these waves do not disappear completely, but could survive for a limited period of time. The persistence of these oscillations in the presence of achaotic regime suggested the use of synchronization to obtain ordered states from the chaotic dynamics. Here we present the results of an experiment using Liquid Crystal in which an external parametric modulation induces a pattern by means a synchronization of standing localized waves.

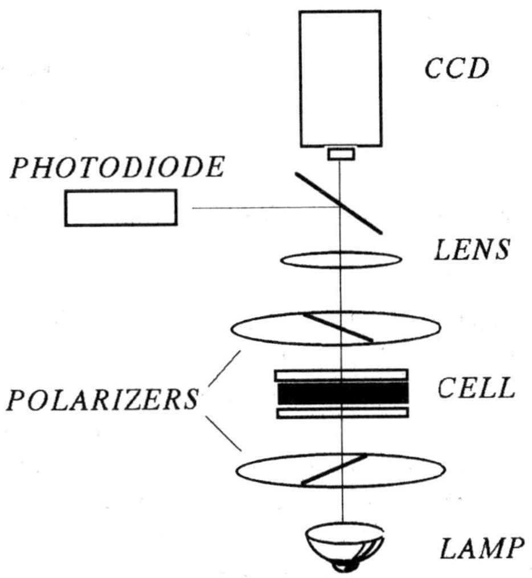

Figure 4. Scheme of the experimental apparatus.

THE EXPERIMENT

We use nematic Liquid Crystal MBBA (methyl-butyl aniline) with planar orientation, in cells 15 microns of thickness and with a square surface with sides of one centimeter. We apply an oscillating voltage of 1 KHz with the temperature of the sample at 25 ^{0}C. The threshold of the Williams' pattern is 15 Volts RMS. The cell temperature is stabilized at 1/10 of a degree. The images of the cells are recorded by means of a CCD camera and a photodiode pick up the light coming from the sample. The output of the photodiode is sent to a digital oscilloscope DATA GENERAL 6100 (DATAPRECISION INC), which computes the Fast Fourier Transform (FFT) of the data. A scheme of the apparatus is shown in Figure 4. At a voltage of 18 V_{RMS} the sample presents a chaotic dynamics. Figure 5 shows the FFT of the photodiode signal in that condition. The decaying shape of the spectrum indicates the presence of chaos while the peaks indicate the presence of oscillations or travelling wave. An increase in the voltage applied between the plates induces a shift in the position of the peak foward higher of frequency values, Figure 6.

Using a second generator we modulate the amplitude of the main generator. In this way a parametric modulation is applied to the system and in order to obtain the synchronization of the standing waves it is important to detune the frequency modulation around twice the frequency of the peak in the spectrum. The results are

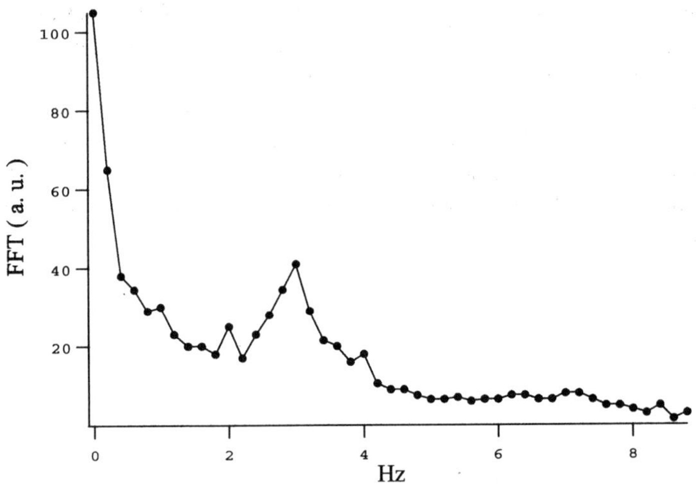

Figure 5. Fast Fourier Transform (FFT) of the photodiode signal in the presence of full chaos.

shown in Figure 7, in which the modulation amplitude at which synchronization is plotted against the modulation frequency. The experimental points indicated in the figure are obtained considering the values at which ordered regions occur in the sample. An example of these localized patterns is shown in Figure 8, late A. The new structure begin to appear in a small region of the sample and subsequently the chaos is suppressed over a wide region of the surface. The extension of this pattern depends on the difference between the frequency modulation and the double of the Hopf frequency. Of course, the pattern is synchronized with the modulation so that, in a generic region, zig-state alternates with the zag-state triggered by the external signal .

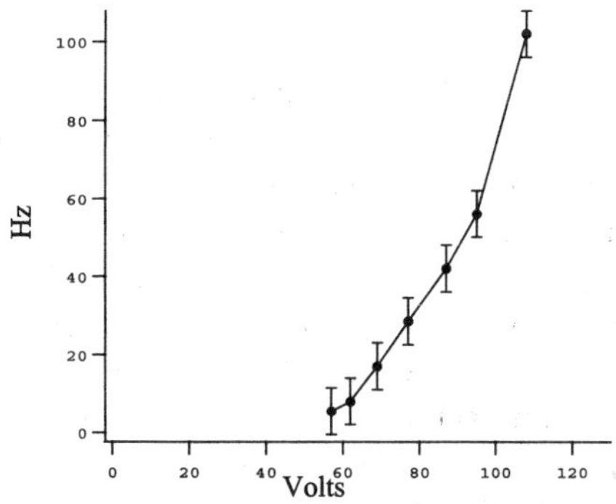

Figure 6. Frequency corresponding to the maximum of the peak in the FFT as a function of the voltage applied to the sample.

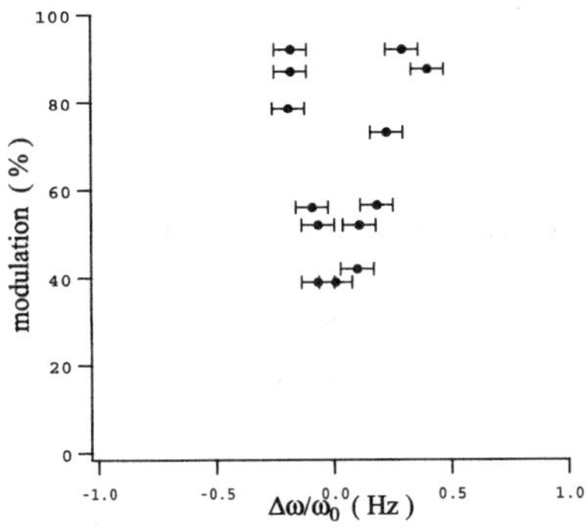

Figure 7. Experimental data corresponding to amplitude modulation versus frequency difference $\Delta\omega$. In the region over the data the cell presents synchronized patterns.

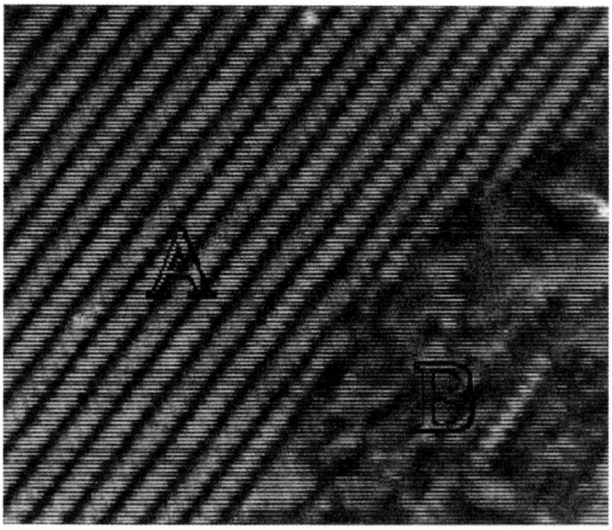

Figure 8. CCD image of the new pattern in the presence of synchronization.

CONCLUSIONS

In this paper we have shown that, using electrohydrodynamic instability in the planar cells of Liquid Crystals is an effective method for studying spatiotemporal chaos. In particular, we present an experiment, in which a new pattern is obtained from the turbulence state by means of a suitable modulation of the electric field. This phenomenon can be explained by a process of synchronisation between the amplitude modulation and travelling wave. The use of Liquid Crystals makes it possible to obtain clear evidence of the phenomenon and the peculiarity of the planar configuration, in the presence of modulation, generates a specific pattern. This peculiarity depends on the fixed initial orientation of the molecules parallel to the plates. This condition is not present in isotropic systems, such as Benard [13] and Faraday [14] instabilities and the pattern assumes any possible direction on the plane. We intend to study isotropic systems in such a way as to extend and generalize our results. It is not necessary to use isotropic liquid. Indeed, nematic Liquid Crystal in cells with molecular orientation perpendicular to the plates, is sufficient to generate patterns with any possible orientation on the plane of the cell.

REFERENCES

1. H.Poincare', *The Foundation of Science: Science and Method*, English. Translation, *The Science Press*, N.Y. (1921).
2. E. Otto, C. Grebogi and J. A. York. *Controlling Chaos Phis. Rev. Lett.*64 ,p.1196, (1990). R. Lima, . and M. Pettini, *Suppression of chaos by resonant parametric perturbation ,Phys.Rev.* A41, pp 726:733 (1990).
3. K. Pyragas, *Continuous control of chaos by self-controlling feedback, Phys.Lett. A* 170, pp. 412:428 (1992).
4. L. Fronzoni, M. Giocondo, and M.Pettini. *Experimental evidence of suppression of chaos by resonant parametric perturbation, Phys.Rev. A 43*, pp. 6483-6487 (1990).
5. P.G. de Gennes, *The Physics of Liquid Crystals, CALENDRON PRESS* , Oxford (1974).
6. E. Dubois-Violette, G Durand, E.Guyon, P. Manneville and P. Pieranski, *Instabilities in Nematic Liquid Crystals In Liquid Crystals Suppl.* 14, 147.(1978).
7. E..Bodenschatz, W Pesch and L. Kramer, *Structure and Dynamics of Dislocations in Anisotropic Pattern-forming System, Physica D* 32, 135.(1998).
8. S. Kai , M. Arakoa, H.Yamazaki and K. Hirakawa, *Phy. Soc. of Japan.* vol 46, 2, p. 401 (1979).
9. B. van der Pol, *On the relaxation Oscillation, Phil. Mag.*2, p.978 (1926).
10. R. Ribotta , A. Joets, and L. Lei, *Oblique Roll Instability in Electroconvective Anisotropic Fluid, Phys. Rev. Lett.* 56, 1595 (1986).
11. M. Triber, and L. Kramer, *Coupled Complex Ginzburg-Landau Equations for the Weak Electrolite Model of Electroconvetion, Phys. Rev. E*, 58, 1973 (1998).
12. S.H. Strogatz, and R.E. Morillo. *Phase Loking and Critical Phenomena in Lattices of Coupled Nonlinear Oscillators with Random Intrisic Frequencies. Physica D 31*, p.143:168 (1988).
13. H. Benard, *Les Tourbillons Cellulaires dans une Nappe Liquide Transportant de la Chaleur par Convection en Regime Permanent, Ann. Chim. Phys.7* (Ser.23),62 (1900).
14 . M. Faraday, *On the forms and states assumed by fluids in contact with vibrating-elastic surfaces, Philos. Trans. R., Sc. London* 52, 319 (1831).

CHAOS AND ORBIT COMPLEXITY

Stefano Galatolo
Dipartimento di Matematica
Università di Pisa
Pisa, Italia

Abstract

In this paper we discuss the relation between entropy, initial data sensitivity and complexity of the orbits in a dynamical system. We expose some classical result holding in the compact case and some new result for non compact dynamical systems where constructivity plays an important role.

INTRODUCTION

When we think of a chaotic system there are two things that we expect to happen:

- the behavior of the system is unpredictable and complex to be described
- small differences on initial conditions leads to big differences in the evolution of the system.

A rigorous measure of initial conditions sensitivity is not difficult to be formalized, it leads to the definition of Lyapunov exponents[1] or to the more general Brin-Katok[2] local entropy.

A measure of the complexity of the behavior of a system is less simple to be formalized, in the case of dynamical systems it was given by Brudno[3]. A set of strings is associated by a certain construction to the orbit of a point and then the complexity of the orbit is defined by the Kolmogorov complexity (second section) of the associated strings. The complexity of an orbit is then a measure of the amount of information that is necessary to describe the orbit.

If the dynamics is ergodic and lies on a compact space it can be proved that the initial conditions sensitivity is almost everywhere equal to orbit complexity.

This implies that in the compact case orbit complexity and instability are both faces of the same coin. This motivates in the case of compact dynamical systems the general Ford's claim[4] that orbit complexity was a synonym of chaos.

The Application of Mathematics to the Sciences of Nature, edited by Cerrai *et al.*
Kluwer Academic/Plenum Publishers, New York, 2002.

If the space is not compact the relation between sensitivity and complexity will wanish, mainly for two reasons. The first is: Brudno's definition is not suitable for the non compact case, it gives infinite complexity to all the divergent orbits, even if the dynamic is very simple and non chaotic; the second reason is: however there are dynamical system acting over spaces that are not compact, having complex orbits and no sensitivity on initial conditions (see example 17). Such dynamical systems will be possible if we use non constructive maps: maps that cannot be defined using finite information.

In this exposition we present a definition of orbit complexity which is (in our opinion) suitable for the non compact case. This definition depends on the choice of some computable structure on the space where the dynamics lies. Computable structures allows the definition of constructive maps. To support our opinion we state results showing that if we replace compactness with constructivity we can have again a relation between chaos and orbit complexity. In the compact case our definition of orbit complexity gives the same value as the Brudno's one. In the case where the space is not compact,under some constructivity hypotesys our definition gives a value that is not greater than the topological entropy of the system.

In the next section we give a short introduction to Kolmogorov complexity.

In the third section we define Brin-Katok local entropy, which we use as a rigorous measure of initial data sensitivity.

In the fourth section we outline our approach to constructivity. We define computable structures and constructive maps.

In the last section we define orbit complexity with respect to a computable structure and discuss its links with initial data sensitivity when the space is compact. We consider the case when the space is not compact and state the results showing the link between constructivity,orbit complexity and entropy.

Almost all results are presented without proofs. For the proof of the new results we refer to[5].

ALGORITHMIC INFORMATION THEORY

Here we give the basic definitions of the concepts coming from algorithmic information theory that will be used in the following sections, a detailed exposition can be found in[6].

Let us consider the function $c : \Sigma \to \mathbf{N}$ from the set $\Sigma = \{0,1\}^*$ of finite (possibly empty) binary strings and the set $\mathbf{N}$ of natural numbers

$$\emptyset \to 0, \quad 0 \to 1, \quad 1 \to 2, \quad 00 \to 3, \quad 01 \to 4, \quad 10 \to 5, ...$$

This function allows us to interpret natural numbers as strings and converse when it is needed. If s is a string we define $|s|$ as the lenght of s, we remark that $|s| \leq log(c(s)) + 1$. [1]

[1] In this paper all the logarithms are in base two.

Let us consider a turing machine C. By the notation $C(p) = s$ we mean that C starting with input p (the program) stops with output s (C defines a partial recursive function $C : \Sigma \to \Sigma$). If the input gives a never ending computation the output (the value of the recursive function) is not defined.

The following definition due to Kolmogorov and Chaitin is the basis of the algorithmic information theory. The definition will be not used directily in the paper but its understanding idea is used in the definition of orbit complexity.

Definition 0.1. *The Kolmogorov complexity or algorithmic information entropy of a string s given C is the length of the smallest program p giving s as the output:*

$$K_C(s) = \min_{C(p)=s} |p| \tag{0.1}$$

if s is not a possible output for the computer C then $K_C(s) = \infty$.

Notations. If s is a string with $|s| = n$ we denote by $s\hat{}$ the string $s(0)s(0)s(1)s(1)...s(n-1)s(n-1)01$. If $a = a(1)...a(n)$ and $b = b(1)...b(m)$ are strings then ab will be the string $a(1)...a(n)b(1)...b(m)$. If a and b are strings then $a\hat{}b$ will be an encoding of the couple (a, b) as a single string.

An universal Turing machine is a general purpose computer that can simulate any other Turing machine if an appropriate input is given. There is a recursive enumeration $A_1, A_2...$ of all the Turing machines.

Definition 0.2. *A Turing machine $\mathcal{U}$ is said to be universal if for all $m \in \mathbf{N}$ and $p \in \Sigma$ then $\mathcal{U}(c^{-1}(m)\hat{}p) = A_m(p)$.*

It can be proved that an universal Turing machine exists.

Definition 0.3. *A Turing machine F is said to be* asimptotically optimal *if for each Turing machine C and each binary string s we have $K_F(s) \leq K_C(s) + c$ where the constant c depends on C and not on s.*

The following proposition can be proved from the definitions

Proposition 0.4. *If $\mathcal{U}$ is an universal Turing machine then $\mathcal{U}$ is asimptotically optimal.*

If we chose an universal Turing machine the complexity of a string will be defined independently of the given Turing machine up to a constant. In the next sections we consider the value of fractions like $\frac{K(s)}{|s|}$, when the lenght of s goes to infinity the value of the constant will becomes irrelevant. For the rest of the paper we will suppose that an universal Turing machine $\mathcal{U}(p)$ is chosen once forever.

LOCAL ENTROPY

We recall the most basic definitions of ergodic theory, and the definition of local entropy[1]. Let (X, d) be a metric space, we suppose that X is also equiped with a probability measure μ defined on the Borel σ-algebra of X. Let T be a measure preserving homeomorphism of (X, μ), that is, a measurable map wich is an homeomorphism on X and such that for each measurable set A we have $\mu(T^{-1}(A)) = \mu(A)$.

A set A is said to be invariant for T if $\mu(T^{-1}(A)\triangle A) = 0$. The measure μ is said to be ergodic for the map T if for each invariant set A either $\mu(A) = 0$ or $\mu(A) = 1$.

If $x \in X$ the Brin-Katok local entropy $h(T, x)$ is a measure of initial data sensitivity of the dynamical system (X, μ, T) at the point x. Local entropy will be a sort of generalization of the Lyapunov exponents, for its definition it is not needed a differential structure but only a metric and a measure.

We start defining the set

$$B_\epsilon(T, n, x) = \{y : d(T^j(x), T^j(y)) \leq \epsilon, 0 \geq j \geq n\}$$

$B_\epsilon(T, n, x)$ is the set of points "following" x at a distance less than ϵ for at least n steps.

Clearly the more the orbit is sensible to initial conditions the set $B_\epsilon(T, n, x)$ is smaller. Moreover if n increases the measure of $B_\epsilon(T, n, x)$ decreases. We expect that if a system is chaotic the rate of decreasing will be exponential, for this reason we consider

$$h_\epsilon(T, x) = limsup\Big(-\tfrac{1}{n}\log\big(\mu(B_\epsilon(T, n, x))\big)\Big).$$

$h_\epsilon(T, x)$ is a monotone function with respect to ϵ, then finally we define the local entropy of x as:

$$h(T, x) = \lim_{\epsilon \to 0} h_\epsilon(T, x).$$

We remark that this function is invariant ($f(T(x)) = f(x)$ a.e.). It follows easily from the definitions that if (X, μ, T) is ergodic an invariant function must be almost everywhere equal to a constant. That is $h(T, x) = h_\mu$ almost everywhere. Brin and Katok[2] proved that this constant is equal to the value of the metric entropy of (X, μ, T). We will not give here the definition of this important concept, the reader could see[7,1] for its definition and its applications. Instead we give the definition of topological entropy for the non compact case that will be used in the following sections. Topological entropy is greater than local entropy ($h(T, x) \leq h_{top}(T)$ a.e.) and it is another important measure of the chaoticity of a dynamical system.

We recall the definition of topological entropy for the non compact case[7]. If $x, y \in X$ let us say that x, y are (n, ϵ) separated if $d(T^k x, T^k y) > \epsilon$ for some $k \in \{0, ..., n-1\}$, if $d(T^k x, T^k y) \leq \epsilon$ for each $k \in \{0, ..., n-1\}$ then x, y are said to be (k, ϵ) near. A set $E \subset X$ is called (n, ϵ) separated if $x, y \in E$ and $x \neq y$ implies x, y are (n, ϵ) separated.

For each compact $K \subset X$ we define

$$s_K(n, \epsilon) = max\{card(E) : E \subset K \; is \; (n, \epsilon) - separated\},$$

$$h_K(T, \epsilon) = \underset{n \to \infty}{limsup} \frac{\log s_K(n, \epsilon)}{n}$$

$$h_K(T) = \lim_{\epsilon \to 0} h_K(T, \epsilon)$$

and finally

$$h_{top}(T) = \sup_K h_K(T)$$

where k ranges over all the compact subsets of X.

COMPUTABLE STRUCTURES

We introduce the notion of computable structure on a metric space. A computable structure on a metric space is a class of dense immersions of the space of finite strings Σ in a metric space (X, d). Computable structures allow us to interpret finite strings as points of the metric space. An interpretation will be computable if the distance between two points in its image is computable with arbitrary precision by some algorithm.

Let us consider (X, d), a separable metric space and Σ is the set of all finite binary strings:

Definition 0.5. *A computable interpretation function on (X, d) is a function $I : \Sigma \to X$ such that $I(\Sigma)$ is dense in X and there is a total recursive function $D : \Sigma \times \Sigma \times \mathbf{N} \to \mathbf{Q}$ such that $\forall s_1, s_2 \in \Sigma, n \in \mathbf{N}$:*

$$|d(I(s_1), I(s_2)) - D(s_1, s_2, n)| \leq \tfrac{1}{2^n}.$$

Definition 0.6. *Let I_1 and I_2 be two interpretations in (X, d); we say that I_1 and I_2 are equivalent if a total recursive function $D^* : \Sigma \times \Sigma \times \mathbf{N} \to \mathbf{Q}$ exists, such that $\forall s_1, s_2 \in \Sigma, n \in \mathbf{N}$:*

$$|d(I_1(s_1), I_2(s_2)) - D^*(s_1, s_2, n)| \leq \tfrac{1}{2^n}.$$

The relation defined by definition 6 is an equivalence relation[5], this motivates the following definition:

Definition 0.7. *A computable structure $\mathcal{I}$ on X is an equivalence class of computable interpretations in X.*

For example if $X = \mathbf{R}$ we consider the interpretation $I : \Sigma \to \mathbf{R}$ defined in the following way: if $s = s(1)...s(n) \in \Sigma$ then $I(s) = \sum_{1 \leq i \leq n} s(i)2^{[n/2]-i}$. I is a computable interpretation. This interpretation is equivalent to all the effective ways to interpretate a string as a rational. The computable structure containing I will be indicated as the standard or natural computable structure over the reals. It is not difficult to imagine how this definition can be extended to $\mathbf{R}^n$.

If $r = r(1)r(2)...$ is an infinite string such that $lim \frac{K_U(r(1)...r(n))}{n} = 1$ then the interpretation I_r defined as $I_r(s) = I(s) + \sum r(i)2^{-i}$ will be computable but not equivalent to I. I and I_r belongs to different computable structures.

Many concrete metric spaces used in analysis or in geometry have a natural choice of a computable structure. In separable function spaces there are structures containing interpretations of strings as polynomials with rational coefficients or step functions, and so on. For example, let us consider $C[0,1]$ with the *sup* norm. This space contains the polynomials with rational coefficients as a dense subset. It is possible to codify such polynomials with strings in a way that the *sup* of each one of them is computable (by an algorithm) up to arbitrary precision. This constructs a computable interpretation function in $C[0,1]$. In a similar way we can use step function to construct a computable interpretation in $L^2[0,1]$.

We give a definition of *morphism* of metric spaces with computable structure, a morphism is heuristically a computable function between metric spaces.

Definition 0.8. *If $(X, d, \mathcal{I})$ and $(Y, d', \mathcal{J})$ are spaces with a computable structure; a function $\Psi : X \to Y$ is said to be a morphism of computable structures if Ψ is uniformly continuous and for each pair $I \in \mathcal{I}, J \in \mathcal{J}$ there exists a total recursive function $D^* : \Sigma \times \Sigma \times \mathbf{N} \to \mathbf{Q}$, such that $\forall s_1, s_2 \in \Sigma, n \in \mathbf{N}$:*

$$|d'(\Psi(I(s_1)), J(s_2)) - D^*(s_1, s_2, n)| \leq \frac{1}{2^n}.$$

We remark that Ψ is not required to have dense image and then $\Psi(I((*))$ is not necessarily an interpretation function equivalent to J.

We remark as an example of the properties of the morphisms that if a map $\Psi : X \to Y$ is a morphism then given a point $x \in I(\Sigma) \subset X$ it is possible to find a point $y \in J(\Sigma) \subset Y$ as near as we want to $\Psi(x)$ the procedure is simple: if $x = I(s)$ and we want to find a point $y = J(z_0)$ such that $d'(\Psi(I(s)), y) \leq 2^{-m}$ then we calculate $D^*(s, z, m+2)$ for each $z \in \Sigma$ until we find z_0 such that $D^*(s, z_0, m+2) < 2^{-m-1}$, clearly $y = J(z_0)$ is such that $d'(\Psi(x), y) \leq 2^{-m}$. The existence of such a z_0 is assured by the density of J in Y.

The definition of a computable structure allow the definition of *constructive* maps. This definition is similar to the definition of Grzegorczyk/Lacombe for maps $\mathbf{R}^n \to \mathbf{R}^n$ [8].

Definition 0.9. *A function $T : (X, \mathcal{I}) \to (Y, \mathcal{J})$ between spaces with computable structure $(X, \mathcal{I}), (Y, \mathcal{J})$ is said to be constructive if T is a morphism between the computable structures and it is effectively uniformly continuous ,i.e. there is a recursive function $f : \mathbf{N} \to \mathbf{N}$ such that for all $x, y \in X$ $d(x, y) < 2^{-f(n)}$ implies $d(T(x), T(y)) < 2^{-n}$*

ORBIT COMPLEXITY

Let us consider a topological dynamical system (X, T) over a separable metric space (X, d). We now give our definition of orbit complexity with respect to a computable structure. If the space is compact and has a computable structure this definition will be equivalent to the original Brudno's one that is defined without the use of computable structures. This implies that in the compact case the choice of the computable structure is irrelevant. While in the non compact case Brudno's definition gives infinite complexity to each divergent orbit[9] making non difference between divergent-simple and divergent-chaotic ones, the use of a computable structure allows us to give a definition that (in our opinion) can see such a difference. We also present some example and theorems to support our opinion and to show the importance of constructivity in the non compact case. We finally remark that even if the physical reality where not constructive the models used to describe such a reality must be to allow calculations. For this reasons we think that in general, the assumption of constructivity is not unnatural.

We want to define real numbers estimating the amount of information needed to describe the orbit of a point x at a given accuracy λ. The first step is to consider a computable structure $\mathcal{I}$ on X and a computable interpretation $I \in \mathcal{I}$. Then we approximate the first n steps of the orbit of x with points in the image of I in a way that the information needed to describe this approximating orbit is as small as possible. This quantity of information is used to define the complexity of an orbit.

We want to interpret the output of a calculation as a finite sequence in X. For this purpose we consider a total recursive surjective $\mathcal{Q} : \Sigma \to \Sigma^*$ where Σ^* is the set of finite sequences in Σ.We also choose an universal Turing machine $\mathcal{U}$.

Composing I, $\mathcal{Q}$ and $\mathcal{U}$ we can interpretate the output of a calculation as a finite sequence in X: for each program p we define $U(p) \in X^*$ (the set of finite sequences in X) as

$$U(p) = I(\mathcal{Q}(\mathcal{U}(p)))$$

where I is extended in the obvious way to a map from the space Σ^* to X^*. $U_i(p) \in X$ is defined as the $i-$th point of $U(p)$. If $x_1, x_2, x_3, ..., x_n$ are points of X and λ is a real number we define the information contained in $x_1, ..., x_n$ with accuracy λ and interpretation I as:

$$\mathcal{E}^I_\lambda(x_1, ..., x_n) = \min \left\{ |p| \ s.t. \ \underset{i}{max} \left(d(U_i(p), x_i)\right) < 2^{-\lambda} \right\}.$$

In other words $\mathcal{E}^I_\lambda(x_1, ..., x_n)$ is the length of the smallest program that can approximate the points $x_1, ..., x_n$ with its output. Now we let the number of steps go to infinity and we consider the average quantity of information that is necessary to describe each step of the orbit (with accuracy λ).

$$\underline{S}_\lambda(T, I, X, x) = \underset{m\to\infty}{minlim} \frac{\mathcal{E}_\lambda(x, T(x), T^2(x), ..., T^m(x))}{m},$$

$$\overline{S}_\lambda(T, I, X, x) = \underset{m\to\infty}{maxlim} \frac{\mathcal{E}_\lambda(x, T(x), T^2(x), ..., T^m(x))}{m}.$$

The functions $\overline{S}_\lambda, \underline{S}_\lambda$ are non decreasing with respect to λ. We remark that the value of $\overline{S}_\lambda$ and $\underline{S}_\lambda$ does not depend on the choice of the universal computer $\mathcal{U}$ and of the function $\mathcal{Q}$. If we change these choices the values of $\mathcal{E}^I_\lambda$ will not change up to a constant which vanishes to the limit. Since $\overline{S}_\lambda$ and $\underline{S}_\lambda$ are non decreasing the limits for $\lambda \to \infty$ are defined.

Definition 0.10. *The upper and lower orbit complexity of x with respect to the interpretation I are defined as:*

$$\overline{S}(T, I, X, x) = \lim_{\lambda\to\infty} \overline{S}_\lambda(T, I, X, x)$$

$$\underline{S}(T, I, X, x) = \lim_{\lambda\to\infty} \underline{S}_\lambda(T, I, X, x)$$

Proposition 0.11. *If the interpretation function I_1 is equivalent to I_2 then $\forall x \in X$*

$$\underline{S}(T, I_1, x) = \underline{S}(T, I_2, x) \ , \ \overline{S}(T, I_1, x) = \overline{S}(T, I_2, x).$$

The last proposition allows the following definition.

Definition 0.12. *If X is a metric space with computable structure $\mathcal{I}$ and $I \in \mathcal{I}$. The orbit complexity of x with respect to the computable structure $\mathcal{I}$ is defined as*

$$\underline{S}(T, \mathcal{I}, x) = \underline{S}(T, I, x)$$

$$\overline{S}(T, \mathcal{I}, x) = \overline{S}(T, I, x).$$

If the metric space is compact the choice of the computable structure is not important as the following theorem states[5].

Theorem 0.13. *If X is compact and $\mathcal{I}$ and $\mathcal{J}$ are different computable structures then for all $x \in X$*

$$\underline{\mathcal{S}}(T,\mathcal{I},x) = \underline{\mathcal{S}}(T,\mathcal{J},x)$$

$$\overline{\mathcal{S}}(T,\mathcal{I},x) = \overline{\mathcal{S}}(T,\mathcal{J},x).$$

If X is a metric space with a computable structure the orbit complexity will be invariant under maps which are morphisms of the computable structures.

Theorem 0.14. *If (X,d,T), (Y,d',T') are topological dynamical systems over metric spaces with computable structures $\mathcal{I},\mathcal{J}$ and f is a morphism between $(X,d,\mathcal{I})$ and $(Y,d',\mathcal{J})$ such that the following diagram*

$$\begin{array}{ccccc} & & f & & \\ & X & \rightarrow & Y & \\ T & \downarrow & & \downarrow & T' \\ & X & \rightarrow & Y & \\ & & f & & \end{array}$$

commutes,if $x \in X$ and $y = f(x) \in Y$ then

$$\overline{\mathcal{S}}(x) \geq \overline{\mathcal{S}}(y).$$

$$\underline{\mathcal{S}}(x) \geq \underline{\mathcal{S}}(y)$$

Example 0.15. *Periodic points are 0-complexity points. If x is a $m-$periodic point for T then $\mathcal{S}_\lambda(x) = 0$ for all λ and I. We have*

$$\lim_{n\to\infty} \frac{\mathcal{E}_\lambda(x,T(x),T^2(x),...,T^n(x))}{n} = 0$$

because given a program p_1 such that $\max_i (d(U_i(p_1),x_i)) < 2^{-\lambda}$ for $i \in \{1,...,m\}$ (p_i approximates the orbit of x for the first m steps) it is possible to write a program p_2 that repeats p_1 n times and approximates the orbit of x for nm steps. Then $|p_2| = |p_1| + log(n) + constant$, this proves the assertion.

Another trivial example is the topological dynamical system over the real line given by $(\mathbf{R},T)$ where T is a translation. This is an evidently non chaotic dynamical system where the orbits are divergent and the orbit complexity (with respect to the natural computable structure of $\mathbf{R}$) of all points is 0. This can be proved directly but also follows from theorem 20 (in the case where the translation is constructive).

When a dynamical system has positive entropy and exponential sensitivity to initial conditions the information required to approximate an orbit for n steps increases linearly with respect to n. Heuristically this fact has to be expected because in such systems the "loss of information" is linear with respect to the time. In one of his papers[6] about orbit complexity Ford said the following:"exponential sensitivity is precisely the point at with, in order to maintain constant calculational accuracy, we must input about as much information as we get out".

The very interesting fact is that in the compact case even the converse is true: ergodic systems with complex orbits must have exponential sensitivity. It can be rigorously proved that the rate of algorithmic information that is required to approximate an orbit (per unit of time) is exactly equal to the local entropy (for almost all points with respect to the invariant measure).

Theorem 0.16. *(Brudno) If X is metric,compact, with a computable structure $\mathcal{I}$ and a probability measure μ, T is an ergodic measure preserving transformation. Then*

$$\underline{S}(x) = \overline{S}(x) = h(T, x), \mu - a.e.$$

We remark that the function $\overline{S}(x)$ is defined without the use of the measure μ, while the value of $h(x, T)$ depends on the choice of the ergodic measure. Topological dynamical systems can have an infinite family of ergodic measures (with different entropy). It it quite surprising that the choice of an ergodic measure sets automatically the set of total measure to be the set with the appropriate orbit complexity.

The relation between entropy and orbit complexity does not exist in the non compact case if we do not assume constructivity. The next example shows a non constructive map, having zero topological entropy and no sensitive dependence on initial conditions for which almost each orbit is algorithmically complex.

Example 0.17. *Let us consider the system (X, T) (figure 1). X is a set of squares in the real plane:*

$$X = \{(x, y) \in \mathbf{R}^2 | -\frac{1}{2} < y < \frac{1}{2}, x = 2n + t, n \in \mathbf{Z}, 0 < t < 1\}$$

it is a metric space with the metric induced by the Euclidean metric in $\mathbf{R}^2$. There is also a natural choice of a computable structure "induced" by the standard computable structure of $\mathbf{R}^2$

Let $a_i, i \in \mathbf{N}$ be an infinite string, such that $\lim_{n\to\infty} \frac{K_U(a(1)...a(n))}{n} = 1$. T is a continuous function $X \to X$ defined as follows.

$$T(x, y) = (x + 1, y * -1^{(a_{|[x/2]|})})$$

where [ ] denotes the integer part. The application T translates the point (x,y) on the coordinate x and flips the y coordinate according to the value of a_i corresponding to the absolute value of the integer part of $x/2$ (to each square it is associated a digit of a).

The map T is not constructive. The complexity of the orbit of all points (x,y) with $|y| > 0$ is 1 because the knowledge of the position of $T^k(x,y)$ with accuracy $2^{-\lambda} < |y|$ implies the knowledge of the k-th digit of a_i.

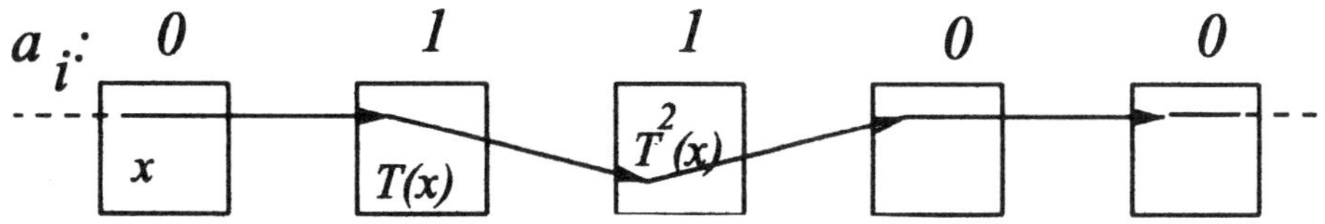

Figure 1 A non compact, 0-entropy dinamical system where all orbits have hight complexity

We now state a result which helps to show how constructivity can recover the link between chaos and orbit complexity in the non compact case: if (X,T) is a "constructive " topological dynamical system the complexity of "much" of the orbits will be not greater than the topological entropy of the system.

A set is recursively totally bounded if for each n there is an algorithmic way to find a finite cover made of balls with radius 2^{-n}. This is done by two functions: the first outputs how many balls are in the cover, and the second outputs the centers of the balls in the cover.

Definition 0.18. *A set A is said to be recursively totally bounded if there are total recursive functions $N(n) : \mathbf{N} \to \mathbf{N}, S(n,i) : \mathbf{N} \times \mathbf{N} \to \Sigma$ such that for each n $A \subset \bigcup_{i \in \{0,\ldots,N(n)\}} B(x_i^n, 2^{-n})$ where $x_i^n = I(s_i^n)$ and $s_i^n = S(n,i)$.*

Definition 0.19. *A set is said to be recursively compact if it is closed and recursively totally bounded.*

Theorem 0.20. *If (X,T) is a topological dynamical system on a metric space with a computable structure $\mathcal{I}$, T is constructive, $h(T)$ is the topological entropy of T, if $x \in K$ recursively compact then*

$$\overline{S}(x,T,\mathcal{I}) \leq h(T)$$

The condition that x is contained in a recursively compact set may look like technical and unnatural, but is automatically verified in many interesting examples.

Example 0.21. *Suppose that M is a topological manifold and a metric space with a computable structure $\mathcal{I}$. In other words M is a manifold where the distance is computable. We also consider $\mathbf{R}^n$ with the standard computable structure $\mathcal{J}$. Suppose that B is an open ball of $\mathbf{R}^n$ with rational center and rational radius. B is a metric space with a computable structure induced by the standard computable structure of $\mathbf{R}^n$.*

Definition 0.22. *The computable structure of M,$\mathcal{I}$ is said to be locally compatible at $x \in X$ with the standard one on $\mathbf{R}^n$ if there is an homeomorphism $\phi : B \to \phi(B)$ of b on a neighborhood of x and ϕ is a constructive map.*

Proposition 0.23. *If the computable structure of $M$is locally compatible with the standard one on $\mathbf{R}^n$ at x then x is contained in a recursive compact.*

Proof. *It is easy to see that B is recursively totally bounded. Let us consider a recursive cover of B with centers x_i^n (and radius 2^{-n}). The map ϕ is recursively uniformly continuous and there is a recursive function f such that $d(x, x') < 2^{-f(m)}$ implies $d(\phi(x), \phi(x')) < 2^{-m}$. To have a recursive cover of $\phi(B)$ with radius 2^{-m} let us consider a cover of B with centers $x_i^{-f(m+1)}$ and radius $2^{-f(m+1)}$. The points $\phi(x_i^{-f(m+1)})$ are centers of a cover of radius $2^{-(m+1)}$ on $\phi(B)$. Since ϕ is a morphism, then there is an algorithm that is able to find strings s_i^m corresponding to points $y_i^m = I(s_i^m)$ as near as we want to the points $\phi(x_i^{f(m+1)})$ (see section 4).Finding points on $\phi(B)$ at a distance less than 2^{-m} from the $\phi(x_i^{-f(m+1)})$ then we have a recursive function giving a recursive cover with radius 2^{-m} of $\phi(B)$.*□

Example 0.24. *If $X = C[0,1]$ (equipped with the sup metric) and $f \in X$ is recursively uniformly continuous then f is contained in a recursive compact.*

Proof. *We show that $x \in K$ where K is a recursive compact, we show how to construct for each m a set of points which is a 2^{-m} net. The construction is similar to the proof of the Ascoli-Arzela theorem*[10]*. Since the map f is recursively uniformly continuous there is a recursive function g such that $d(x, x') < 2^{-g(m)}$ implies $d(f(x), f(x')) < 2^{-m}$. If $M \in \mathbf{Z} > ||f||$ then the graph of the function f is contained in the rectangle $[0,1] \times [-M, M]$. The compact K will be the set of all continuous functions $\phi : [0,1] \to [-M, M]$ satisfying the same continuity relation as $f : d(x, x') < 2^{-g(m)}$ implies $d(f(x), f(x')) < 2^{-m}$. Let us consider a partition $y_1, ..., y_l$ of the segment $[-M, M]$ into segments of length not greater than $\frac{2^{-m}}{5}$. Let also consider a partition $x_1, ..., x_j$ of $[a, b]$ into segments of length not greater than $2^{-g(m)}$. It is possible to find such partitions by an algorithm. Next we consider as centers of the recursive cover of radius 2^{-m} all the piecewise linear functions ϕ_i which are continuous on $[0,1]$, linear inside each interval $[x_i, x_j]$ and at the points x_i:$\phi_h(x_i) = y_j$ for some y_j. This set can be constructed by an algorithm and is a 2^{-m} cover of K.*□

Example 17 shows that if the system is constructive then hight orbit complexity will not imply initial data sensitivity or hight entropy, theorem 20 and the last examples shows that if the system is constructive the implication comes back true. Replacing compactness with constructivity we recover a link between chaos and orbit complexity even in the case where the space X is not compact.

REFERENCES

1. Mane R. *Ergodic theory and differentiable dynamics.* Springer-Verlag, Berlin-New York, 1987.
2. Brin M., Katok A. *On local entropy* in Geometric dynamics (vol 1007,Lecture notes in Matematics)(Springer,Berlin 1983).
3. Brudno A.A. *Entropy and the complexity of the trajectories of a dyn amical system* Trans. Moscow Math. Soc. **2** 127-151 (1983)
4. Ford J. *Directions in classical chaos* in Directions in chaos, Vol. 1, 1–16, World Sci. Publishing, Singapore, 1987.
5. Galatolo S. *Orbit Complexity by Computable Structures* Preprint of Dipartimento di Matematica, Universitá di Pisa.
6. Zvorkin A.K., Levin L.A. *The complexity of finite objects and the algorithmic theoretic foundations of the notion of information and randomness* Russ. Math. Surv. 25 (1970)
7. Petersen K. *Ergodic Theory* (Cambrigdge University Press,Cambridge 1983)
8. M.B. Pour-El and J.I. Richards *Computability in Analysis and Phisics* (Springer Berlin,1988).
9. Batterman R.; White H. *Chaos and algorithmic complexity* Found. Phys. 26 (1996) no. 3, 307–336.
10. Kolmogorov, A. N.; Fomin, S. V. *Elements of the theory of functions and functional analysis* With a supplement, "Banach algebras", by V. M. Tikhomirov. Sixth edition. (“Nauka”, Moscow 1989).

ON THE RIEMANN-MANGOLDT CONSTANT

Danilo Merlini and Luca Rusconi

Cerfim and Issi
Via F.Rusca 1
6600 Locarno-CH

ABSTRACT

We study numerically the thermodynamic limit for the zeros of the Riemann Zeta function and propose a possible general law of the asymptotics for their real part in the critical strip. The experimental results are then used to control the numerical behavior of the asymptotics of the Riemann-Mangoldt function, giving some support for the existence of the Riemann-Mangoldt constant, which would ensure the truth of the Riemann hypothesis.

INTRODUCTION

It is known that a very important and very fundamental problem in mathematics and related fields is connected with a proof or disproof of the Riemann Hypothesis (RH) [1,2,3] (expected to be solved in the near future in the international scientific comunity). Some scientists are working on the problem with very advanced mathematical structures, while others are of the idea that a "solution" could possibly emerge by using refined computer techniques.

If $z=\rho+iE$ is a complex variable, the RH says that the Riemann Zeta function (to be defined below) has all its zeros on two lines in the complex plane: the so called "trivial zeros" lie on the real axis $E=0$ at $z=\rho=-2n$, $n\in N^*$ (the set of integers), while all the others, which are called "non trivial zeros", lie on the line $\rho=\frac{1}{2}$, or better $z=\frac{1}{2}+iE_n$, $n\in N^*$ (i.e. their real part is exactly $\frac{1}{2}$). The problem has connections with many areas of mathematics and physics: arithmetics, phase transitions, quantum theory, quantum spectral fluctuations and cryptography [4-13].

In $z=\rho+iE$, E is called the energy [5] and for many millions of zeros, the value of E_n are known [2,3,12]. We now recall that for $Re\{z\}=\rho>1$, the Riemann Zeta function $\zeta(z)$, is given by

$$\zeta(z)=\sum_{n=1}^{\infty}\frac{1}{n^z}=\prod_{p\ prime}\frac{1}{1-p^z}\qquad \rho>1$$

where the last term is called the Euler product and involves all primes.

It is known [4] that all non trivial zeros are located in the so called "critical strip" of the complex plane given by $0<\rho<1$, so that in investigating some properties of the zeros, it is more convenient to use an expression of $\zeta(z)$ which is convergent in the critical strip. The first considered here is given by [2,3]

$$\zeta(z)=\frac{1}{1-2^{1-z}}\sum_{n=1}^{\infty}\frac{(-1)^{n-1}}{n^z}\qquad \rho>0$$

Another one is given by the asympthotic expansion [4]

$$\zeta(z)=\lim_{N\to\infty}\left(\sum_{n=1}^{N}\frac{1}{n^z}-\frac{N^{1-z}}{1-z}\right)\qquad \rho>0$$

We use

$$\zeta(z)=\sum_{n=1}^{N}\frac{1}{n^z}-\frac{N^{1-z}}{1-z}+O(N^{-\rho})\qquad \rho>0$$

where $O(N^{-\rho})\approx c_0N^{-\rho}$, c_0 constant. Moreover, $\zeta(z)$ satisfies the functional equation [1,4]

$$\pi^{-z/2}\Gamma(z/2)\zeta(z)=\pi^{-(1-z)/2}\Gamma((1-z)/2)\zeta(1-z)$$

which says, in particular, that the nontrivial zeros are symmetric with respect to the critical line. So if z is a zero, then 1-z is also a zero. Moreover, taking into consideration the new variable $\omega=\exp(z-½)$ (other transformations are of course possible), then if ω is a zero, $1/\omega$ is also a zero so that if the RH is true, then all non trivial zeros would lie on the unit circle $|\omega|=1$. This situation bares some analogies with facts encountered in statistical mechanics for partitions functions of classical spin models in two as well as in more dimensions. For example, for the well known 2-dimensional Ising model in zero field on a square lattice with nearest neighbor interaction K=J/kT (where T is the absolute temperature), the partition function is a polynomial (up to some trivial factors) in the variable S=sinh(2K) of order N (the number of spin variables on the square), which in the thermodynamic limit $N\to\infty$ has all its zeros on the unit circle $|\omega|=|\sinh(2K)|=1$. This happens only in the thermodynamic limit and also for finite systems with special boundary conditions. Another situation is offered in considering the d-dimensional Ising model in external magnetic field H=h/kT with ferromagnetic two body interaction between nearest neighbors spins with variable K=J/kT, where it is known that, independent of the strength J, the zeros in the variable $\omega=\exp(-2H)$ lie on the unit circle even for finite systems and this says that a phase transition (with a symmetry breakdown of the internal symmetry group) take place at zero magnetic field H=0 (Lee-Yang theorem). Regarding this above, it should be said that fundamental phase transitions connected with the Zeta function have recently been found and rigorously proven by others [5,8,9,10]. In connection with the critical strip and for the Riemann problem, the energies E_n are unobtainable since the boundary conditions are still lacking [6,7].

Concerning the RH, the most well known numerical evidence that this should be true, is offered by extensive computations for the first few millions of zeros of the Zeta function which are found to lie all on the critical line. On the other hand, by a theorem of Hardy and Selberg [4], it is known that at least 30% of the zeros lie on the critical line. In working with the Riemann-Mangoldt function (to be defined below), it should be anticipated here that the RH is equivalent to the assertion that as $x\to\infty$ the Riemann-Mangoldt function behaves as

$\psi(x) \approx x + O(\sqrt{x}\log^2 x)$, where the term $\sqrt{x}$ is typical of fluctuations in one dimensional random walks.

THERMODYNAMIC LIMIT

In the context of the RH and working with the Zeta function, here the thermodynamic limit means that in one of the above forms of the Zeta function, which is convergent in the critical strip, the function is truncated at some integer N (analogous to the number of particles in a statistical system) and then some properties for $N\to\infty$ are investigated. In a preliminary work [15], using the first of the above expansions (the alternating one), we studied the real and the imaginary part of the first zero as a function of N up to N=200 (the first zero of the Zeta function is located at z= ½+14.134725i).

The values of ρ(N) as well as that of E(N), behaves as a function of N in the same way and approaches the asymptotical values with slow decaying oscillations described by a Bessel function of argument N whose analytical form may be given explicitely [15]. A work with the second form of the function given above is in progress [16]. Here we just present some values we have found up to N=200 (with precise calculations), because the law of approaches to the thermodynamic limit (which we believe does not depend on which zero we are dealing with), is of interest to the study of the Riemann-Mangoldt function presented below. We give the values for the real part only.

N	ρ(N)
10	0.491
20	0.520
30	0.484
40	0.486
50	0.517
60	0.509
70	0.468
80	0.519
90	0.519
100	0.483
110	0.476
120	0.505
130	0.520
140	0.516
150	0.499
160	0.481
170	0.479
180	0.492
190	0.507

Table 1

As can be seen, the mean value is given by ρ=0.49847. The asymptotic of the computations was found to be given by $\frac{1}{2} \pm \frac{k}{\sqrt{N}}$ where k is of the order of unity. We now report some other values for bigger N obtained directly from a program in Maple V (still for the first zero).

N	ρ(N)
500	0.511
600	0.492
700	0.4993
800	0.509
900	0.498
1000	0.4915
1500	0.4943
2000	0.505

Table 2

In the same way, for the second zero which is located at E=21.022039, we have

N	ρ(N)
100	0.48
200	0.506
300	0.5072
400	0.5075
500	0.5065
600	0.491
700	0.5076
800	0.4925
900	0.504
1000	0.503
1500	0.4946
2000	0.495

Table 3

For the third zero at N=2000, we obtain ρ=0.503.

The results of our computations concerning the way in which the thermodynamic limit is approached for some zeros in the critical strip indicate that the general law should be the same for all zeros. This law is, as stated before, important for the discussion of the Riemann-Mangoldt "constant" to be discussed below.

THE RIEMANN-MANGOLDT FUNCTION

It is one of the important functions in the field of arithmetics and it is suitable for investigating the RH in a numerical context. It is defined as follows

$$\Psi(x) = \sum_{n \leq x} \Lambda(n)$$

with

$$\Lambda(n) = \begin{cases} \log p & if\ n = p^m\ (p\ prime, m \geq 1) \\ 0 & otherwise \end{cases}$$

Futhermore, if

$$\Lambda(n) = \begin{cases} \log p & if\ n = p\ (p\ prime) \\ 0 & otherwise \end{cases}$$

the function is called the Tschebysceff function [14], and is usually indicated by Θ(x). The prime number theorem states that π(x), which is the number of primes at most equal to x , has the leading term given by $\frac{x}{\log x}$ and is expected that the following is true

$$\pi(x) \approx \frac{x}{\log x} + O(\sqrt{x}\log x) = Li(x) + O(\sqrt{x}\log x)$$

where $Li(x) = \int_2^x \frac{1}{\log(t)} dt$ is the logarithmic integral. The prime number theorem may be stated in the form

$$\prod_{p \leq x, p\ prime} p = e^x \quad x \to \infty$$

or

$$\sum_{p \leq x, p\ prime} \log p = x \quad x \to \infty$$

Now, the truth of the RH [4] is connected with the fluctuations in the above equation and it is claimed that

$$\psi(x) = x + O(x^{\theta}\log^2 x) \quad x \to \infty$$

The point is that it is not possible to have a substantially better error in the prime number theorem than the one in the above expression where θ is still unknown; θ=½ realizes the best possible error term, and this is equivalent to the RH which is thus formulated, using the Riemann-Mangoldt function, as

$$\psi(x) = x + c\sqrt{x}\log^2 x \quad x \to \infty$$

A theorem shows that θ=½ is determined precisely by the dimension of the critical strip along the real axis in the complex plane where the non trivial zeros are located i.e |ρ-½|<½. More precisely: suppose that θ<1 exists, so that the set of non trivial zeros is in the strip given by θ, i.e that all non trivial zeros have |ρ-½|<|θ-½|, then as x→∞,

$\psi(x) = x + O(x^{\theta}\log^2 x)$. Now, in numerically studies of the RH with the function ψ(x) it is also possible to restrict to the function Θ(x). The reason is that it can easily be shown that

$$\psi(x) = \Theta(x) + k\sqrt{x}\log x \quad k > 0, x \to \infty$$

Thus

$$f(x) = \lim_{x\to\infty}\left|\frac{\psi(x)-x}{\sqrt{x}\log^2 x}\right| = \lim_{x\to\infty}\left|\frac{\Theta(x)-x}{\sqrt{x}\log^2 x}\right| + O(\frac{1}{\log x})$$

In connection with the above theorem and expansion, we have numerically investigated the function f(x) which is expected to behave like $cx^{\theta-\frac{1}{2}}$.

In fact, from our computation [15] on the thermodynamic limit for the first zero, we have seen that $\theta \approx \frac{1}{2} \pm \frac{k}{\sqrt{N}}$ as x→∞ (asympthotic expansion for the Bessel function discussed above). So we espect that

$$f(x) = \lim_{x\to\infty}\left|\frac{\psi(x)-x}{\sqrt{x}\log^2 x}\right| = \lim_{x\to\infty}\left|\frac{x + cx^{\theta}\log^2 x - x}{\sqrt{x}\log^2 x}\right| =$$

$$= \lim_{x\to\infty}\left|cx^{\theta-\frac{1}{2}}\right| = \lim_{x\to\infty}\left|c\right|e^{(\theta-\frac{1}{2})\log x}$$

Using the above numerical result, we have that

$$f(x) = \lim_{x\to\infty}\left|c\right|e^{(\theta-\frac{1}{2})\log x} = \lim_{x\to\infty}\left|c\right|e^{k\frac{\log x}{\sqrt{x}}} = \left|c\right|$$

i.e. the existence of an upper bound c, which we call the Riemann-Mangoldt constant.

Evidently the deviation of Θ from ψ is $\frac{const.}{\log x}$. So if the constant exists and is small, there may be some discrepance between the computations with ψ and Θ and the constant k may be only meanigsfull for the computation with ψ. In any cases

$$\psi(x) = \Theta(x) + \varepsilon \quad \varepsilon > 0$$

so that f(x) computed with the Θ function should be an upper bound for c.

COMPUTATIONS AND CONCLUSIONS

Our computations have been carried out for Θ(x) up to x=1.2 10^6, which is a very small number in an arithmetical context. Nevertheless we are presently working on a more substantial computation. At the moment we are reporting the numerical results of our preliminary computations involving Θ(x) in y=f(x) up to x=1.2 10^6. In Table 4 we give some values of the maxima in the above mentionned range of x, and in Table 5 the corrisponding values of c using our constant k=0.365905.

During our computations we encountered pairs of twin primes (p, p+2) near the local maxima of f(x) but also around the absolute maxima in bigger intervals (we remember that a proof that the set of twin primes is infinite is still lacking).

x	f(x)
302831	0.0113612
463219	0.0102959
643847	0.008668
757993	0.0085972
993821	0.0083838
1062869	0.0088649
1090697	0.0089638

Table 4

Maximum	c
302831	0.0113516
463219	0.0102886
643847	0.0086627
757993	0.0085923
993821	0.0083795
1062869	0.0088605
1090697	0.0089594

Table 5

In conclusion, up to $x=1.2\ 10^6$, we have found an upper bound c for the Riemann-Mangoldt constant situated in the range **c=0.00896**. The results in the computations up to many millions will be reported in the future .

REFERENCES

1. Bernhard Riemann, *Oeuvres Mathèmatiques*, Ed. Jacques Gabay, pp. 165-176 (1990).
2. D.B.Zagier, *Zetafunktionen und Quadratische Körper*, Springer Verlag (1981).
3. D.B.Zagier, Die erste 50 milionen Primzahlen, Beilage Nr. 15 *Elemente der Mathematik* pp 1-24, Birkhäuser Verlag (1977).
4. S.J.Patterson, *An Introduction to the Theory of the Riemann Zeta-Function*, Cambridge Univ. Press (1995).
5. B.Julia, On the 'statistics' of primes, *J. Phys France*, Vol 50., pp 1371-1375 (1989).
6. M.V.Berry, J.P.Keating, *The H=xp and the Riemann-Zeros*, preprint (1997).
7. M.V.Berry, J.P.Keating, *The Riemann-Zeros and Eigenvalues Asymptotics*, preprint (1998).
8. A.Connes, *Trace formula in noncommutatives geometry and the zeros of the Riemann Zeta-Function*, preprint (1997)
9. P.Cohen, Sur la mécanique statistique d'après les travaux de Bost-Connes, *Actes des Journées Academiquées de Lille*, to appear (1998).
10. P.Cohen, Dedekind Zeta Functions and quantum statistical mechanics, *Atti del Colloquio di Matematica*, CERFIM vol. 9, pp 99-107
11. M.Katz, P.Sarnak, Zeroes of Zeta Functions and symmetry, *Bull. (new series) of the A.M.S.*, Vol 36, Nr. 1, , pp 1-26 (1999).
12. A.Odlyzko, On the distribution of spacings between Zeros of Zeta Functions, *Math. Comp.* Vol 48, pp 273-308 (1987).

13. K.S.Alexander, K.Baglawski, G.C.Rota, A stochastic interpretation of the Riemann Zeta function, *Proc. Nat. Acad. Sci.* USA, Vol 90, , pp 697-699 (1993).
14. K.Prachar, *Primzahlverteilung*, Springer Verlag (1957).
15 D.Merlini, L.Rusconi, *Analisi Asintotica Degli Zeri della Funzione Zeta di Riemann*, ed. ISSI , Vol. 5, pp 103-127 (1999).
16. D.Merlini, L.Rusconi et al., in preparation (1999).

LONG-TERM STABILITY IN CIRCULAR ACCELERATORS

Walter Scandale

CERN
CH-1211
Geneva 23
Switzerland

ABSTRACT

In the modern hadron colliders, the stability of the single-particle motion is basically determined by the field-shape imperfections of the superconducting magnets, especially during the injection plateau, when the effect of the persistent currents is maximum and the transverse size of the beam is large. The non-linear fields are at the origin of two effects: the betatron tunes change with the amplitude and the momentum of the circulating particles, and, for certain combinations of the horizontal, vertical, and synchrotron tunes, non-linear resonances are excited. These phenomena have a destabilizing influence on the particle motion, over a time-scale extending up to several million turns. Some precautions can make the motion of the particles less sensitive to the non-linear components of the guiding fields. Correcting multipoles can be foreseen in the regular cells, to reduce the non-linear tune-shift caused by the systematic components of the field errors. The variations of the orbit functions can be limited along the insertions. The closed orbit and the linear coupling can be corrected sufficiently well. Finally the ripple of the power supplies can be reduced as much as possible.

INTRODUCTION

The motion of charged particles in circular accelerators is basically governed by the magnetic field of the guiding dipoles and the focusing quadrupoles. Intentional and non-intentional non-linear fields are in general also present, the side-effect of which is to induce losses at large amplitude. Sextupoles are used to reduce the chromaticity and octupoles make the tune dependent on the amplitude, which is sometime exploited to improve the current-dependent behavior. In hadron accelerators, the destabilizing action of chromaticity sextupoles is self-compensated to a large extent due to the regularity of the lattice. However, usually, a strongly focused lattice is necessary to reduce the sensitivity to field errors, and this in turn produces two adverse effects: the strength of chromaticity sextupoles has to be increased, therefore the beam stability is reduced, and the space available for dipoles along the ring is shortened with a consequent decrease of the maximum beam

The Application of Mathematics to the Sciences of Nature, edited by Cerrai *et al.*
Kluwer Academic/Plenum Publishers, New York, 2002.

energy. Unintentional multipoles due to unavoidable imperfections of the guiding and focusing fields introduce additional non-linearities, which represent the greatest hazard. However, compromises are to be found between making magnetic fields as uniform as possible and keeping the magnet cost low. This is a difficult achievement especially for superconducting magnets, whose quality depends on the mechanical tolerances of the coil geometry, rather than on the shape of the poles. In the magnets of the Tevatron (FNAL, Chicago), of RHIC (BNL, N.Y.) and of Hera (Desy, Hamburg), typical deviations from uniformity have been limited to about one part in ten thousand at 2.5 centimetres from the magnetic axis. Similar values, properly extrapolated with the inner coil diameter and the superconducting filament size, are expected to be reached in the magnets of the LHC.

The single-particle approach provides a sufficiently simple, reliable and coherent model of the real accelerator to investigate performances related to non-linear dynamics. The key issue is to estimate the stability of the motion over the operational cycle of the accelerator. In a linear machine with irrational tunes the motion is stable and regular all around the closed reference orbit near the magnetic axis. The non-linear fields add a tune dependence with the amplitude, which shifts tunes to rational values, provoking resonant phenomena accompanied in the phase space by islands of finite area surrounded by thin chaotic layers. The islands and the chaotic layers exist through the entire phase space. However, at small amplitude, trajectories follow invariant surfaces, the KAM tori [1], and remain stable for indefinite time. As amplitude increases, the islands become larger until they overlap. Beyond that point, the chaotic layers become interconnected and the particle motion is no longer bounded. This is the domain in which the non-linear forces are able to provoke particle losses, that sometimes may occur after millions of turns. The border between regular and chaotic motion is called dynamic aperture. This is analytically well defined for 12 dimensional (12-D) systems only [2]. With more degrees of freedom, particles in stochastic layers, even close to the origin, may escape through the entire phase space, due to the so-called Arnold diffusion. However, for all practical purposes, the border between mostly regular and mostly chaotic trajectories can be used as the dynamic aperture.

Both analytical and numerical tools are used to estimate the dynamic aperture as a function of various machine parameters. Improvements of the linear lattice and correction schemes are studied to reduce the influence of the non-linear forces, and to specify upper limits for the magnet imperfections. The final confirmation is in general performed with numerical simulations in which the particle position is tracked element by element around the machine for large numbers of turns.

Simpler dynamical systems, such as the Hénon map [3], are advantageously used to investigate mechanisms of long-term losses.

Machine experiments with existing accelerators, in which non-linear perturbations are deliberately introduced, allow comparisons with predictions of numerical simulations.

In the following sections three subjects are reviewed, with special emphasis for the case of the LHC: the tools by which predictions on beam stability are formulated, the particular applications on accelerator design, and finally the outcome of the experiments performed in the CERN-SPS and in the FNAL-TEVATRON.

TOOLS FOR DYNAMIC APERTURE ESTIMATES

Tracking Simulations

Simulations with thin-lens approximation and symplectic integrators are considered as the master tool for quantitative estimates of particle behavior in the LHC with non-linear elements. They provide exact solutions of the equation of motion for a dynamical system approximating the accelerator. A sequence of linear transfer matrices interleaved with localized polynomial non-linearities is to be computed. Reliable results are easily obtained, since computer rounding errors can be kept under control [4]. However a vast computing power is required to get reliable estimates of the dynamic aperture as a function of various lattice and beam parameters. A fully realistic description of the accelerator structure is difficult if not impossible. Simplifications are also imposed by limitations in computing power. Thin-lens description of guiding and focusing fields is used, and do not imply relevant changes to orbit functions. Ignoring fringing fields, and representing non-linearities with equivalent thin-lenses is also considered acceptable.

Two computer codes are routinely used to describe LHC lattice models and compute particle trajectories of given initial conditions: MAD [5], developed at CERN, and SIXTRACK[6], developed at DESY. Both of them have scalar versions to be processed in the modern farms of workstations as well as vectorized versions to make use of modern parallel processors.

Maps

In linear lattices, particle coordinates can be propagated along the accelerator azimuth by Twiss transfer matrices.

The use of maps can be extended to non-linear dynamics with some precautions. This extension, originally motivated by the need to speed-up long-term tracking simulations in hadron colliders, in fact provides a powerful tool to handle dynamical quantities, like the tune dependence with the amplitude and the momentum, the distortion functions and the smear, the high-order non-linear invariants, and finally the Fourier harmonic coefficients of the resonance driving terms. Non-linear matrices can be constructed very efficiently with Differential Algebra techniques using Taylor expansion to some high order of algebraic operators [7]. One-turn Taylor maps resulting from the composition of all the linear and non-linear elements in the accelerator, are inherently not symplectic because of the high order truncation, therefore inappropriate to preserve volume in phase space. A way to restore symplecticity, whose physical meaning, however, is not fully understood, is to replace the truncated map with a Normal Form [8,9], that is an integrable map, represented by a rotation of an angle depending on the amplitude of the orbit. In practice this is performed by a local change of coordinates in the phase space, which brings intricate orbits into circles. However, the normalized map has an optimized order of accuracy. Above it, the approximation is improved at lower amplitude and worsened at higher amplitude. The domain of convergency is limited by resonances of low order that are allowed by the truncated Taylor map. There are ways to handle the first limiting resonance, with Resonant Normal Forms, which are not yet made of practical use.

In the LHC, the mapping approach based on Taylor expansion and Normal Form is used to evaluate the dependence of tune-shift on the amplitude and the momentum due to systematic field-shape imperfections [10]. By this it is possible to identify the multipoles that are more dangerous for the stability of the motion, taking into account the quite strong high order cross terms, and to define and optimize the most suited scheme of multipolar correctors.

High order Taylor maps are also used to estimate the dynamic aperture in a faster way than with the usual element by element tracking [11]: however, in the LHC, this approach is non-controversial only for simulations up to few 10^{4} turns. By increasing the order of the map the violation of area preserving transport can be made arbitrarily small, but the map size grows exponentially and the computing speed decreases accordingly.

Early Indicators of Chaos

Early indicators of chaotic motion have been used to speed up the estimate of the dynamic aperture in the LHC. The exponential divergence of two initially very close trajectories is a criterion for chaos, a linear growth indicating regular motion. The exponential coefficient, called Lyapunov exponent [12], can be used to localize stochastic layers in the phase space and eventually to identify the stability border below which its value is zero. The routine way to evaluate the Lyapunov coefficient is to track simultaneously two particles with a slightly different initial amplitude, and to compute periodically and plot their mutual distance[13]. An equivalent method is based on the analytical evaluation of the Jacobian in the phase space domain of interest [14]. A third possibility is to compute the slow change of an action invariant[15]. The predictability of all three methods is enhanced when the non-linear deformation of the phase space is removed by Lie algebraic or Normal Forms type change of coordinates. It is currently admitted that through early indicators of chaos a conservative estimate of the dynamic aperture can be obtained with about ten times less computing power than for standard tracking.

Figure of Merit and Data Processing

The linear aperture, based on smear and tuneshift with the amplitude, and the short-term dynamic aperture were widely used in the past [16] to estimate non-linear effects, since threshold values for detuning and amplitude distortion were considered sufficient to ensure long-term stability. However, the validity of this extrapolation has not been confirmed by deeper studies. Therefore, intensive tracking and sophisticated data processing are preferred nowadays to estimate the dynamic aperture, after a preliminary selection of rather few significant cases, on the basis of short-term simulations. Results are presented in graphical form: survival plots depict the maximum number of stable turns as a function of starting amplitude [17]. These plots and early indicators of chaos provide a practical estimate of the stable region. Dense survival plots are ragged and show a large spread in the survival time close to the chaotic border, rapidly decreasing at larger amplitudes. Such an irregular shape reflects the local origin of the particle instability: at moderate amplitude in presence of weaker perturbations, the escape time is largely influenced by microscopic changes of initial coordinates; at large amplitude, instead there are only fast losses. Under the influence of non-linearities, particles migrate across different nests of resonances, which can be at least phenomenologically correlated to average lifetime. The loss mechanism is in general sudden: the particle may stay confined even for millions of turns and then diverge in a few thousand turns.

APPLICATIONS IN ACCELERATOR DESIGN

The LHC must operate with negligible loss for long periods, up to 10^8 turns, in spite of the unavoidable field shape imperfections. An upper limit to unintentional multipoles and practical compensation strategies have to be devised for a safe operation. This implies a thorough understanding of the influence of the non-linearities on the long-term behavior of particle trajectories. Analytical methods available are not yet fully exploited. Numerical simulations are too heavy and time consuming for an exhaustive overview of all the possible situations. Nevertheless, remarkable progress has been made through heuristic approaches proposed in the recent past, based on the investigation of simplified non-linear models of the LHC lattice and on the use of empiric criteria for beam stability.

Too crude simplifications of the lattice structure itself have dramatic effects on non-linear performances. Cell lattice models with only regular cells and no interaction regions show a regular azimuthal pattern of the orbit functions and in particular of the betatronic phase advance leading to unrealistic enhancement of the particle stability. They are in general used for numerical studies of simple dynamical systems as the Hénon map. A more realistic way to drop the insertions is to replace them with equivalent rotation matrices. Part of the chromatic aberration and some unintentional field errors are disregarded in this way. However, relevant informations can be gained with less computing power and complexity, especially during the injection plateau, where the stability of motion is mainly determined by non-linear perturbations in the arcs. This approach was used to determine the optimum value for the length and the phase advance of the LHC cell [18], fixed to about 100 m and 90 degrees, respectively. More advanced studies are based on models with realistic descriptions of the insertions.

The field-shape imperfections are equivalent to multipoles up to large order, which can be expressed as the sum of two parts, one systematic and the other random. The general agreement is to stop at order 11th in the multipolar expansion and to neglect correlations between random multipoles of different orders. Systematic errors are larger at injection due to persistent currents. Large low-order (3rd and 5th) values provoke a sizeable detuning with the amplitude and the momentum, which can be corrected either locally [19], as in the LHC lattice version 2, or using a clever cancellation of the detuning terms by means of Sympson rules [20], as in the LHC lattice version 1. In the latter case, multipolar correctors are to be located near the main quadrupoles as well as at about the middle of each half cell [9]. Octupolar imperfections have, in the LHC, a particular behavior related to the symmetry of the magnetic field in two-in-one magnets: the integrated value along the azimuth is expected to be zero, therefore the detuning with the amplitude is expected to be self-compensated without specific correctors. Large high-order (7th and 9th) systematic multipoles destabilize off-momentum particles and have to be minimized by design.

Random imperfections, which vary from magnet to magnet due to manufacturing tolerances, are the main source of non-linear resonances and distortion functions. Statistical distributions can be easily predicted, but are insufficient for a complete knowledge of the non-linear optics, since resonance strengths depend on the specific sequence of the random errors around the ring rather than on statistical properties. Therefore, criteria for magnet design are to be studied on several non-linear lattices, with different sequences of random multipoles. In fact, there are many parameters that limit the stability of the particle motion in the LHC, therefore the first task is to identify the most important ones, in order to reduce to a reasonable amount the enormous need of computing time required for an exhaustive set of simulations. Parameters routinely considered in the accelerator models are residual closed orbit, linear coupling due to imperfections, synchrotron motion, and main power supplies ripple. Chromaticity and non-linear detuning are corrected with suited set of correctors. Special cases with some residual uncompensated chromaticity are considered as well. With this strategy [21], beam stability has been found to be strongly influenced by linear lattice parameters like tune, residual linear coupling, and peak-ß values in the insertion quadrupoles as well as by a residual chromaticity of a few units. Instead, residual closed orbit associated to magnet misalignment and tune ripple of a few 10^{-4} units showed a weak interference with beam stability.

Particles above the stable region are expected to diffuse towards the vacuum pipe at a speed strongly increasing with the transverse amplitude. A set of collimators [22] is used to absorb them before they hit the magnets and provoke an unwanted deposition of energy in the superconducting coils. The transverse position of the primary collimator defines the mechanical aperture of the accelerator. It is basically fixed taking into account the mechanical tolerances of the cold bore and of the thermal screen of the main magnets, the expected peak-values of the closed-orbit, of the dispersion, and of the ß-function modulation, and, of course, the optimum value of the separation between the primary and the secondary collimators. For a safe operation, careful matching of physical aperture and stability border is to be performed[21]. With a small mechanical aperture, trajectories with small amplitude oscillations only are allowed, which are weakly perturbed even in presence of large field-shape imperfections, whilst, with larger mechanical apertures, and larger amplitude oscillations, the non-linear perturbation becomes larger and the size of the magnetic errors start to play a leading role for the dynamic aperture. On the other hand, we believe that particles with amplitudes up to the chaotic boundary are stable, although the non-linear perturbations induce a finite smear of their trajectories, whilst particles with larger amplitudes may become unstable after a sufficiently large number of turns. Ideally, the chaotic limit should be equal to the dynamic aperture evaluated in presence of collimators, in which case only the unstable particles will be intercepted by the collimators. In this respect the mechanical size and the field quality of the LHC at injection have been found to be well matched to a value of 6 σ, i.e. of 7.2 mm, which is considered a wise choice for the needed dynamic and physical aperture.

Strategies of magnet sorting have been invented, by which the magnets are installed in such a sequence in the machines as to minimize the combined non-linear effects. For practical and theoretical reasons, the sorting scheme should be as local as possible and must refer to a limited kind of multipoles. Different solutions have been proposed [23,24,25]. By introducing a quasi-periodicity of multipoles every two betatronic wavelengths, the harmonic content of non-linearities can be shifted away from harmful frequencies. Alternatively, small groups of magnets, typically ten, are ordered in such a way to minimize a broad band of non-linear driving terms computed to 2nd perturbative order, contributing to resonances up to order 12th. The first method is used in the LHC and the SSC, the last method is used for HERA and is still under investigation for the LHC.

Diffusion with steady state increase of the amplitude has never been detected in numerical simulations, even in presence of external tune modulation, contrary to what is usually observed with beam-beam interaction. Due to resonance crossing and non-linear coupling, migration of particles in the tune diagram and mixing of horizontal and vertical oscillations are well visible in long-term tracking results. However large increases in amplitude and particle losses are sudden and unpredictable, even if they occur after a large number of turns.

EXPERIMENTS

In the last decade, experiments have been devised to study the effect of high-order resonances under controlled conditions. The hope is that by comparing theoretical or numerical results with experimental ones, it may be possible to define suitable criteria for beam stability and evaluate their predictive power. This is an ambitious goal, since real machines are much more complicated than models used in tracking codes or analytical evaluations. There are many phenomena, like collective instabilities, synchro-betatron resonances, linear imperfections affecting the orbit functions, the linear coupling, the closed orbit, non-linear imperfections due to fringing fields and saturations, which may take a long time to be quantified in order to disentangle single-particle effects from measurements. However, analysis of operating situations provides a wealth of informations which can be exploited to bridge the gap between models and reality. These experiments have been performed in the CERN-SPS and the FNAL-TEVATRON which are already well understood, so that clear conditions could be defined, spurious effect eliminated and phenomena under study carefully isolated, with reasonably small effort. Having repeated similar measurements in different accelerators is invaluable to help in distinguishing results of general interest from those which are just a property of the machine used. Common motivations of the two experiments are related to the refinement of aperture and field quality criteria for future large hadron accelerators, like LHC or RHIC. A common procedure consists in exciting already existing sextupoles in order to introduce in a controlled fashion non-linearities in an otherwise linear lattice. To probe large amplitudes, a pencil beam with small emittance and momentum spread is used, to which a large enough coherent deflection is applied. In a few hundred turns, a 'hollow' distribution of charges is created around the central orbit due to nonlinear filamentation, whose behavior is observed with several instruments: current transformers record lifetime, Schottky noise detectors give tune and tune-spread, flying wires provide transverse profile, and orthogonal pairs of position monitors are able to produce a phase space portrait.

In the experiment E778 at FNAL [26], sextupoles were exciting strongly the third integer resonance together with the higher-order ones. Smear, injection efficiency and short-term dynamic aperture were measured and compared with tracking. The agreement is good, however long-term losses could not be quantitatively reproduced. The existence of stable nonlinear resonance islands was demonstrated experimentally by observing coherent persistent signals of particles captured into them. Tune modulation effects were explored and compared with those of 1-D forced pendulum.

In the experiments at the SPS [27,28], the sextupolar excitation chosen such as to minimize the third integer resonance. Detuning compensation was experimentally tested by using existing octupoles. A 30% increase of dynamic aperture resulted from a factor ten reduction of tune-spread. This provides experimental guidance in devising correction schemes for large hadron accelerators. However most of the emphasis was put on the study of slow diffusion induced by power supply ripple, and controlled modulation of a special quadrupole. The diffusion coefficient was measured as a function of the amplitude, the modulation frequency and depth, and the tune. It was obtained by scraping the beam tail with horizontal and vertical collimators, retracting them suddenly by a few mm, and observing the beam lifetime to estimate the time taken by the particles to fill the gap created by the retraction. Diffusion immediately sets in when tune modulation is turned on, and there is evidence that a ripple which leads to tune modulation of 10^{-3} cannot be tolerated in a machine with strong non-linearities. A tune modulation at two frequencies is more destructive than a modulation at a single frequency for the same overall depth. The agreement with numerical simulations is of the order of 20%, however the strong dependence of diffusion on modulation depth and the dependence on frequency are not yet understood.

Recent experiments at FNAL [29] have also addressed the problem of ripple induced diffusion. Intensity and transverse profile were recorded and used to deduce a phenomenological dependence of the diffusion coefficient on amplitude. The proposed model assumes a threshold amplitude, below which there is no diffusion, and above which the diffusion speed increases as a polynomial of the amplitude. A steady reduction of beam size was pointed out which was never observed at the SPS. This is likely to be typical of the regime of large losses explored at FNAL.

The phenomenological model of FNAL for diffusion has been found to be in contradiction with some results of the CERN experiment as well as with the sudden manifestation of fast amplitude growth in tracking [30]. More general models of diffusion based on Markov process with jumps, using master equations on status transition probability are under investigation.

REFERENCES

1. V.I. Arnold, *Mathematical Methods of Classical Mechanics*, Springer-Verlag, (1978)
2. B.V. Chirikov, *Physics Report* **52**:263 (N.5)(1979).
3. M. Q. Hénon, *Quarterly of Appl. Math.* **27**:291 (1969).
4. E. Lohrmann, P. Wilhelm, *Particle Acc.* **19** :99 (1986).
5. H. Grote, C. Iselin, *CERN/SL*/90-13 *(AP) Rev*.2 (1991).
6. F. Schmidt, *CERN/SL/AP* 91-52 (1991).
7. M. Berz, *Particle Acc.* **24**:109 (1989).
8. A. Bazzani, et al., *Il Nuovo Cimento*, **102 B**, N. 1 (1988).
9. E. Forrest, J. Irwin, M. Berz, *Particle Acc.* **24**:91 (1989).
10. W. Scandale, et al., *Particle Acc.* **35**:53 (1991).
11. R. Kleiss, F. Schmidt, F. Zimmermann, *Particle Acc.***41**, (N.2) (1993).
12. A. M. Lyapunov, Stability of Motion, *Ac. Press* (1966).
13. F.Schmidt, F. Willeke, F. Zimmermann, *Particle Acc.* **35**:249 (1991).
14. F. Zimmermann, *Dipl. Thesis, DESY-HERA* 91-03 (1991).
15. J.Irwin, Y. Yan, *IEEE PAC 89*:1340 (1989).
16. A. Chao et al., *CERN 88-04*:152 (1988).

17. F. Galluccio, F. Schmidt, *EPAC 90:*640 (1990).
18. M. Bassetti et al., *IEEE PAC 87:*1284 (1987).
19. M. Giovannozzi, F.Schmidt, General Normal Forms procedure to correct Tune-Shift and Non-Linear Chromaticity for Large Accelerators like the LHC, *PAC* 1993
20. D. Neuffer, *CERN* 88-04:179 (1988).
21. Z. Guo, T. Risselada, W. Scandale, *EPAC* 92:637 (1992).
22. Design Study of the LHC, *CERN* 91-03:115 (1991).
23. R. Gluckstern, S. Ohnuma, *IEEE PAC* 85:2315 (1985).
24. F. Willeke, *DESY-HERA* 87-12 (1987).
25. W. Scandale, L. Wang, *CERN SPS* 89-41 (AMS) (1989).
26. L. Merminga et al., *IEEE PAC* 89:1429 (1989).
27. D. Brandt et al., *EPAC* 90:1438 (1990),
28. X. Altuna et al., *CERN SL*/91-41 (AP) (1991).
29. T. Chen et al, *Phys.Rev.Lett.* **68**:33 (N.1) (1992).
30. A. Gerasimov, *FERMILAB-Pub*-92/185 (1992).

MATHEMATICAL MODELS IN BEAM DYNAMICS

Giorgio Turchetti

Dipartimento di Fisica
Università di Bologna, Via Irnerio 46
40126 Bologna (ITALIA)
turchetti@bo.infn.it

INTRODUCTION

The high energy accelerators are the basic tool to investigate the fundamental interactions among elementary particles. Indeed a particle accelerated up to a momentum $p \gg 1$ Gev/c is a probe of wavelength $\lambda = h/p$ capable of exploring distances much smaller than the proton size ($r_p \sim 10^{-13}$ cm corresponding to $p = 1$ Gev/c). The very high energies are reached with a multi-stage structure formed by an injector, a linear accelerator (Linac) and a sequence of rings (Synchrotrons) of increasing radius. At the CERN facility the Linac energy is 0.05 Gev and 4 rings are needed to accelerate a proton up to 8000 Gev, thc final energy of the 30 Km long Large Hadron Collider [1].

The high intensity ion accelerators at intermediate energies (1-10 Gev) allow to explore the nuclear matter at extreme density conditions and offer serious perspectives for the energy production. In particular the transmutation of nuclear wastes and the heavy ion inertial fusion are being considered as possible candidates for a safe and clean nuclear energy production for the second half of this century [2,3] With an estimated population doubling from 6 to 12 billions within 2050 and a tripling of the energy consumption from 10 to 30 GT (giga oil tons) the energy production by nuclear technologies will be at least 10% of the total, in order to keep the level of CO_2 to a tolerable level. By rising the safety standards and solving the wastes problem, the nuclear energy production will become socially more acceptable, due to its low environmental impact.

Inertial fusion

A fusion reaction such as

$$D + T \rightarrow He^4 + n + (17.6 \text{ Mev})$$

The Application of Mathematics to the Sciences of Nature, edited by Cerrai *et al.*
Kluwer Academic/Plenum Publishers, New York, 2002.

occurs if a DT pellet is compressed up to density n for a time τ which satisfies the Lawson criterion $n\tau \sim 10^{15}\text{cm}^{-3}\text{s}$. With the rapid compression $\tau = 10^{-8}s$ produced by the ion bunches, a density of $n = 10^{23}$ particles per cm^3 must be reached, corresponding to 10^3 the density of the solid DT. The pellet must be irradiated very uniformly either with a large number of isotropic beamlets (direct drive) or by x rays produced when a converter hit by the primary ion beams (indirect drive).

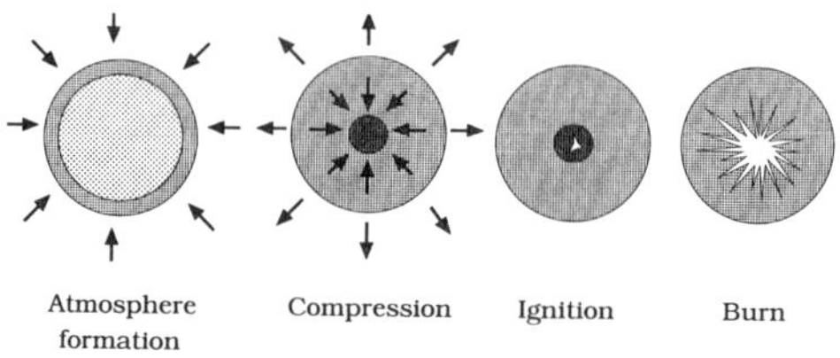

Figure 1. Pellet ignition

Typically Bi^+ ions would be accelerated by a Linac up to 10Gev, stored and bunched in storage rings and further compressed before the final focusing [3,4]

Space travel

The inertial fusion opens long term perspectives for the space travel too. Indeed if the dimension of the driver can be reduced substantially to allow the installation into a space vehicle, the achievable speeds would be increased enormously, allowing very fast travel in our solar system and even missions to nearby stars lasting just a few centuries [5,6]. Letting Δm be the mass defect of a fusion reaction and v the spaceship velocity and assuming the energy produced is converted into kinetic energy of the spaceship we have

$$\frac{v}{c} = \left(2\frac{\Delta m}{m}\right)^{1/2}$$

Since $\Delta m\, c^2$ is of the order of 1 ev in chemical reactions and of 1 Mev in nuclear reactions an increase of speed by 3 orders of magnitude from $v \sim 10$ Km/s to $v = 10^4$ Km/s is possible. Moreover velocities of the exhaust comparable with the cruise velocity are possible with a reasonable ratio of the payload to fuel ratio for orbital or return missions.

Mathematical Models

The accelerators are complex systems consisting of a large number of constituents for the focalization of particles on the design orbit and their acceleration up to the design energy. However there are only a few basic ingredients like the focusing FODO cell, the thin sextupoles and RFQ cavities exerting transverse and longitudinal forces required for focusing, chromatic corrections, acceleration and bunching. Symmetries reduce the complexity and allow to formulate simple models such as the Hénon map. These models present strict analogies with the basic models of celestial mechanics such as the restricted three body problem, whereas a realistic accelerator looks more like the solar system with its thousands of components if one includes planets satellites and asteroids.

The basic questions are also similar. The solar system is known to be stable since a few billion years namely namely one billion of revolutions of Jupiter. Stability of the beam on a day time, corresponding to one billion of revolutions, is required for the

Large Hadron Collider. In a very high energy accelerator the particles of the beam are subject to the non linear forces of the lattice but do not interact with each other because the Coulomb repulsion and the magnetic attraction between two charges moving with a speed close to c are almost equal (the ratio is v^2/c^2) and cancel each other.

For a very high intensity low energy accelerator on the contrary the Coulomb interaction among the particles is very important. In this case the number of turns is is much less but the space charge effect becomes relevant. The evolution of a large number of interacting particles has to be faced using the strategies of plasma physics, statistical mechanics and dynamical systems The halo problem concerning the formation and evolution of of small clouds of particles out of the beam core can be investigated by formulating simple dynamical models, where the electric field of a given equilibrium distribution of the core is assigned. As a consequence accelerators are also a laboratory of non linear dynamics, where the transition from ordered to chaotic motions can be observed experimentally and theoretically analyzed.

GEOMETRICAL OPTICS AND BEAM DYNAMICS

In a circular accelerator the design trajectory is a closed orbit obtained by joining straight lines and arcs of circle in the regions where a magnetic dipole field is present. Given a charge q with momentum p_0, the radius ρ of the arc it describes in a magnetic field of intensity B_0 is

$$\frac{1}{\rho} = \frac{qB_0}{cp_0}$$

In a Linac the design orbit is just a straight line.

The particles of the beam move close to the reference orbit and their trajectories are determined by the Maupertuis principle. The functional

$$W = \int \mathbf{p} \cdot \mathbf{v}\, dt = \sqrt{2mE} \int n\, ds$$

defined on the isoenergetic trajectories with fixed end points is stationary on the physical trajectory. The same functional $\int n\, ds$ appears in the Fermat's principle for light-rays trajectories, if we interpret n as the refraction index [7,8]. Choosing a Cartesian system, where the z axis is tangent to the reference trajectory, the element of arc length ds can be written as dz times a function of $x' = dx/dz$ and $y' = dy/dz$. In the simplest case of a rectilinear reference trajectory we have $ds = (1+x'^2+y'^2)^{1/2}\, dz$. If the reference trajectory is a plane curve, choosing a coordinate system with the z axis tangent to the curve, and the x axis in the plane with orientation opposite to the center of curvature we have instead $ds = ((1 + x/\rho)^2 + x'^2 + y'^2)^{1/2}\, dz$, where ρ is the radius of curvature and z denotes the arc length along the curve. Referring to the former case, the physical trajectory satisfies the Euler-Lagrange equations for the Lagrangian $\mathcal{L} = n\,(1 + x'^2 + y'^2)^{1/2}$ and the corresponding Hamiltonian is

$$H = -(n^2 - p_x^2 - p_y^2)^{1/2}, \qquad p_x = n\frac{x'}{(1 + x'^2 + y'^2)^{1/2}} = n\frac{dx}{ds}, \qquad p_y = n\frac{dy}{ds}$$

If the forces are conservative $n = (1 - V/E)^{1/2}$ and the particle orbit is the same as for a ray moving in a medium of refraction index n. †. If n depends only on

† In the analysis of the the slow transverse motion the use of a non relativistic kinematics is justified

z the conservation of $p_x = n\sin\theta\cos\phi$ and $p_y = n\sin\theta\sin\phi$ implies that the ray trajectory is planar and that $n\sin\theta$ is constant (Snell's law), see figure 2. If n is slowly varying, namely if $|V| \ll E$, the trajectory is close to the reference orbit. In this case $n \simeq 1 - V/(2E)$ and $p_x \simeq x'$, $p_y \simeq y'$. The orbit in the paraxial approximation is defined by the Hamiltonian

$$H = \frac{{p_x}^2 + {p_y}^2}{2} + \frac{V}{2E}, \qquad\qquad p_x = \frac{dx}{dz}, \quad p_y = \frac{dy}{dz}$$

The equations of the orbit read

$$\frac{d^2x}{dz^2} = -\frac{\partial}{\partial x}\left(\frac{V}{2E}\right) \qquad\qquad \frac{d^2y}{dz^2} = -\frac{\partial}{\partial y}\left(\frac{V}{2E}\right)$$

where $2E = v_0\, p_0$. If no force acts in the transverse plane x, y then the orbit has a linear divergence $x(z) = x(0) + p_x(0)z$, $\quad y(z) = y(0) + p_y(0)z$

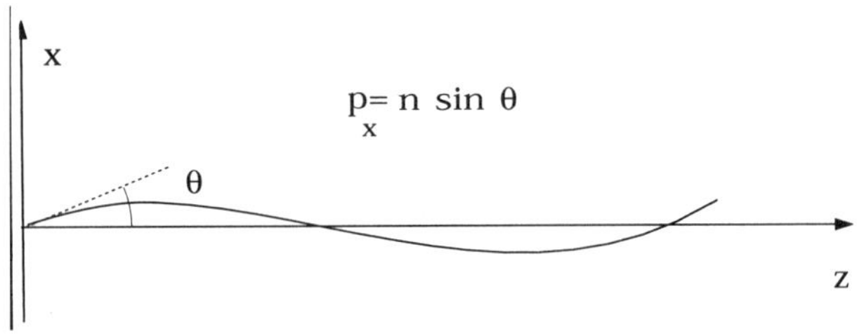

Figure 2. Ray trajectory

For small deflections we can interchange the arc ds on the orbit with the arc dz on the reference orbit. Choosing the angles as indicated in figure 2 we have

$$\frac{dx}{ds} = \sin\overline{\theta}_x, \qquad \frac{dx}{dz} = \tan\theta_x, \qquad\qquad \theta_x \simeq \overline{\theta}_x$$

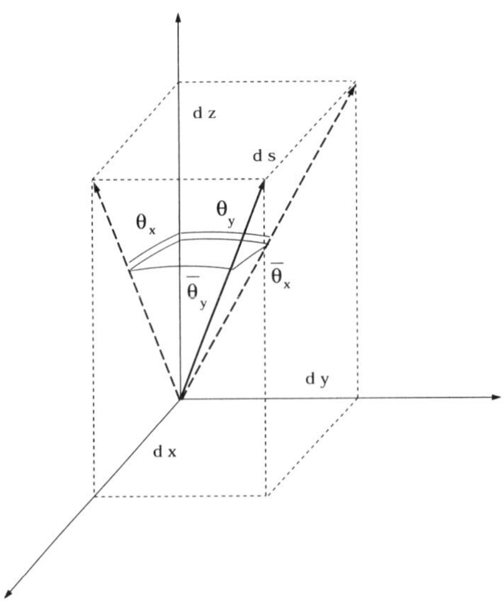

Figure 3. Angles for a ray trajectory

In the paraxial approximation the coordinate z and the arc length s are equivalent. We shall use s, as customary, from now on.
The basic elements of a lattice are the vacuum tube (drift), the solenoid, the FODO cell and the sextupole.

The *solenoid* provides a constant focusing due to the edge effect. If the magnetic field is $\mathbf{B} = B\mathbf{e}_z$ then the potential of the force exerted on a charge q is

$$\frac{V}{2E} = \frac{k}{2}(x^2 + y^2), \qquad k = \left(\frac{\omega_L}{v_0}\right)^2 = \left(\frac{qB}{2p_0 c}\right)^2$$

The *FODO cell* is the basic focusing system. A focusing quadrupole is formed by 4 coils whose field is $\mathbf{B} = k\,\rho\,B_0(y\mathbf{e}_x + x\mathbf{e}_y)$.

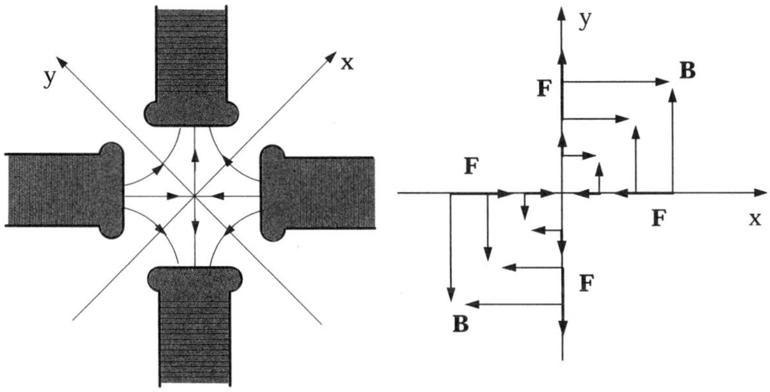

Figure 4. Quadrupole and its force field

The potential of the corresponding force is

$$\frac{V}{2E} = k\frac{x^2 - y^2}{2}$$

The force is attractive along the x axis and repulsive along the y axis and its equipotential lines are hyperbolae. A defocusing quadrupole is obtained by a $\pi/2$ rotation which interchanges the x and y axis. The FODO cell is formed by a drift a defocusing quadrupole another drift and a focusing quadrupole, and the potential has the same expression where k depends on s and is a piecewise constant function. The focusing effect of this system is very similar to the stabilization of a reversed pendulum with a vertically oscillating suspension point.

The *sextupole* is used to correct the chromatic aberrations namely the effect due to a longitudinal momentum $p \neq p_0$. The potential is

$$\frac{V}{2E} = k_2\,\frac{3xy^2 - x^3}{6}$$

If the sextupole is thin $\ell_s \ll 1$, the impulsive approximation is justified and the effect of the sextupole is to change the momenta according to

$$\Delta p_x = \frac{k_2}{2}\,\ell_s\,(x^2 - y^2) \qquad \Delta p_y = -\,k_2\,\ell_s\,x\,y$$

LINEAR DYNAMICS

The propagation through a magnetic element or a fraction of it is given by a symplectic transfer map

$$\begin{pmatrix} x' \\ p'_x \end{pmatrix} = \mathsf{L}_x \begin{pmatrix} x \\ p_x \end{pmatrix}, \qquad \begin{pmatrix} y' \\ p'_y \end{pmatrix} = \mathsf{L}_y \begin{pmatrix} y \\ p_y \end{pmatrix}$$

The matrix elements are related to the focal distances f_1, f_2 and the distances d_1, d_2 of the principal planes as for any optical system.

Neutral elements

In a straight section (drift) of length ℓ the propagation is rectilinear and $\mathsf{L}_x = \mathsf{L}_y = \mathsf{O}$

$$\mathsf{O} = \begin{pmatrix} 1 & \ell \\ 0 & 1 \end{pmatrix}$$

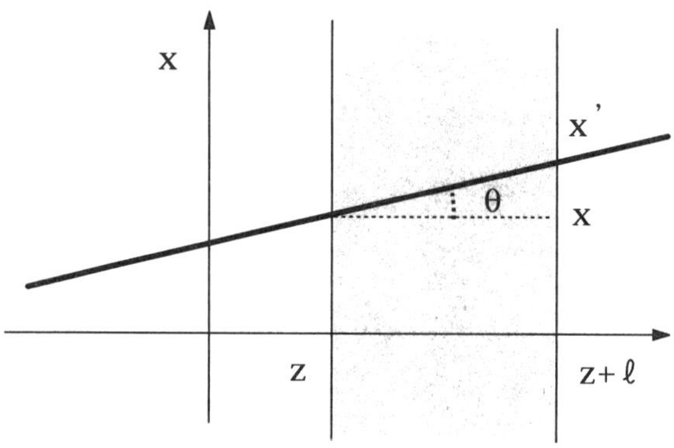

Figure 5. Rectilinear ray propagation $p_x = dx/dz = \tan\theta$ and $x' - x = \ell \tan\theta$. In the paraxial approximation s and z can be interchanged and $\sin\theta \sim \tan\theta \sim \theta$

Active elements

For a thick converging lens of length ℓ, by solving $\dot{x} = p_x$, $\dot{p}_x = -kx$, we determine the transfer matrix

$$\mathsf{F} = \begin{pmatrix} \cos(k^{1/2}\ell) & k^{-1/2}\sin(k^{1/2}\ell) \\ -k^{1/2}\sin(k^{1/2}\ell) & \cos(k^{1/2}\ell) \end{pmatrix}$$

For a thick diverging lens, where the force is reversed $\dot{x} = p_x$, $\dot{p}_x = kx$

$$\mathsf{D} = \begin{pmatrix} \text{ch } (k^{1/2}\ell) & k^{-1/2}\text{sh } (k^{1/2}\ell) \\ k^{1/2}\text{sh } (k^{1/2}\ell) & \text{ch } (k^{1/2}\ell) \end{pmatrix}$$

The magnetic elements are a combination of converging and diverging lenses on the x and y axis.

Focusing quadrupole	$\mathsf{L}_x = \mathsf{F}$	$\mathsf{L}_y = \mathsf{D}$
Defocusing quadrupole	$\mathsf{L}_x = \mathsf{D}$	$\mathsf{L}_y = \mathsf{F}$
Solenoid	$\mathsf{L}_x = \mathsf{F}$	$\mathsf{L}_y = \mathsf{F}$

Thin Lenses

In the limit $\ell \to 0$ and $\ell k \to 1/f$ one obtains the thin lens approximation

$$\mathsf{F} = \begin{pmatrix} 1 & 0 \\ -\frac{1}{f} & 1 \end{pmatrix}, \qquad \mathsf{D} = \begin{pmatrix} 1 & 0 \\ \frac{1}{f} & 1 \end{pmatrix}$$

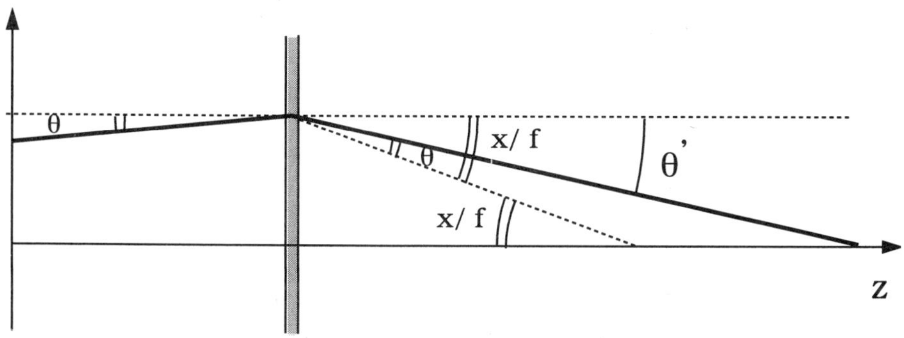

Figure 6. Geometric relations for a thin converging lens. If θ=0 one has θ'=$-x/f$. In general θ'=$\theta-x/f$ in the paraxial approximation

The transfer matrix for a FODO cell in the thin lens approximation is readily computed and reads

$$\mathsf{FODO} = \begin{pmatrix} 1+\frac{\ell}{f} & 2\ell+\frac{\ell^2}{f} \\ -\frac{\ell}{f^2} & 1-\frac{\ell}{f}-\frac{\ell^2}{f^2} \end{pmatrix}$$

Since one half of its trace is $1-\ell^2/(2f^2)$ the cell has a focusing effect on both axis.

The one turn map

By composing the maps L_k of the elements which form the accelerator one obtains the one turn map $\mathsf{L}(s)$. It depends on s namely on the point of the lattice we consider. If $|\mathrm{Tr}\ \mathsf{L}_x\,| < 2$ the eigenvalues are $e^{\pm\, i\,\omega_x}$ and the map is conjugated to a rotation [9,10].

$$\mathsf{L}_x(s) = \mathsf{W}_x(s)\,\mathsf{R}(\omega)\,\mathsf{W}_x^{-1}(s)$$

The map $\mathsf{W}_x(s)$ is fixed uniquely by choosing $\mathsf{W}_{12} = 0$ and reads

$$\mathsf{W}_x = \begin{pmatrix} \beta_x^{1/2} & 0 \\ -\alpha_x\,\beta_x^{-1/2} & \beta_x^{-1/2} \end{pmatrix}$$

An invariant for the orbit of the particle is given by the action

$$\jmath_x = \frac{1}{2\beta_x(s)} \left(x^2 + (\beta_x(s)\, p_x + \alpha_x(s)\, x)^2 \right),$$

After each turn the particle does not return to the initial position but moves on an ellipse of area $2\pi\jmath_x$. The amplitude is given by $A_x = (2j_x\beta_x)^{1/2}$. At each turn the phase is increased by $\omega = \int_0^L \beta^{-1}(s)\, ds$.

Using *normal coordinates*

$$\begin{pmatrix} X \\ P_x \end{pmatrix} = \mathsf{W}^{-1} \begin{pmatrix} x \\ p_x \end{pmatrix} \qquad\qquad \jmath_x = \frac{1}{2}(X^2 + P_x^2)$$

the orbit becomes a circle of radius $(2\jmath_x)^{1/2}$. When s is varied the ellipse changes but its area remains unchanged.

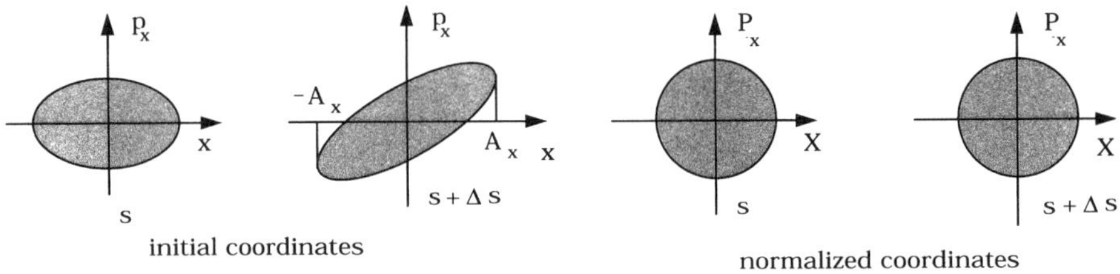

Figure 7. Elliptical beam at s, $s{+}\Delta s$ and its image in normal coordinates

The transfer matrix from s_0 to s reads

$$\mathsf{L}_x(s, s_0) = \mathsf{W}_x(s)\; \mathsf{R}\; (\phi_x(s_0, s))\; \mathsf{W}_x^{-1}(s_0)$$

where the phase advance is the sum of a linear function plus a periodic one $\phi_x(s_0, s) = \omega\frac{s-s_0}{L} + \hat{\phi}_x(s - s_0)$ where $\hat{\phi}_x(0) = 0$. It can be shown that $\phi_x = \int_{s_0}^{s} \beta_x^{-1}(t)dt$. Consider a section of the machine where $\alpha_x = 0$, and a uniform distribution of particles on an ellipse with semi-axis along x and p_x and ratio $\beta(s_0)$. Such an ellipse changes while the beam advances, as shown by figure 7, but its density remains constant since the transformation is area preserving. Using normal coordinates the distribution is uniform on a circle which remains unchanged.

NON LINEAR DYNAMICS

The non linear effects on a high energy accelerator are due to the the chromatic sextupoles and also to the multipolar errors in the superconducting dipoles. The potential of a sextupole of length ℓ_S is given by $V = 2E\, k_2\, (3xy^2 - x^3)/6$ and in the thin lens approximation, where the sextupolar force acts impulsively, the one turn map reads, letting $K_2 = \ell_S\, k_2$ be the integrated gradient

$$M = \mathsf{L} \begin{pmatrix} x \\ p + \frac{1}{2}K_2\,(x^2 - y^2) \\ y \\ p_y - K_2\, xy \end{pmatrix}$$

Referring for simplicity to a flat beam $y = p_y = 0$ and performing the change of coordinates $\frac{1}{2} K_2 \beta_x^{3/2} \mathsf{W}_x^{-1}\mathbf{x} \rightarrow \mathbf{x}$ where $\mathsf{L}_x = \mathsf{W}_x \mathsf{R}(\omega)\mathsf{W}_x^{-1}$, we obtain the two dimensional Hénon map

$$\begin{pmatrix} x' \\ p'_x \end{pmatrix} = \mathsf{R}(\omega) \begin{pmatrix} x \\ p_x + x^2 \end{pmatrix}$$

For $\omega \rightarrow 0$ this map is the symplectic integrator of the cubic Hamiltonian $H = \omega \frac{p_x^2 + x^2}{2} - \frac{x^3}{3}$ with step $\Delta s = 1$ and its phase space portrait exhibits and integrable behavior, see figure 8. By increasing the linear frequency ω, closed curves (tori) around the origin are still present but new structures such as chains of islands and chaotic orbits appear. The frequency Ω of an orbit is defined as average phase advance and specifies its nature. The tune $\nu = \Omega/2\pi$ is irrational on a torus, it has the same rational value p/q on all the orbits of a chain with q islands and varies randomly for a chaotic orbit [11,12]

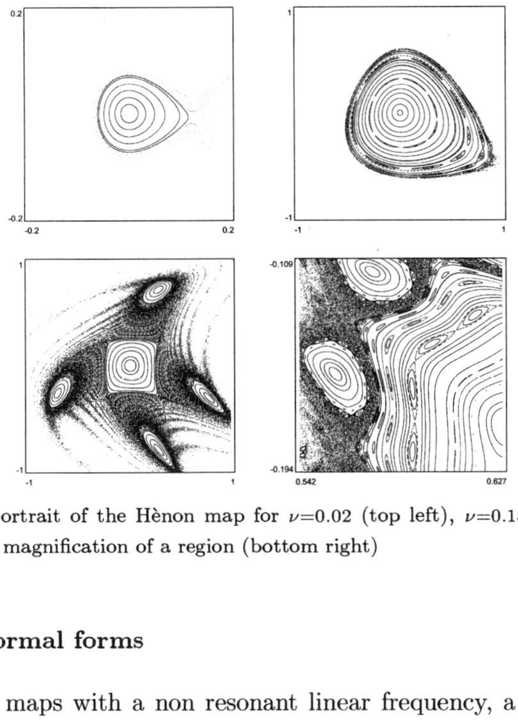

Figure 8. Phase portrait of the Hènon map for ν=0.02 (top left), ν=0.15 (top right), ν=0.255 (bottom left) and the magnification of a region (bottom right)

Non resonant normal forms

For nonlinear maps with a non resonant linear frequency, a coordinate change can be found such that all the orbits near the origin become circles

$$M = \Phi \circ U \circ \Phi^{-1}$$

The new map U is just a rotation of a angle Ω which depends on the radius and Φ is a symplectic map close to the identity. The map U is invariant with respect to the group generated by the linear part $R(\omega)$ of the map M. This group is $O(2)$ if ω is non resonant. Letting $(X, P_x) = \Phi^{-1}(x, p_x)$ be the normal coordinates and (Θ, J) be the action angle variables defined by $X = \sqrt{2J}\cos\Theta, \quad P_x = -\sqrt{2J}\sin\Theta$ the frequency is a function of J which is invariant. One introduces an interpolating Hamiltonian $\hat{H}(J)$, whose flow at integer times $t = s/L$ (normalized to the revolution period if L

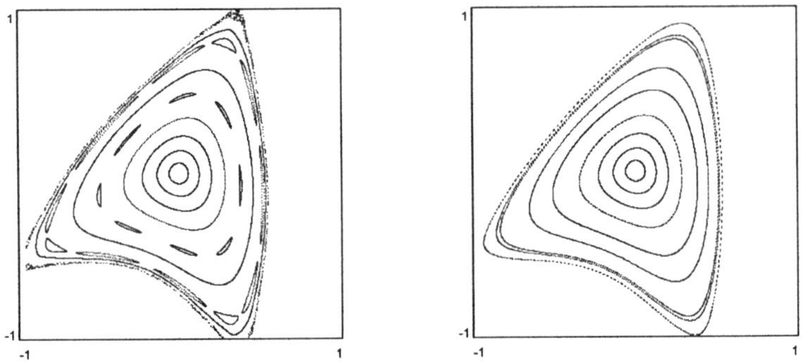

Figure 9. Hènon map with $\omega/2\pi{=}(\sqrt{5}{-}1)/2$ (left). Level curves of $H{=}\hat{H}(\Phi^{-1})$ (right)

is the accelerator length) is the orbit of the one turn map.

$$U = R(\Omega), \qquad \Omega(J) = \frac{\partial H}{\partial J}, \qquad J = \frac{P^2 + X^2}{2}$$

Resonant Normal Forms

When the linear frequency is resonant $\omega_R = 2\pi\ p/q$ or close to a resonance $\omega = \omega_R + \epsilon$, we construct a map U in normal form with respect to the discrete group generated by $R(\omega_R)$. We introduce another map $\hat{U}$

$$U = R(\omega_R)\,\hat{U}$$

which is the time one flow of an interpolating Hamiltonian in normal form $\hat{H} = \hat{H}(R(\omega_R))$. As a consequence the orbits of U itself also belong to the level curves of $\hat{H}$ since $\hat{H}(U) = \hat{H}(R(\omega_R)\hat{U}) = \hat{H}(\hat{U}) = \hat{H}$. Using the Lie derivative $D_{\hat{H}}$ we can write $U = R(\omega_R)\exp\left(D_{\hat{H}}\right)$. In action angle coordinates the Hamiltonian $\hat{H}$ has only the harmonics kq in its Fourier expansion and its pendulum like orbits exhibit a chain of q islands, see figure 10. The rotation $R(\omega_R)$ takes into account the jumps from one island to another which occur when U is iterated [13].

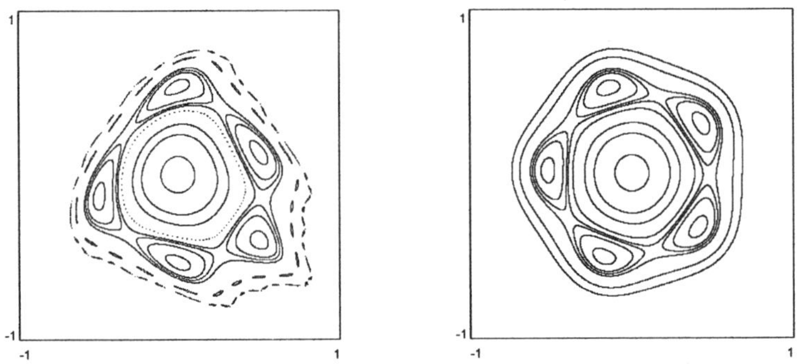

Figure 10. Hènon map $\omega/2\pi{=}0.21$ (left). Level curves of $\hat{H}$ with $\omega_R/2\pi{=}1/5$ (right) exhibit five-fold symmetry.

Perturbative Expansion

This method allows to compute the normal forms. Letting U_N and Φ_N be order N polynomials and E_N a remainder defined by a series of polynomials of order larger than N one has

$$M = \Phi_N \circ (U_N + E_N) \circ \Phi_N^{-1}, \qquad \qquad \|E_N\| \le (CrN^\alpha)^N$$

where the coefficients of the polynomials are determined recursively order by order, by imposing the additional constraint that Φ_N is the truncation of a symplectic map. Letting $N \to \infty$ the error norm does not vanish and the series defining U and Φ diverge. However their asymptotic nature allows to define an optimal truncation order. Having fixed the radius r of a disc in the phase plane, the error decreases until a minimum is reached for $N \sim r^{-1/\alpha}$. For a symplectic map of $\mathbb{R}^{2d}$ the orbit remains stable for an exponentially long number of iterations

$$n \ge n_* \exp\left[\left(\frac{r_*}{r}\right)^{\frac{2}{d+1}}\right]$$

where $\alpha = (d+1)/2$ is the optimal exponent [14,15]. The one turn map of a beam has $d = 2$ and supposing that the orbits within a disc of radius r are stable for $n = 10^4$ turns, in a disc of radius $r/4$ they are stable for $10^{4\times 4^{2/3}} = 10^{10}$ turns, assuming $n_* = 1$. A similar scaling law has been found for the Hénon map in 4D and the LHC map [16,17].

The Fourier Analysis

The frequency of a given orbit is obtained by Fourier analysis. The accuracy can reach n^{-4}, with appropriate filters, for an orbit of length n.

$$(x_n - ip_{x,n},\ y_n - ip_{y,n}) = \sum_{k_1,k_2} e^{in(\Omega_1 k_1 + \Omega_2 k_2)}(a_{k_1,k_2},\ b_{k_1,k_2})$$

The actions of non resonant tori are also determined, since the interpolated orbits are dense, by integrating over the basic cycles Γ_1, Γ_2.

$$J_\ell = \frac{1}{2\pi}\int_{\Gamma_\ell}(p_x dx + p_y dy) = \frac{1}{2}\sum_{k_1,k_2} k_\ell(|a^2_{k1,k_2} + |b_{k_1,k_2}|^2), \quad \ell{=}1{,}2$$

As consequence the action-frequency map is numerically accessible [18], see next figure.

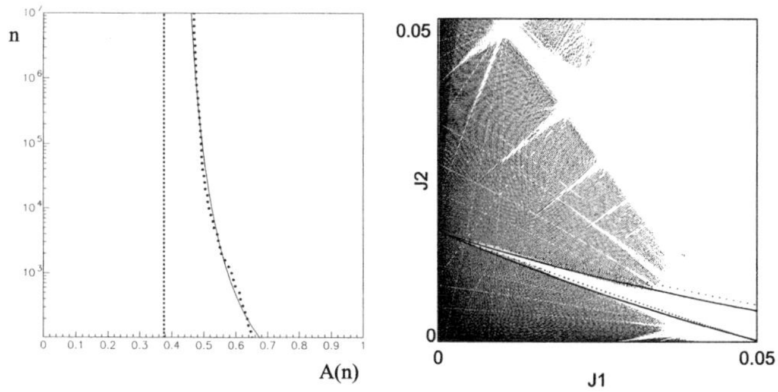

Figure 11. Dynamic aperture $A(n)$ for the Hénon map with linear frequencies 0.168, 0.201 and exponential fit (left). Non resonant invariant tori in action space for the Hénon map with linear frequencies 0.205, $\frac{1}{2}(\sqrt{5}-1)$ and boundary of the resonance 5 from resonant normal forms (right)

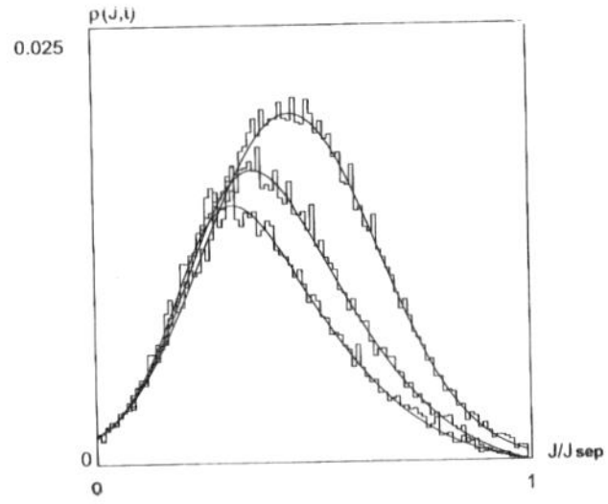

Figure 12. Action distribution for $H=\frac{1}{2}(x^2+p^2)+\frac{1}{3}x^3+\frac{1}{2}\,\xi(t)x^2$ with $\epsilon=0.0025$ at $t=0,2\,10^5,4\,10^5$

NOISE EFFECTS

Fluctuations of the lattice parameters may arise from mechanical vibrations or misalignments, from current ripples or from random emission of photons and determine a linear growth of the beam size. The simplest model is obtained by combining a random rotation with a FODO cell plus a thin sextupole.

$$M_{\text{stoc}} = \mathsf{R}(\epsilon\xi)\,M, \qquad\qquad M = \mathsf{R}(\omega)\begin{pmatrix} x \\ p+\ x^2 \end{pmatrix}$$

where ξ is a random variable with zero mean and variance σ. The deterministic orbits of the map M are interpolated by a Hamiltonian $H_0(x,p)$, while the stochastic map is interpolated by the quadratic Hamiltonian

$$H_{\text{stoc}} = \epsilon\,\Xi\left(\frac{s}{L}\right)\,\frac{x^2+p^2}{2}$$

where $\Xi(t)$ is a non stationary process, whose value in the unit length intervals $[n, n+1[$ is constant and equal to a random variable ξ. The noise gives rise to a diffusion process.

In order to determine the diffusion rate and the equation satisfied by the probability density $\rho(x,p;t)$, we assume that the linear frequency is non resonant and introduce a set of normal coordinates (X,P) such that the level curves of H_0 become circles and we denote by V the quadratic Hamiltonian $(x^2+p^2)/2$ written in normal coordinates. For $\epsilon\to 0$ it can be proved that, denoting by $\xi(t)$ a white noise, the stochastic Hamiltonian is equivalent to

$$H = H_0\left(\frac{P^2+X^2}{2}\right) + \epsilon\xi(t)V(X,P) = H_0(J) + \epsilon\xi(t)\;V(\Theta,J)$$

where Θ, J are the normal action-angle coordinates defined by $X = \sqrt{2J}\cos\Theta$ and $P = -\sqrt{2J}\sin\Theta$. The probability density satisfies the Fokker-Planck equation [19]

$$\frac{\partial\rho}{\partial t} + \Omega\,\frac{\partial\rho}{\partial\Theta} = \frac{\epsilon^2\sigma^2}{2}\,[[\rho,V],V]$$

If Ω varies with J the angle diffusion rate is much faster with respect to the action diffusion according to $\langle(\Theta-\Omega t-\Theta(0)\,)^2\rangle \propto \epsilon^2t^3$, $\langle(J-J(0))^2\rangle \propto \epsilon^2 t$ and angle averaging it is justified. The average density distribution $\rho(J,t)$ satisfies the equation [19]

$$\frac{\partial\rho}{\partial t} = \frac{\partial}{\partial J}D\frac{\partial\rho}{\partial J}, \qquad\qquad D(J) = \frac{\epsilon^2\sigma^2}{4\pi}\int_0^{2\pi}\left(\frac{\partial V}{\partial\Theta}\right)^2\,d\Theta$$

whose solutions are in excellent agreement with the simulations [20] as shown by the figure 12. A similar agreement is found for the stochastically perturbed 4D Hénon map.

The theory extends to the case of quasi resonant frequencies, where a chain of islands appears in the phase space of the unperturbed system [21]. By imposing that the diffusion coefficient is infinite $D = \infty$ within the islands, the Fokker-Planck solutions are in agreement with the simulations as shown by figure 13.

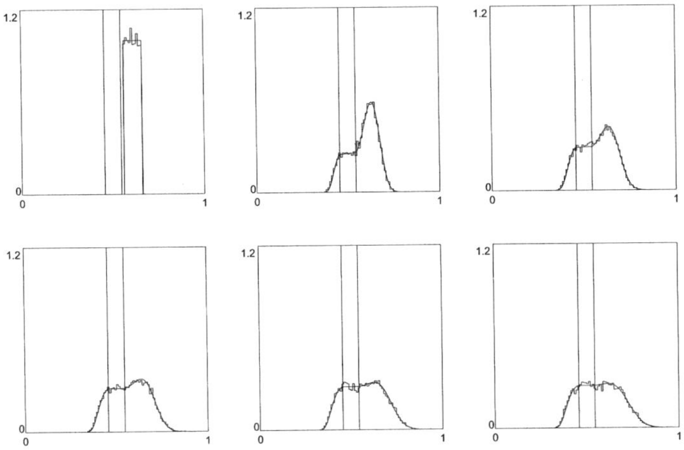

Figure 13. Action distribution for the Hamiltonian $H_0{=}2\,j\,(1{+}\epsilon\,\xi(t)){+}A\,j^2\,(1{+}\epsilon\cos(2\theta))$ with 4 islands where $A{=}10$, $\epsilon{=}0.01$. At times $t{=}3k\,10^3$ with $k{=}0,1,\ldots,5$ the simulation (histograms) and Fokker-Planck solution (curves) are shown; vertical lines delimit the island.

SPACE CHARGE EFFECTS

For low energy high intensity beams the Coulomb repulsion among the charged particles cannot be neglected, since it is no longer compensated by the magnetic attraction. Indeed letting qE_r be the radial force on a particle of charge q due to the Coulomb repulsion, the magnetic attraction force is $-qE_r\,v_0^2/c^2$, where v_0 is the longitudinal velocity. For a uniform distribution of charges in a cylinder of radius R with a velocity v_0 along the cylinder axis, the electric field is $E_r = 2Q\,r/R^2$ where $Q = I/v_0$ is the charge per unit length and I is the beam current. As a consequence the force acting on a particle is proportional to $I\,(1 - v_0^2/c^2)$; if it is not small, the collective effects are important and must be taken into account [22]. In this case the Hamiltonian $H_{\,\rm tot}$ is no longer the sum of single particle Hamiltonians without interaction terms and it is hard to describe the transport with the Liouville equation for the density $\rho_{\,\rm tot}$ in phase space, because its dimension is very high. The use of numerical solutions with symplectic integrators is limited by the computational complexity $\sim N^2$ and by the presence of collisions. The standard approach consists in using the Vlasov equation, valid whenever the collisions can be neglected and distribution function $\rho_{\,\rm tot}$ can be written as a product of single particle distributions $\rho(\mathbf{r}_1, \mathbf{p}_1, s)\cdots\rho(\mathbf{r}_N, \mathbf{p}_N, s)$. The Vlasov equation reads

$$\frac{\partial\rho}{\partial s} + [\rho, H] = 0$$

where H is the single particle Hamiltonian. This is defined by

$$H = \frac{p_x^2 + p_y^2}{2} + k_{0\,x}(s)\frac{x^2}{2} + k_{0\,y}(s)\frac{y^2}{2} + \frac{q\,Q}{v_0 p_0}\left(1 - \frac{v_0^2}{c^2}\right)\,V(x, y)$$

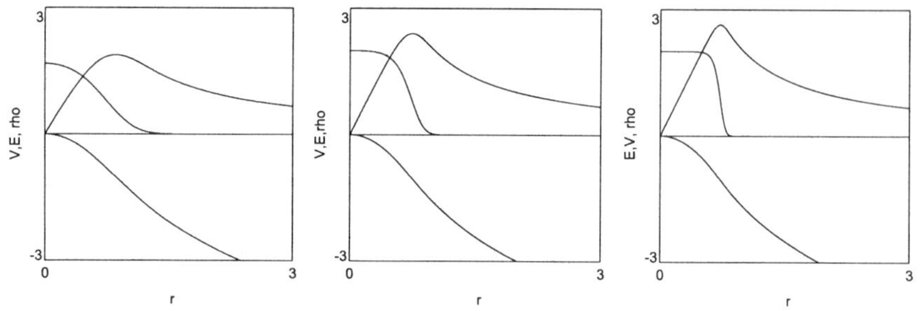

Figure 14. Self consistent Maxwell-Boltzmann distribution and corresponding electric field E and potential V for $\xi/(k_0R^2)$=0.5 and KT=0.1 (left), KT=0.02 (center) KT=0.005 (right)

where V is the self-consistent potential for a unit linear charge density, which satisfies Poisson's equation

$$\Delta V = -4\pi\rho_s(x,y,s), \qquad \rho_s = \int \rho(x,y,p_x,p_y,s)\,dp_x\,dp_y$$

and Q is the linear charge density related to the beam current by $Q = I/v_0$ Vlasov equation can be simulated because the computational complexity at each step Δt is of order N and some stationary solutions are analytically known. This is the case for an axially symmetric distribution on a constant focusing channel where $k_{0\,x} = k_{0\,y} = k_0$. In this case an equilibrium distribution $\rho = \rho(H)$ with support on $H \leq E_T$ (here E_T denotes a transverse energy whereas $E_L = v_0p_0/2$ is the longitudinal energy). Denoting by $W(r)$ the overall potential and by R the distribution radius $W(R) = E_T$, the space distribution is given by

$$\rho_s(r) = \int_0^{2\pi} d\theta \int_0^{E-W(r)} \rho(H)\,d\frac{p^2}{2} = 2\pi \int_{W(r)}^{W(R)} \rho(H)\,dH, \quad W(r) = \frac{k_0^2}{2}r^2 + \frac{q}{mv_0^2}\,V(r)$$

The radial Poisson equation becomes a nonlinear differential equation for V to be satisfied with the additional constraint that ρ_s is normalized

$$\frac{1}{r}\frac{d}{dr}r\frac{dV}{dr} = -4\pi\rho_s(r), \qquad 2\pi\int_0^R \rho_s(r)r\,dr = 1$$

The behavior of $V(r)$ at the origin follows from Gauss law $E_r = 4\pi\,r^{-1}\int_0^r \rho_s(r')\,r'\,dr'$ where $E_r = -V'(r)$ is the electric field and reads $V(r) = -\pi\,\rho_s(0)\,r^2 + O(r^3)$. The equation for V has to be solved with vanishing initial conditions and $r^{-1}V'(r)|_{r=0} = -2\pi\rho_s(0)$. The normalization of ρ follows from the asymptotic behavior of the solution

$$2\pi\int_0^R \rho_s(r)\,r\,dr = \lim_{r\to\infty} \frac{r\,E_r(r)}{2}$$

Starting with an arbitrary normalization, the correct normalization and the corresponding potential are determined recursively. In two special cases $\rho = \rho_0\delta(H - E)$ and $\rho = \rho_0\theta(H - E)$ there are analytical solutions known as KV and waterbag distributions. For the Maxwell-Boltzmann distribution $\rho = \rho\,e^{-H/\alpha}$, to which a generic distribution should relax if collisional effects were present choosing $\alpha = kT/(v_0p_0)$, the self consistent potential must be determined numerically. In the laminar limit $T \to 0$ the space distribution is uniform for $r < R$, whereas in the limit $T \to \infty$ it becomes Gaussian.

The KV Distribution

The normalized space density corresponding to $\rho = \rho_0 \delta(H - E)$ is uniform within a cylinder namely $\rho_S = (\pi R^2)^{-1}\theta(R - r)$ where the electric potential is $V = -r^2/R^2$ for $r < R$. The Hamiltonian reads

$$H = \frac{p_x^2 + p_y^2}{2} + \frac{k}{2}r^2, \qquad k = k_0 - \frac{\xi}{R^2}, \qquad \xi = \frac{2qQ}{mv_0^2}\left(1 - \frac{v_0^2}{c^2}\right)$$

where k is the strength of the elastic force depressed by the Coulomb repulsion, whose strength is measured by the perveance ξ. The radius is fixed by $E = \sqrt{k}\, \jmath_0$, where $\jmath_0$ is the action, and from $E = kR^2/2$ one obtains $R^2 = 2\jmath_0/\sqrt{k} = 2\jmath_0\,(k_0 - \xi/R^2)^{-1/2}$. The space charge effect is measured by the ratio between the tune $\nu = L\sqrt{k}/(2\pi)$ and the bare tune $\nu_0 = L\sqrt{k_0}/(2\pi)$ known as tune depression $\nu/\nu_0 = (k/k_0)^{1/2}$.
In the general case the KV self consistent solution can be obtained for a distribution $\rho = \rho_0\, \delta(F)$ whose support $F = 0$ is an ellipsoid [22]. Indeed its projection $\rho_S(x, y)$ is uniform on an ellipse and the corresponding electric field is linear. As a consequence the ellipse semi-axis can be self-consistently determined. The Hamiltonian in the initial coordinates reads

$$h = \frac{p_x^2 + p_y^2}{2} + k_{0\,x}(s)\frac{x^2}{2} + k_{0\,y}(s)\frac{x^2}{2} + \frac{q\,Q}{v_0 p_0}V = \frac{p_x^2 + p_y^2}{2} + k_x(s)\frac{x^2}{2} + k_y(s)\frac{x^2}{2}$$

and the normalized coordinates it becomes

$$H = \frac{P_x^2 + X^2}{2\beta_x(s)} + \frac{P_y^2 + Y^2}{2\beta_y(s)} \equiv \frac{\jmath_x}{\beta_x(s)} + \frac{\jmath_y}{\beta_y(s)}$$

where $H \neq h$ since the transformation depends on s. The actions $\jmath_x$, $\jmath_y$ are invariant. Any function $F(\jmath_x, \jmath_y)$ is invariant, but the Hamiltonian H is not since it depends explicitly on s. Choosing F as a linear combination of the invariants with constant coefficients we have the normalized distribution

$$\rho = \frac{1}{4\pi^2 \jmath_{0x} \jmath_{0y}}\, \delta\left(\frac{\jmath_x}{\jmath_{0x}} + \frac{\jmath_y}{\jmath_{0y}} - 1\right)$$

which is an equilibrium distribution in the normalized coordinates system. With respect to the initial phase space coordinates F is invariant $dF/ds = 0$ but exhibits an explicit dependence on s such that $\partial F/\partial s = -[h, F] \neq 0$. Since F is periodic in s with period L we say it is stationary. The support of ρ is a fixed ellipsoid $F(X, P_x, Y, P_y) = 0$ in the normal coordinates, a variable ellipsoid $F(x, p_x, y, p_y, s) = 0$ in the initial coordinates. The ellipsoids are foliated by a one parameter family of invariant 2D tori defined as products of circles (ellipses) in the normal (initial) phase planes $\jmath_x = c$, $\jmath_y = \jmath_{0\,y}(1 - c/\jmath_{0\,x})$. The largest ellipses in the initial coordinates are given by

$$x^2 + (x\,\alpha_x + \beta_x\, p_x)^2 = A_x^2 \qquad y^2 + (y\,\alpha_y + \beta_y\, p_y)^2 = A_y^2$$

$A_x = (2\beta_x \jmath_{0x})^{1/2}$ and $A_y = (2\beta_y \jmath_{0y})^{1/2}$ are the x, y amplitudes. The projection of the ellipsoid in the configuration plane is a fixed ellipse in normal coordinates, a variable ellipse in the initial coordinates. The distribution is uniform and can be written as

$$\rho_S(x, y, s) = \int \rho dp_x dp_y = \frac{1}{\pi A_x(s) A_y(s)}\, \vartheta\left(1 - \frac{x^2}{A_x^2(s)} + \frac{y^2}{A_y^2(s)}\right)$$

The electric field generated by a charge distribution ρ_s is

$$E_x = \begin{cases} \dfrac{4\,x}{A_x(A_x + A_y)} & \text{if} \quad t < 1 \\ \\ \dfrac{4\,x}{A_x'(A_x' + A_y')} & \text{if} \quad t > 1 \end{cases} \qquad\qquad t = \frac{x^2}{A_x^2(s)} + \frac{y^2}{A_y^2(s)}$$

${A_x'}^2 = A_x^2 + \chi$, ${A_y'}^2 = A_y^2 + \chi$ are semi-axis of the confocal ellipse $x^2/{A_x'}^2 + y^2/{A_y'}^2 = 1$.

Envelope Equations

The equations of motion within the core

$$\frac{d^2x}{ds^2} + k_{0\,x}(s)x = \frac{\xi}{2}\,E_x \equiv \frac{2\xi\;x}{A_x(A_x + A_y)}, \qquad \frac{d^2y}{ds^2} + k_{0\,y}(s)y = \frac{\xi}{2}\,E_y \equiv \frac{2\xi\;y}{A_y(A_x + A_y)}$$

where $\xi = 2qQ/(v_0 p_0\,\gamma^2)$ is the *perveance* and $\gamma^{-2} = 1 - v_0^2/c^2 \;\rightarrow\; 1$ in the non-relativistic limit, have to be solved with a self consistent electric field. An iterative procedure can be used starting with zero electric field the beam is distributed uniformly in an ellipse of semi-axis $A_x^{(0)} = (2\jmath_{0\,x}\,\beta_x^{(0)})^{1/2}$ and $A_y^{(0)} = (2\jmath_{0\,y}\,\beta_y^{(0)})^{1/2}$. At the next step the electric field is computed for a charge distributed in this ellipse and the particles, due to repulsion, accommodate into a bigger ellipse of semi-axis $A_x^{(1)} = (2\jmath_{0\,x}\,\beta^{(1)})^{1/2}$ and $A_y^{(1)}$, see figure 15. The sequences $A_{x,y}^{(n)}$ and $\beta_{x,y}^{(n)}$ converge for moderate space charges corresponding to a tune depression ν/ν_0 above 0.6.

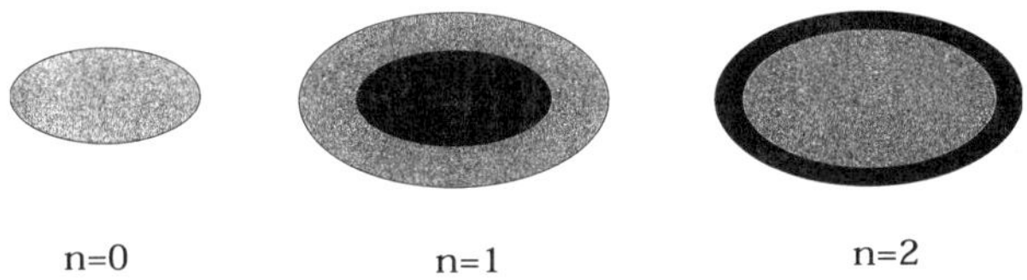

Figure 15. Ellipse without charge (left). Ellipse at the first iteration (center) where the previous ellipse is uniformly charged. Ellipse at the second iteration (right). The charge distribution is darker.

The limit amplitudes prove to be the periodic solutions of the self consistent envelope equations. Given a uniform charge distribution on an ellipse of semi-axis B_x, B_y, the equation of motion is linear within the ellipse and the amplitudes A_x, A_y of the betatronic oscillations satisfy the envelope equation

$$\begin{aligned} \frac{d^2A_x}{ds^2} - \frac{4\jmath_{0\,x}^2}{A_x^3} + k_{0\,x}(s)\,A_x &= \frac{2\xi}{B_x(B_x + B_y)}\,A_x, \\ \frac{d^2A_y}{ds^2} - \frac{4\jmath_{0\,y}^2}{A_y^3} + k_{0\,y}(s)\,A_y &= \frac{2\xi}{B_y(B_x + B_y)}\,A_y \end{aligned}$$

The iterative procedure corresponds to solve for the amplitudes $A_{x,y}^{(0)}$ without space charge, then to choose $B_{x,y} = A_{x,y}^{(0)}$ and to solve for $A_{x,y}^{(1)}$ and so on, requiring that the solutions are periodic. The limit $A_{x,y}(s)$ is the periodic solution of the above equation where B_x, B_y are replaced with A_x, A_y. The solution of this nonlinear equation exists for arbitrary space charge.

In the constant focusing case the periodic solutions become constant and correspond to the the critical point of the above equation, which proves to be stable. In the symmetric case $k_{0\,x} = k_{y_0} = k_0$ and $j_{0\,x} = j_{0\,y} = j_0$ the amplitude is solution of

$$\frac{4j_0^2}{A^4} = k_0 - \frac{\xi}{A^2}$$

and setting $k = k_0 - \frac{\xi}{A^2}$ we have $\beta = k^{-1/2}$ and $A^2 = 2j_0\beta$.

Beyond the periodic solutions the envelope equations has mismatched solutions, which are are quasi-periodic. The periodic solutions are stationary (constant in normal coordinates) and represent equilibria of the Vlasov equations, whereas the mismatched solutions describe the non equilibrium states. To find the mismatched solutions we consider a small perturbation η_x, η_y to a periodic solution and linearize.

$$\ddot{\eta}_x + k_1\,\eta_x + \tau\,(\eta_x + \eta_y) = 0 \qquad \ddot{\eta}_y + k_2\,\eta_y + \tau\,(\eta_x + \eta_y) = 0$$

where

$$k_1 = k_{0x} + 3k_x \quad k_2 = k_{0y} + 3k_y \quad k_{x,y} = k_{0x,y} - \frac{2\xi}{A_{x,y}(A_x + A_y)} \qquad \tau = \frac{2\xi}{(A_x + A_y)^2}$$

The eigenfrequencies of the system can then be evaluated † and in the symmetric case $k_{0x} = k_{0y} = k_0$ and $j_{0x} = j_{0y} = j_0$ are given by $\omega_1^2 = k_0 + 3k$ and $\omega_2^2 = 2(k + k_0)$ and the corresponding eigenvectors are $\mathbf{e}_1 = (1, -1)$ and $\mathbf{e}_2 = (1, 1)$. Since unperturbed beam is round, exciting the first mode corresponds a deformation into an ellipse of semi-axis $A_x = R + \eta_0 \cos(\omega_1 t)$ and $A_y = R - \eta_0 \cos(\omega_1 t)$, the second mode to a breathing disc $A_x = A_y = R + \eta_0 \cos(\omega_2 t)$, as shown by figure 16. The perturbation to the potential is in this case is of normal quadrupole type $V = c_1 x^2 + c_2 y^2$.

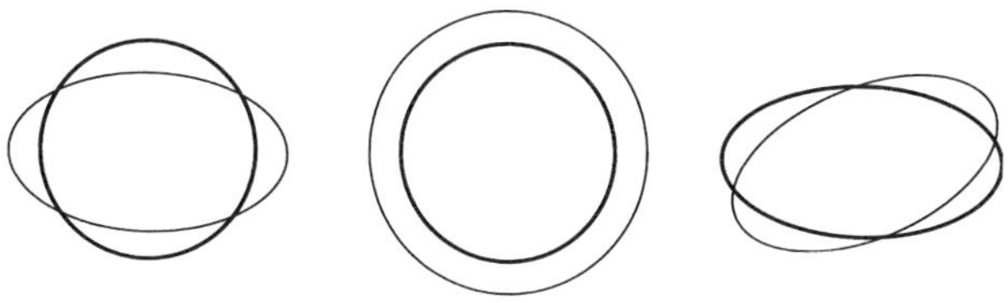

Figure 16. Quadrupolar modes normal slow mode (left side), normal fast mode (center) for a round beam. Skew quadrupolar mode for an elliptic beam (right)

By perturbing the solution of the Vlasov equation one finds a skew quadrupolar mode (and higher multipolar modes) perturbation to the potential like $c\,xy$. The frequency of this *tilting* mode is the same as the previous slow mode $\omega_1^2 = k_0 + 3k$. Breaking

† In the asymmetric case the mismatch frequencies are given by

$$\omega_\pm^2 = \frac{k_1 + k_2}{2} + \tau \pm \sqrt{\left(\frac{k_1 - k_2}{2}\right)^2 + \tau^2}$$

and can be written as a function of the bare $\omega_{0x,y} = L\,k_{0x,y}^{1/2}$ and depressed betatron frequencies $\omega_{x,y} = L\,k_{x,y}^{1/2}$ accounting for $k_{1,2} = (\omega_{0x,y}^2 + 3\omega_{x,y}^2)\,L^{-2}$ and $\tau = \left((\omega_{0x}^2 - \omega_x^2)^{-1} + (\omega_{0y}^2 - \omega_y^2)^{-1} \right)^{-1} L^{-2}$

the axial symmetry makes this mode becomes unstable [23] if $k_{0y} > k_{0x}$, $k_x > k_y$ and $\jmath_{0x}/\jmath_{0y} > (k_{0y}/k_{0x})^2\,(k_x/k_y)^{1/2}$.

Particle in Core models and Simulations

Given the electric field $\mathbf{E}$, corresponding to to the solution of the Vlasov-Poisson for a density distribution ρ, the equations of motion for a test particle

$$\frac{d^2x}{ds^2} + k_{0\,x}(s)x = \frac{\xi}{2}\,E_x(x,y,s), \qquad \frac{d^2y}{ds^2} + k_{0\,y}(s)y = \frac{\xi}{2}\,E_y(x,y,s)$$

are integrated by computing a sequence of micro-maps with the impulsive approximation for $\mathbf{E}$. It is possible also to include the effect of lattice nonlinearities, such as a thin sextupole, if the size of the beam is small enough with respect to the dynamic aperture so that it does not influence the charge distribution of the core [24,25]. For a matched beam, the electric field has the same periodicity as the lattice $\mathbf{E}(s+L) = \mathbf{E}(s)$ and the Poincaré section is suitable to examine the dynamical behavior of the system. In figure 17 the phase space for low and high charge distribution corresponding to a matched KV is shown.

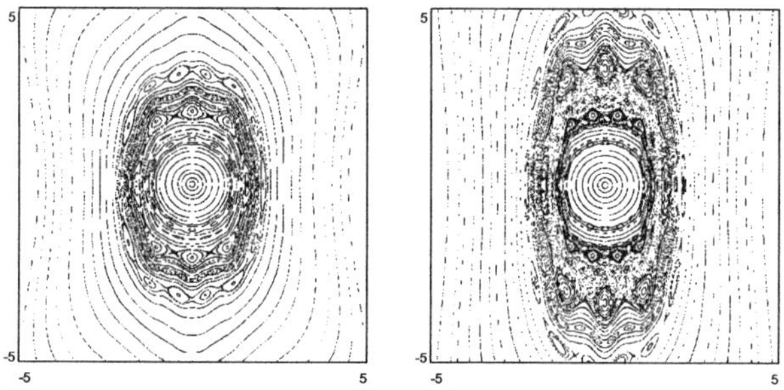

Figure 17. Orbits of matched KV beam in a FODO cell for small (left), large (right) space charge

It is possible to simulate the space charge effects numerically by replacing the particles with macroparticles of longitudinal extension Δs and charge q_{macro} interacting with a smoothed Coulomb force. This is achieved by solving the Poisson equation with a spectral method, namely by a Fourier expansion with $K \times K$ harmonics. If Q is the charge per unit length, the number of macroparticles in each micro-cell is $N = Q\Delta s/q_{\text{macro}}$. Choosing in the x, y plane a square of of area $A_0 = D_0^2$ and a basis $e^{2\pi i(x\kappa_x + y\kappa_y)/D}$, the smallest wavelength is $\delta = D_0/K$. The spectral method amounts to partitioning the square into cells of side δ. If $A = D^2$ is the area occupied by the beam the number of filled cells is $n_A = A/\delta^2 = K^2 A/A_0$ and the average number of macroparticles per cell is $n_P = N/n_A$, see figure 18. Requiring that the statistical fluctuation $(n_P)^{-1/2}$ of macroparticles per cell is equal to the space position uncertainty within the core $\delta/D = K^{-1}D_0/D = K^{-1}\,A_0^{1/2}/A^{1/2}$ the scaling law $N = K^4\,A^2/A_0^2$, is obtained [26,27].

Accurate simulations require a fine grid and consequently a high number of particles. The KV distribution being very singular is not reproduced very accurately and mismatch oscillations of the beam are observed. The actions which should be conserved do not remain exactly constant. However the field out of the the beam core is well reproduced and the dynamical evolution of test particles is close to the results obtained with an exact KV distribution. In figure 19 we compare these orbits for a typical

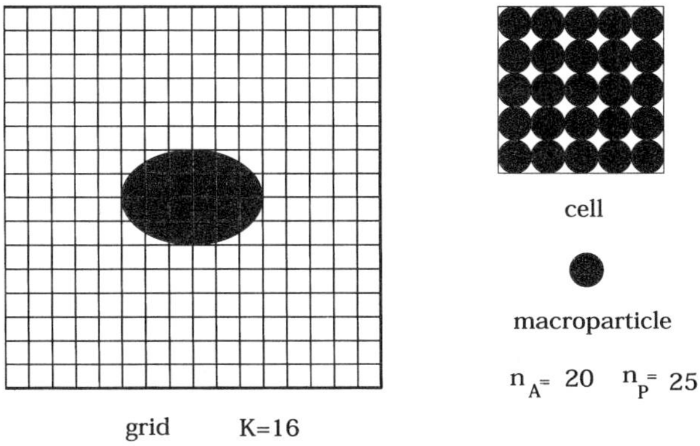

Figure 18. Grid for the numerical solution of the Poisson equation and macroparticles. The square has side 1 and the beam is on an ellipse of sides A_x=3/16, A_y=2/16 the grid size is 1/16. The beam is covered by n_A~20 cells, with n_p=25. In each cell there are n_p=25 macroparticles

FODO cell and a moderate space charge. The orbits obtained from the simulation exhibit a noise effect which is a typical effect of the numerical scheme. The noise is responsible of a slow linear increase of the actions. When the beam is mismatched, namely the initial distribution is not a KV or another stationary solution of the Poisson-Vlasov equations, the beam oscillates with a period different from the lattice length. These coherent oscillations of the beam can resonate with the betatron frequencies which vary out the beam core for the KV distribution, because the field is no longer linear.

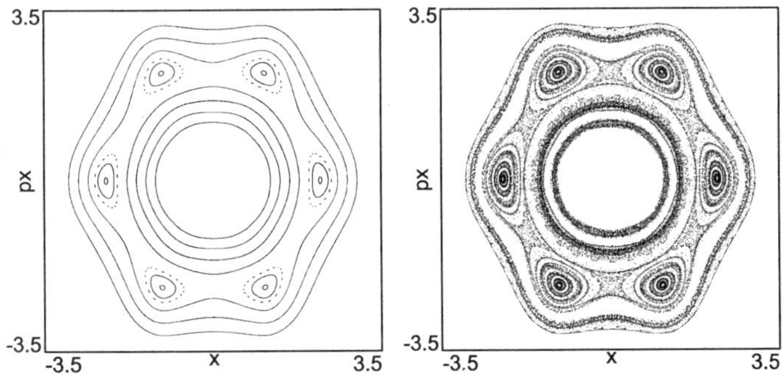

Figure 19. Poincaré map of test particles in a FODO cell with bare tunes ν_{0x}=0.1785, ν_{0y}=0.1645, depressed by space charge to ν_x=0.1438, ν_y=0.1285, Each orbit is traced twice for the particle in core model and the simulation, with the same initial point (right). Tune ν_x computed on the orbits given by the particle in core model and the simulation (left). In both cases the differences are hardly distinguishable. The simulation refers to n_P=50000 macroparticles, K=128 Fourier components in a box of size D=20. Small mismatch ($<10^{-2}$) are present since n_P is finite and their frequency agrees with the theoretical value $\nu_{\pm}$−=0.3176, 0.2837,. The emittance varies within $\sim 10^{-3}$.

In this case the one turn map analysis cannot be adopted. By approximating the mismatch frequency with a rational number m/n, the stroboscopic map iterated n times still provides a picture of the orbits behavior. A more quantitative analysis is provided by the frequency map, see figure 19, which allows to detect the resonances and the chaotic regions where the halo spreads out.

REFERENCES

1. The LHC Study group *Design study of LHC* CERN 91-03 (1991)
2. J. Perkins *The role of inertial fusion in the energy marketplace in the 21 century and beyond* Nucl. Instr. & Methods **A 415**, 44 (1988)
3. Jurgen Meyer-ter-Vehn, S. Atzeni, R. Ramis *Inertial confinment fusion* Europhysics News **29**, 202 (1998)
4. I. Hofmann, G. Plass *The HIDIF study*, GSI-98-06, Darmstadt (1998)
5. E. Mallove, G. Matloff *The starflight handbook*, Wiley and Sons, N.Y. (1989)
6. R. J. Halyard Journ. British Interplanetary Soc. **50**, 129 (1997) and **52**, 429 (1999)
7. H. Wollnik *Optics of charged particles* Academic Press N. Y. (1987)
8. A. Nussbaum *Contemporary optics for physicists and engineers* Prentice Hall (1976)
9. A. Bazzani, E. Todesco, G. Turchetti, G. Servizi *A normal form approach to the theory of nonlinear betatronic motion* CERN Yellow Reports **94-02** (1994).
10. G. Turchetti *Dinamica Classica dei sistemi fisici* Ed Zanichelli, Bologna (1998)
11. V. I. Arnold Russ. Math. Surveys **18**, 85 (1963)
12. A. J. Lichtenberg, K. A. Lieberman *Regular and stochastic motion*, Springer Verlag, Berlin (1983)
13. A. Bazzani, M. Giovannozzi, G. Servizi, G. Turchetti, E. Todesco Physica **D64**, 66 (1993)
14. A Bazzani, S. Marmi, G. Turchetti Celestial Mechanics **47**, 333 (1990)
15. A. Bazzani, G. Turchetti *Order and Chaos theory and practice* E. T. Bountis, Springer Verlag, pag. 59 (1992)
16. R. Bartolini, A. Bazzani, M. Giovannozzi, A. Scandale Part. Accel. **52**, 147 (1996)
17. M. Giovannozzi, W. Scandale, E. Todesco Part. Accel. **56**, 195 (1997)
18. A. Bazzani, L. Bongini, G. Turchetti Phys. Rev. **E 57**, 3432 (1998)
19. A. Bazzani, S. Siboni, G. Turchetti Physica **D 76**, 8 (1994)
20. A. Bazzani, S. Siboni, G. Turchetti Journal of Physics **A30**, 27 (1997)
21. G. Turchetti, F. Davico Bonino Journal de Physique IV, **8**, 173 (1998)
22. M. Reiser *Theory and design of charged particle beams* Wiley and Sons N. Y. (1994)
23. I. Hofmann Phys. Rev. **E57**, 4713 (1998)
24. A. Bazzani, M. Communian, A. Pisent Particle Accelerators **63**, 79 (1999)
25. J. M. Lagniel Nucl. Instr. Meth **A 345**, 405 (1994)
26. G. Turchetti et al. *Frequency map analysis of resonances in a non linear lattice with space charge* Nucl. Instr. Meth. to appear.
27. G. Turchetti et al *3D solution of the Poisson-Vlasov equations for a line of FODO cells* Phys. Rev. Special Topics to appear.

BIOLOGY

AN AXIOMATIC APPROACH TO SOME BIOLOGICAL THEMES

Marco Forti*, Paolo Freguglia†, and Lodovico Galleni§

*Dip. Matematica Applicata "U. Dini", Università di Pisa
Via Bonanno 25 B
56100 Pisa (Italia)
forti@dma.unipi.it

†Dip. Matematica Pura e Applicata, Università dell'Aquila
Via Vetoio snc, Loc. Coppito
67010 L'Aquila (Italia)
paolo.freguglia@gr.technet.it

§Centro Interd. per lo Studio dei Sistemi Complessi, Università di Pisa
Via Bonanno 25 B
56100 Pisa (Italia)
LGalleni@agr.unipi.it

INTRODUCTION

In this paper we continue the enterprise of elaborating an axiomatic framework suitable for modern Biology, started in [1,2,3]. We isolate and propose to the attention of the scientific community a few axioms expressing some basic biological facts and their theoretical interpretations. We hope we can foster criticism and contributions from researchers of any field of the Natural Sciences. In fact, we conceive this paper as part of a general research programme on the Foundations of Science, originated in the Eighties by E. De Giorgi at the Scuola Normale Superiore, Pisa and carried on by several researchers after his death in 1996 (see [4,5,6,7,8]). The aim of this programme is not to provide *safe and unquestionable grounds* to scientific activity, but rather to develop *conceptual environments* where this activity can be carried out naturally and without artificial constraints, and the notions considered in various disciplines can be presented in a simple and clear-cut way, so as to allow for both a deeper analysis by researchers in the field, and a proficuous interdisciplinary debate.

The main features of any axiomatic framework stemming from the foundational programme of E. De Giorgi are the following:

- *Non-reductionism*: there are many kinds of "qualitatively different" objects and concepts which are studied by science. Mathematical models are often fruitful in providing

The Application of Mathematics to the Sciences of Nature, edited by Cerrai *et al.*
Kluwer Academic/Plenum Publishers, New York, 2002.

a rigorous analysis of these concepts, but reducing natural scientific notions to their mathematical codes can undermine the conceptual clarity of the notions themselves. Moreover the possibility of using *qualitatively* (and not only quantitatively) different kinds of objects simplifies the enterprise of engrafting new notions, especially in natural sciences. In this paper, for instance, *living objects* are defined *per se*, and not on the basis of a decomposition into simpler components. The analysis of complex objects into appropriate simpler components is essential to any theory of these objects, but any decomposition returns only a fragment of the intended properties of the objects themselves. In fact, any living object has properties and is involved in relations which refer to it as a whole (*in toto*), and these are not adequately analysed in terms of the various components. Last but not least, in Biology any object is also a component of higher-order entities; also the consideration of these entities can be essential in the study of basic characteristics of the object itself. For instance, any animal has to be studied as a member of some species, and species cannot be accounted to simply as *sets of individuals*. In fact any species has properties and relations involving it as a whole, independently of the specific set of individuals belonging to it at a given time: a species has to be considered in relation with the whole ecosystem, and also in a historical context including its origin and evolution.

- *Open-endedness*: the introduction of new notions should be always possible. Any sufficiently analysed and clarified concept should be engraftable (*innestabile*) in a natural way. To this end, the axioms and postulates involving the objects under consideration have to be formulated so as to allow for the introduction of new objects and concepts which can emerge from theoretical and experimental developments of the considered disciplines, and this aim has to be achieved without having to revise the formulation of the main properties previously axiomatized. This feature is of primary importance in Biology, since here one needs conceptual instruments suitable to systematize the new data which are produced continuously by practical research on the field. As a matter of fact, it is often difficult to integrate this huge quantity of new information within the current conceptual frameworks.

- *Semiformal axiomatization*: the theory should be presented in a clear and rigorous axiomatic form, so as to allow for critical analysis by scholars of different fields of research. To this end a rigorous presentation using the *natural language* seems preferable to a *complete formalization* using some artificial formal language. This seems of particular importance in Biology, because this science has not yet developed formal languages of its own for expressing the different theories (see [9,10]). Of course, one has to render the statements as unambiguous as possible, by giving a precise axiomatic determination of the basic technical terms, and this can be achieved by giving an axiomatization in the style of traditional Mathematics. It is worth mentioning that Classical Mathematics itself was intended as a scientific theory of the natural world!

- *Selfdescription*: the main properties, relations and operations involving the objects under consideration are themselves first class objects in the framework.

Engrafting Biology into this framework amounts to isolating few objects and concepts studied by some branch of Biology, investigating the way they are connected and described through the basic notions of the general theory. We focus here on the morphological characters of living objects and species, in relation with the processes of generation and speciation. The paper is organized as follows:

- in the first section we outline the *general axiomatic framework*, based on the notions of *quality, relation, operation* and *collection* (Axioms G.1-9);
- in the second section we introduce *biological* and *living objects*, and we sketch an axiomatization of *variable* objects and operations, suitable for dealing with the exper-

imental sciences (Axioms V.1-5);
- in the third Section we adapt to this enriched framework the axioms on *generation* and *species* already proposed in [2,1] (Axioms S.1-4);
- in the final section, after a short review of the classical model of Schiaparelli [11] (according to the reconstruction of [12,13]), we embody the main ideas of the model in three axioms *à la* De Giorgi (Axioms M.1-3).

We hope that this axiomatization may clarify, at least in part, some of the problematic aspects of the biological notions considered here, and in particular that it may facilitate a critical dialogue among researchers of various scientific disciplines, more or less related to Biology.

The topics of this paper have been discussed during workshops held in 1998-99 at Arcidosso and at Domus Galilæana (Pisa): the authors are grateful to the organizers and to all participants for useful discussions. The authors are also thankful to Furio Honsell for suggestions and comments.

THE GENERAL AXIOMATIC FRAMEWORK

Most disciplines in the Sciences and Humanities deal, ultimately, with *several qualitatively different* objects and study *relations* over them. It seems therefore appropriate to isolate as *primitive*, i.e. not reducible to preceding concepts, the notions of *having a given quality* (or *property)* and of *being in a given relation with something.* Following [7], we propose, as starting point, a general axiomatic framework consisting of a few fundamental *qualities* and *relations.* Additional qualities and relations, peculiar to each specific field, can be naturally introduced later on within this framework. More precisely, we assume as *primitive*, i.e. not reducible to preceding concepts, the following notions:

- *the object q is a quality;*
- *the object r is a binary relation;*
- *the object s is a ternary relation;*
- *the object t is a quaternary relation;*
- *the object x has the quality q* (written as $q\ x$);
- *the objects x, y are in the binary relation r* (written as $r\ x, y$);
- *the objects x, y, z are in the ternary relation s* (written as $s\ x, y, z$);
- *the objects x, y, z, w are in the quaternary relation t* (written as $t\ x, y, z, w$).

A first form of *selfdescription* is obtained by introducing four distinguished qualities corresponding to the primitive kinds of objects considered above, i.e. the *classifying qualities Qqual, Qrelb, Qrelt, Qrelq.*[a] Hence we postulate:

Axiom G.1. *Qqual, Qrelb, Qrelt, Qrelq are qualities:*
1. no object has simultaneously two of the above qualities;
2. x is a quality if and only if Qqual x;
3. x is a binary relation if and only if Qrelb x;
4. x is a ternary relation if and only if Qrelt x;
5. x is a quaternary relation if and only if Qrelq x.

[a] As an aid to recalling the intended meaning of the distinguished objects, we use *suggestive names* containing fragments of the corresponding *Latin* word. E.g. qualities have initial Q, relations R, qualities and relations concerning qualities or relations usually have suffix or infix *qual* and *rel* respectively. Moreover basic operations or collections have upper case initials and objects involving them have suffix or infix *op* and *coll*, respectively.

We introduce, next, the *fundamental relations Rqual, Rrelb, Rrelt,* which describe the behaviour of qualities and relations. Hence we postulate:

Axiom G.2. *Rqual is a binary relation:*
1. if Rqual x, y *then Qqual* x*;*
2. if Qqual q *then Rqual* $q, x \Longleftrightarrow q\ x$.

Axiom G.3. *Rrelb is a ternary relation:*
1. if Rrelb x, y, z *then Qrelb* x*;*
2. if Qrelb r *then Rrelb* $r, x, y \Longleftrightarrow r\ x, y$.

Axiom G.4. *Rrelt is a quaternary relation:*
1. if Rrelt x, y, z, w *then Qrelt* x*;*
2. if Qrelt s *then Rrelt* $s, x, y, z \Longleftrightarrow s\ x, y, z$.

Qualities and relations have an essentially *intentional* character, hence we do *not* postulate *extensionality* for them. We do not exclude that there are different qualities which are enjoied by exactly the same objects, and different relations which involve exactly the same objects.[b]

Most axiomatizations of scientific theories make use of two other primitive kinds of objects, namely *operations* and *collections* (or sets). According to the basic pattern for engrafting new concepts in our general axiomatic framework, we introduce suitable *qualities*, which classify the objects under consideration, and suitable *relations*, which describe their behaviour and connect them with the objects introduced previously. Operations and collections themselves will be used in the sequel, together with qualities and relations, for axiomatizing new theories.

The concept of *operation* builds solely on the intuition of an operation as an object which *acts* (operates) on one, or two, objects and possibly produces a *result.*[c] We introduce the qualities *Qops* and *Qopb* of being respectively a *simple* (unary) and a *binary operation* and the relations *Rops* and *Ropb*, which connect any operation with the imputs and the corresponding outputs . The *functionality* of operations is expressed by postulating that the relations *Rops* and *Ropb* are "univalent". Hence we postulate:

Axiom G.5. *Qops is a quality, Rops is a ternary relation.*
1. If Rops x, y, z *then Qops* x*;*
2. If Rops f, x, y *and Rops* f, x, z*, then* $y = z$.

Axiom G.6. *Qopb is a quality, Ropb is a quaternary relation.*
1. If Ropb x, y, z, w *then Qopb* x*;*
2. If Ropb g, x, y, z *and Ropb* g, x, y, w *then* $z = w$.

The concept of *collection* aims to capture the most general notion of "aggregation into a whole of different objects of any kind", which underlies the definitions of *class* and *set*, given by Frege and Cantor. These notions are *extensional* in nature, in the

[b]Apparently the action of *quaternary relations* is left here without internal description: it would require a *quinary* relation *Rrelq*, thus giving rise to an infinite sequence of fundamental relations of increasing arities, as in [4]. We do not exclude, of course, that there are relations of any arity, but we prefer to stop at a low level of complexity, since only binary or ternary relations appear naturally in the applications. Quaternary relations, which are used only for giving a natural description of ternary relations and binary operations, can equally well be internally described in various ways, see e.g. [7,6].

[c]As in the case of relations, we do not exclude that there are operations which act on more than two objects at once; we simply don't introduce them here for the sake of simplicity.

sense that collections having the same members are identical.[d] We introduce the quality *Qcoll*, of being a collection, together with the binary relations *Rcoll* (membership) and *Rincl* (inclusion), and we postulate:

Axiom G.7. *Qcoll is a quality; Rcoll and Rincl are binary relations:*
1. if Rcoll x, y *then Qcoll* x*, i.e.* x *is a collection;*
2. if C, D *are collections, then Rincl* C, D *holds if and only if Rcoll* D, x *implies Rcoll* C, x *for any* x*;*
3. if C, D *are collections and both Rincl* C, D *and Rincl* D, C *hold, then* $C = D$*.*

When X, Y are collections, we adopt the standard notation $x \in C$ and $D \subseteq C$ for *Rcoll* C, x and *Rincl* C, D, respectively. Similarly, when f, g are a simple and a binary operation, we write $fx = y$ and $gxy = z$ for *Rops* f, x, y and *Ropb* g, x, y, z, respectively.

We postulate the existence of four fundamental collections, namely the *universal* collection V, the *empty* collection $\emptyset$, the collection *Coll* of all collections, and the collection *Ins* of all *sets*.[e] We postulate that the *difference* (and hence also the *union* and the *intersection*) of two collections always exist, and that the *singleton* of any object, the *union of two sets* and the *intersection of a set and a collection* are always sets.

Axiom G.8. $V, \emptyset,$ *Coll are collections; Compl is a binary operation:*
1. for any object x*,* $x \in V$ *and* $x \notin \emptyset$*;*
2. $x \in$ *Coll if and only if Qcoll* x*;*
3. if $X, Y \in$ *Coll, then Compl* $XY = X \setminus Y$ *exists and is a collection containing all and only those elements of* X *which do not belong to* Y*.*

Axiom G.9. *Ins is a subcollection of Coll, whose elements are the sets.*
1. Sing is a simple operation: for any object x*, Sing* $x = \{x\}$ *exists and is a set whose unique element is* x*.*
2. The union of two sets and the intersection of a set and a collection are sets.

A thorough development of the theory of operations and collections is carried out in [7,14,6]. We do not introduce other operations or collections here, since our aim is only that of presenting a *minimal* axiomatic framework, suitable for our purposes in axiomatizing Biology. In particular all sets considered in this paper are finite, and the existence of any finite set is provided by the axioms above. To be sure, we shall consider also collections which, although not necessarily infinite, are nevertheless very large and rather indeterminate, so as to appear different from the well-determined sets we are introducing as such. However also these collections are granted by the axioms above.

BIOLOGICAL, VARIABLE, AND LIVING OBJECTS

[d]On the contrary, we have *not* postulated extensionality for operations, since we do not want to rule out the possibility that there exist operations, acting on the same objects and giving the same result on each object, which are nonetheless different, for they operate according to different procedures.

[e]The word *Ins* has been chosen in accordance with the Latin *insimul* (meaning simultaneously), which is the root of the word for sets in some Neo-Latin languages, e.g. *insieme* (Italian), *ensemble* (French). Sets are isolated as "small collections which can be totally controlled and freely manipulated", differently from general collections, which are "too large and complicate". They comprehend most collections commonly considered in the applied Sciences, and in general all (intuitively) *finite* collections.

We are interested here in axiomatizing Biology. Following [1], we ground on the tautological statement that "Biology is the Science which studies Biological Objects", and we start by introducing the quality *Qbiol* of being a *biological object*. According to the general principles of our nonreductionist foundational theory, we do not intend to *define* once for all the notion of biological object in terms of *more fundamental* notions; rather we try to find out some of the main features of biological objects and we state them in form of axioms. First of all, we notice that any biological object is always involved in three fundamental relations: it has been *generated* by something, it *exchanges energy* with something, it is *composed* by parts and is itself part (*component*) of complex entities. Therefore we introduce three binary relations

- *Rgen*: the most general relation between *generator* and *generated*;
- *Ren*: the most general relation involving *exchange of energy*;
- *Rcomp*: the most general relation between *complex* and *component*.

We formulate a first axiom concerning these relations and biological objects:

Axiom V.1.
1. Qbiol is a quality. An object x has the quality Qbiol if and only if it is a biological object.
2. Rgen is a binary relation. If Qbiol x then there exists y such that Rgen x, y.
3. Ren is a binary relation. If Qbiol x then there exists y such that Ren x, y.
4. Rcomp is a binary relation. If Qbiol x then there exist y, z such that Rcomp x, y and Rcomp z, x.

In the axiom above, the intended meaning of *Rgen* x, y is that y has been generated by x, while *Ren* x, y means that there is an exchange of energy between x and y, and *Rcomp* x, y means that y is a component of x, i.e. that x is a complex object and y is one of its parts. Here *Rgen* is conceived as the *most general* relation of generation, hence no particular "process of generation" is specified. It may be that x is itself a biological object, but it is also possible that this is not the case (e.g. in photosynthesis CO_2 and H_2O generate sugar). Several specifications of this general relation can be considered in different contexts: we shall deal in the sequel with the *operations of generation* which instantiate this relation when *living objects* are concerned (we shall also introduce *operations of speciation* in order to generate species). Similarly, also the relations *Ren* and *Rcomp* are intended to represent the *most general* way of exchanging energy and constituting a part, respectively. Therefore these relations provide only very generic information about the effective way the involved objects are correlated. In further developments, these relations have to be made more precise by introducing relations and operations specific to the various kinds of objects under consideration. Notice that being involved in the three general relations *Rcomp*, *Ren* and *Rgen* is only a *necessary* condition for having the quality *Qbiol*. It is clearly not *sufficient*: in fact it is easily seen that any atom is involved in all of them. Many less trivial examples could be given against any attempt of characterizing biological objects through general relations involving also non-biological objects. This fact confirms the convenience of introducing axiomatically the quality *Qbiol*.

A serious problem arises in dealing with *experimental sciences*, and especially Biology, within a foundational framework *à la* De Giorgi, like that outlined in the previous section: dew to its origin in Mathematics, Logic and Computer Science, this framework accommodates naturally only abstract, "eternally invariant" objects. A few attemps have been made of engrafting "variation", at the prize of some (perhaps cumbersome) parametrizations. We refer only to the introduction of the concept of *variable* of classical Mathematical Physics suggested in [15], and to the treatment of *contingency* and *modalities* outlined in [16]. However much more is needed for superseding the obstacles met in [1] in axiomatizing biological notions, in particular when dealing with *living objects*: for instance, a lion is a living object until it dies, it is in the relation *Rcomp* with its stomach until it is removed, etc. To be sure, the very primitive notion of biological object seems to bring about a kind of dependence on *time* and *space* parameters, thus leading to a general notion of *variable object*. Obviously, we cannot go deeply into the question here, and we restrict ourselves to presenting a couple of axioms, arising from the investigations of [8]. We start from the naive idea that the objects of natural sciences (say the *Earth*, or the famous policedog *Rex*) have an *individual* nature wich goes beyond the *singular concrete things* with which they are identified at each instant (usually a suitable finite aggregate of particles). Of course, depending on the objects under consideration, not only *time* instants or intervals come into play, but also *spatial* determinations and more generally any kind of relevant data suitable for characterizing a given situation. Therefore we introduce a very large and unspecified collection $\mathcal{S}$ of parameters or "sorts" (the *states* of the universe), conceived to encompass all "phase spaces" considered by any scientific theory under consideration, in particular here Biology. At suitable values of the parameters, any natural object has a "concrete representative" (the *Earth* is an aggregate of determined molecules, and *Rex* of determined cells). Depending on the state, these concrete representatives *change* (some atoms of the Earth are eliminated or substituted, some old cells disappear in generating new ones inside Rex, etc.), while the "abstact object" is still considered to remain the *same*. It seems therefore appropriate to introduce the quality *Qvar* of being a variable object together with a general relation *Rinst* connecting each such object with the states at which it is *instantiated* and with its actual representative at each state. Hence we postulate

Axiom V.2. *Qvar is the quality of being a variable object, $\mathcal{S}$ is the collection of all states, and Rinst is the ternary relation of instantiation:*
1. if Rinst x, y, z then Qvar x and $y \in \mathcal{S}$;
2. if Rinst x, y, z and Rinst x, y, w, then $w = z$.

If X is a variable object, $s \in \mathcal{S}$ is a state, and z is an object such that *Rinst* X, s, z, then z is uniquely determined, by the second clause above. Therefore z will be denoted by X_s and called the *instantiation of X at the state s*. Whenever X has an instantiation X_s at the state s we simply say that X *exists at s*, and that X_s *is X at s*. We cannot investigate here the connections between a variable object and its instantiations, nor the structure of the "phase space" $\mathcal{S}$ (as a matter of fact, this topic is still largely unexplored). We consider only one example, typical in Biology, namely the case of an operation acting on a variable object and producing another variable object. It seems convenient to assume that any such (abstract) operation can be "decomposed" into several (concrete) operations which, at suitable states, act on suitable (instantiations of) objects, and produce other (instantiations of) objects, possibly in new states. The first object has to be (a component of) the argument and the last one (a component of) the result of the abstract operation. Therefore we propose the following axiom

Axiom V.3. *If F is an operation and X, Y are variable objects such that $FX = Y$, then there exist operations $f_1, \ldots, f_n$, variable objects $V^1, \ldots, V^n$, and states $s_1, \ldots, s_n \in \mathcal{S}$ such that $f_i V^i_{s_i} = V^{i+1}_{s_i}$ for $i = 1, \ldots, n-1$. Moreover Rcomp X, V^1 and Rcomp Y, V^n.*

The operations $f_1, \ldots, f_n$, are called the *components* of the variable operation F. We have considered only simple operations for sake of brevity, but clearly we intend to postulate a correspondent axiom for binary operations as well. Moreover binary operations can actually appear as components of simple variable operations and *vice versa.* (Alternatively, we could consider binary operations as acting on *pairs* of objects.) Similar axioms should be given also for relations, qualities, or collections involving variables objects. Finally it should be appropriate to specify suitable relations among the states involved in such axioms. But this task exceedes by far the scope of this paper.

We are now able to consider the *living objects*, which are in fact *essentially variable.* We have already underlined that it seems inconvenient to try to define the living objects by means of previously introduced concepts. Hence we introduce a quality *Qviv* (Lat. *vivens*) which characterizes all and only living objects. Obviously all living objects are biological objects, and we postulate

Axiom V.4. *Qviv is the quality of being a living object. If Qviv x, then also Qvar x and Qbiol x.*

By the axioms above, any living object is involved in the general relations *Rcomp*, *Ren*, and *Rgen*. In this case we can be more specific about *Rgen*, and we consider as a basic characteristic of living objects the fact of being involved in *operations of generation.* Therefore we introduce a collection *Opgen* of operations of generation, with the intended meaning that any $G \in Opgen$ is a *variable operation* which applies to one or two suitable living objects and produces a (possibly empty) set of living objects. Hence we postulate

Axiom V.5. *Opgen is a collection of variable operations (the operations of generation):*
1. if X, Y are living objects, $G \in Opgen$, and GX (or GXY) esists, then it is a set of living objects;
2. if $GX = \{Z_1, ..., Z_n\}$ (or respectively $GXY = \{Z_1, ..., Z_n\}$) then Rgen X, Z_i (and respectively also Rgen Y, Z_i).

We have taken a set of living objects as result of an operation of generation since usually a single such operation can produce more than one individual. Roughly speaking, any operation of generation corresponds to a "specific way" of reproduction. If $g_1, \ldots, g_n$ are the components of $G \in Opgen$, then the operations g_i represent the different concrete actions which, at suitable states, are made by one or two living objects in order to generate other living objects. Thus one could start an analysis of the various forms of reproduction by simply considering the structure of the different components of any operation of reproduction. Such an analysis, as well as a study of the connections of these operations with the fundamental relations *Ren*, *Rgen*, *Rcomp*, is only called for in [1], and clearly it falls outside of the scope of this paper.

THE NOTION OF SPECIES

The concept of species is a very discussed one in contemporary Biology, as can be seen in [17,18,19]. We gave in [1] a first axiomatization of this concept, which we adapt here to the present framework. It seems natural to introduce first the different species as qualities of living objects, together with the quality *Qspec* of being a species. In this way we do not need to introduce a specific relation connecting any living object with its species, since, being any species a quality, the general relation *Rqual* makes the job.

Axiom S.1. *Qspec is the quality of being a species. If Qspec S holds, then S is a quality of living objects. Moreover S is a biological object, i.e. Qbiol S.*

A first nontrivial axiom postulates that any two individuals of the same species always have a common ancestor of the same species. In order to allow for the process of speciation, the converse property (namely that the outspring is always of the same species of the parents) has to be postulated cautiously, so as to admit exceptions at suitable states.

Axiom S.2. *Let X, Y be living objects of the same species S; then:*
1. there are living objects $X_1, \ldots, X_n, Y_1, \ldots, Y_m, Z$, all having the quality S, such that $Rgen\ X_1, X$, $Rgen\ X_2, X_1, \ldots$, $Rgen\ Z, X_n$, $Rgen\ Z, Y_m$, $Rgen\ Y_m, Y_{m-1}, \ldots$, $Rgen\ Y_1, Y_2$, $Rgen\ Y_1, Y$;
2. there exists a collection of exceptional states $\mathcal{E}_S \subset S$ such that, if G is an operation of generation, $GXY = \{Z_1, \ldots, Z_n\}$, and all the components of G act at states $s_i \notin \mathcal{E}_S$, then all the Z_i belong to the species S.

The collection $\mathcal{E}_S$ considered in clause 2 of the axiom above is introduced so as to take into account the exceptional states corresponding to an *operation of speciation*, to be introduced later. Namely, if during the time interval corresponding to $\mathcal{E}_S$ an operation of speciation produced a new species S', then some (or possibly all) descendants of S-individuals generated after the end of $\mathcal{E}_S$ will belong to the new species S'. This seemingly paradoxical situation originates from the contrast between the theory of evolution and the concept of species, born inside of a fixistic paradigm. This contraposition can be composed, in our opinion, by postulating that the individuals generated when an operation of speciation takes place may belong simultaneously to more than one species. On the other hand, following [1], we have not admitted exceptional states in the first clause of the axiom. Hence any two individuals of a new species have necessarily a common ancestor during the speciation period. One could have assumed instead the existence of another special time interval (situated immediately after that of speciation) at which the validity of clause 1 is suspended: then the common ancestor might be found in the old species S.

We assumed in [1] that all processes of speciation considered up to now in Biology can be uniformly described by a collection *Opspec* of *operations of speciation.* These operations apply to one (or two) suitable species and produce one or two species. Like the operations of generation, also those of speciation are variable. Hence we can decompose any $F \in Opspec$ into its components f_i, acting at suitable states $s_i \in S$. It is natural to assume that an operation of speciation operates through a number of generations. During this process, the original species is still present, and the new species begins to appear: this is the reason why we introduced the "exceptional states" $\mathcal{E}_S$ in the Axiom S.2. An operation of speciation takes place during a time interval much larger than those requested for any single operation of generation. It is therefore natural to assume that the components of an operation of speciation applied to some species X act at states that in some sense *include* all exceptinal states of the set $\mathcal{E}_X$. Of course, our states are not, in general, *pure time* intervals. Thus we use the relation of *general* inclusion, *Rincl*, and we postulate

Axiom S.3. *Opspec is a collection of variable operations (the operations of speciation):*
1. if X, Y are species, $F \in Opspec$, and $FX = Z$ (or $FXY = Z$), then Z is a set of species, which in general contains at most two species;
2. if z is an individual of a species $S \in Z$, then z has an ancestor of the species X

(and also one of Y, when $FXY = Z$).
3. Let $f_1, \ldots, f_n$ be the components of $F \in Opspec$, and let X, Y be species such that $FX = Y$. Let $g_1, \ldots, g_m$ be the components of $G \in Opgen$, and let x, y be individuals of the species X such that $Gx = Z$ (or $Gxy = Z$). Assume that x is involved by some f_i at the state t_i, and that in the process of applying G to x (or to x, y) it appears a component $g_j V^j_{s_j} = V^{j+1}_{s_j}$: then Rincl t_i, s_j and $s_j \in \mathcal{E}_X$. Moreover any $z \in Z$ may belong also to the species Y.

In general the operations of speciation are simple. Binary operations occur in the case of speciation by hybridization, which is extremely rare in animals but somewhat more frequent in plants. We take the result of an operation of speciation to be a *set* of species, rather than a single species, so as to comprehend all cases occurring in practice: after the operation a new species can replace the old one, or it can add itself to the old one, or else the old species disappears and two new species take place instead.

The concept of species is a fundamental one in Biology and, even in a first attempt, it is worth a deeper inspection. Here we focus mainly on morphological characteristics, and thus we leave outside many important topics (for more on the origin and the evolution of species, as well as for a few axioms on the problem of hybrids, see [1]). The most effective techniques for characterizing species (without using cross reproduction) are based on morphological criteria. Any species is identified by isolating some morphological characters which identify and distinguish it. Following [1], we introduce the notion of *morphological quality*, intended in a rather wide sense, as some physical characteristic feature, which can be observed and described, even if not necessarily connected with "how the object looks like".

We fix a large collection *Morph*, intended to contain all morphological qualities which are considered at present together with all such qualities which can possibly be considered in future developments. Clearly the whole collection *Morph* is never conceived as completely defined, since it is continuously enriched, according to the advances of experimental Biology. This openendedness can help in finding ever new characteristic qualities. In fact, we are ready to accept morphological qualities corresponding to any visible (or better *describable*) aspect, such as the number and shape of chromosomes, or the presence/absence of fixed sequences of DNA (this can be helpful in meeting the case of the so called "sibling species", see [1]). For every living object we can find a number of qualities from the collection *Morph*, so as to completely identify it. Some of these qualities are shared by all the objects of the same species. Therefore it seems reasonable to postulate:

Axiom S.4. *Morph is a collection of qualities. For any species S there exists a finite family $Morph_S$ of subsets of Morph such that an individual belongs to the species S if and only if it has all the qualities of some set $M \in Morph_S$.*

This axiom is unproblematic in many cases, and in fact very often *just one* set of morphological qualities suffices, and $Morph_S$ can be taken to be the single set $M_S = \bigcup Morph_S$. The set M_S, containing all qualities which are needed for identifying all elements of S, is called the set of the *characteristic qualities* of the species S. However associating to each species a single set of morphological qualities may be unsuitable in some cases, e.g. in presence of strong sexual polymorphism, and it may happen that no individual shares all the characteristics of the species. Keeping together a family of sets, which in most cases share many common qualities, allows for taking into account such morphological differences inside a single species. Even so, given a species S, the problem of actually isolating a characteristic set M_S out from the large indeterminate collection *Morph* of all possible morphological qualities appears very difficult, even impossible at times.

THE SCHIAPARELLI MODEL VIEWED *À LA* DE GIORGI

In a recent paper [13], the second author recapitulates the classical essay [20] of the great italian astronomer Giovanni Virginio Schiaparelli, published in 1898, and warmly recommended by V. Volterra few years later.[f] In this paper, a geometric model of the morphology of all animals and plants is proposed, and the existence of a "fundamental formula" describing and controlling the evolution of any form of life is confidently affirmed.[g] More precisely, Schiaparelli assumes that the collection of all living objects is represented by a finite (albeit very large) number of points of a space $\mathbf{R}^n$. The dimension n corresponds to the number of biological characteristics which are sufficient to determinate uniquely each individual living object (past, present and future), the coordinates of the corresponding point being given by the values of these characteristics at any instant. (It is subsumed that these characters are passible of *measurement with real values*, a hypothesis not far from our idea that morphological qualities are physically describable.) This model fits very well with the morphological characterization of species we outlined in the axiom S.4, therefore we summarize below the account of Sciaparelli's essay given in [12]. In the final subsection we shall engraft onto our axiomatic framework the main ideas laying at the basis of the model, so as to obtain a more accurate description of the organization and evolution of the morphology of living objects.

4.1 A Reconstruction of Schiaparelli's Model

Since the original paper [11] is written in an argomentative-assertive style, we summarize its content in five *propositions*. First of all, imagine a fixed number n of (measurable) properties, and associate to any living object the n-tuple of the values assumed by these properties at a given instant. If the selected properties are adequately descriptive, one can identify each living objects with the corrisponding n-tuple, and also consider its "trajectory" depending on time. Since the living objects exising at any time are in finite number, we obtain immediately:

Proposition 1 (Law of discreteness)[h] *The complex of all living objects is (at any moment) a finite, hence discrete set of points $L \subset \mathbf{R}^n$. In particular there is a minimal distance between different points of L.*

Being an astronome, Schiaparelli conceived a similarity between the distribution of the stars in the sky and that of the points of his model. Like stars are organized

[f]Si deve dunque accogliere l'ardito tentativo del piú illustre astronomo dei nostri giorni, lo Schiaparelli, di costruire un modello geometrico atto allo studio delle forme organiche e della loro evoluzione, con quel medesimo interesse con cui sono accettati e studiati i modelli meccanici di Maxwell e di Boltzmann della induzione elettrica e dei cicli termici; tanto piú che non gravi difficoltà si opporrebbero a trasformare il modello stesso dello Schiaparelli da geometrico a meccanico, rendendolo cosí ancor piú intuitivo. ([11], pp. 3-28).

[g]Noi proponiamo l'ipotesi ... che un sistema di organismi naturali dipenda da una formula fondamentale unica, rappresentante i loro caratteri comuni; nella quale i parametri, ossia elementi discriminatori, colla diversità dei loro valori determinino le differenze di vario ordine e la classificazione gerarchica in varietà, specie, generi, ecc. ([11], p. 335).

[h]In natura il continuo geometrico non esiste. ... la materia forma dunque un sistema discontinuo ([11], pp. 306-307).

in relatively dense galaxies, separated by a large amount of almost empty space, thus living objects are contained in mutually disjoint regions, and they are dense in the middle and very rare near the boundary of each region. Thus the species are naturally identifiable exactly as *galaxies* are:

Proposition 2 (Law of distinction of species)[i] *There exists a cell decomposition of the space* $\mathbf{R}^n$ *such that in each* n*-dimentional cell the points of* L *located near the boundary are relatively few, and become fewer and fewer as the distance from the boundary decreases.*

It seems natural to extend the astronomic analogy to larger aggregates: thus one could imagine complexes of cells corresponding to *classes, phyla, etc.*, and state the coresponding propositions. But Schiaparelli goes also the other way on, and he isolates one point at the center of each cell, to be taken as the "normal type" of that cell (note that this point itself may not belong to L). He claims that a great distance from the normal type is very unlikely in any process of generation, and in most cases this distance has an *external* reason). It follows that a statistical law distributes the points of L in decreasing density, depending on the distance from the normal type.

Proposition 3 (Law of the distribution around normal types)[j] *The center of any cell is called its normal type. A point rather far from the normal type can be produced inside the cell only under the influence of unlikely combinations. In any cell the density of* L *decreases with the distance from the normal type.*

In order to give a better justification and a possible quantification of the statistical distribution stated above, Schiaparelli introduces an *attractive force* which is inversely proportional to the distance from the normal type:

Proposition 4 (Law of attraction of normal types)[k] *Points far from the normal type tend to produce new points closer to it: in fact, there is an attractive force increasing as much as the distance from the normal type increases.*

Proposition 4 provides a great stability of any species, and in fact even the possibility of *speciation* seems excluded. But now Schiaparelli introduces his *fundamental law*, allowing for the persistency of *oriented perturbations.* Each generation has corresponding vectors connecting the parents with the offspring: although the length of these vectors are usually small (according to the precending laws),[l] (according to the preceding laws), nevertheless, if a persistent perturbation contrasts the attraction of the normal type and maintains the vectors pointing in the outside direction, then after several generations part of a population can leave its original cell.

---

[i]Tra due specie distinte e confinanti, cioè che si "assomigliano", gli individui dell'una specie che hanno caratteristiche vicine a quelle degli individui dell'altra specie sono rari, e quanto piú si assomigliano tanto piú sono rari ([11], *p.*318).

[j]Sono determinabili, al di là della loro esistenza in natura, individui ciascuno dei quali è rappresentativo di un tipo. Tale individuo si chiamerà "tipo normale". ... Una digressione (che si ha passando da un individuo ad un altro per generazione) alquanto sensibile da un tipo normale M non potrà aver luogo che a causa di improbabili (cioè alquanto difficili da accadere) combinazioni ([11], p. 319).

Sono via via sempre piú numerosi gli individui, generati nella varietà determinata da un tipo normale, che assomigliano al tipo normale. E ciò per una ben nota legge statistica ([11], p. 317).

[k]Quando un tipo individuale, cioè quando un individuo M si è allontanato (generazione dopo generazione) dal suo tipo normale X per effetto di una qualunque causa perturbatrice, ad ogni ulteriore allontanamento da X si opporrà una forza tanto piú grande quanto piú grande è l'allontanamento precedente MX. E di quanto l'allontanamento di M da X è impedito, altrettanto sarà favorito l'avvicinamento ([11], p. 328).

[l]La direzione (ed il verso) [del vettore] indica la qualità della deviazione [dal genitore] e la lunghezza ne indica la grandezza ([11], p. 319).

Proposition 5 (Fundamental law of speciation)[m] *If some persistent (natural or atificial) perturbation maintains its direction through many generations, then the differences from the normal type can add one to another, so as to reach eventually a previously not inhabited region.*

We conclude this short review by an important remark: the evolution considered by Schiaparelli is different from that of Darwin in a crucial point. The transformations of living objects are not completely casual in the large: the small, casual modifications have to be summed up in such a way that the cell (type) reached at the end of the process is not really new and indeterminate, but is chosen among a *restricted range of possible types* (and it is, in principle, the *most convenient one*).[n]

4.2 An axiomatic engrafting of Schiaparelli's ideas

We are now able to engraft onto our axiomatic framework the main ideas embodied in the model of G. V. Schiaparelli. We propose three axioms aimed to give, within our framework, a more precise and quantifiable account of the organization and evolution of living objects, on the basis of morphological qualities.

We suppose that the collection $Morph$ is finite, and we assume that to each morphological quality $q \in Morph$ it is associated a fixed physical characteristic, passible of evaluation in a suitable metric space R_q (which we construe as subspace of an euclidean space $\mathbf{R}^{n_q}$, for sake of simplicity).[o] To any "concrete" object x is associated a q-value $\eta_q(x) \in M_q$ (the *measure* of x w.r.t. the quality q). Similarly, to any variable object X is associated a *q-trajectory* $\eta_q^X = \{(s, \eta_q(X_s)) \mid s \in \mathcal{S}\}$, which describes the variation of the q-values of (the instantiations of) X, depending on the states of the universe. The idea is that a living object X *has* the morphological quality q if and only if, for a suitable set of states $\mathcal{N}(X)$, the set of the q-values of its instantiations, i.e.

$$N_q^X = \eta_q^X(\mathcal{N}(X)) = \{\eta_q(X_s) \mid s \in \mathcal{N}(X)\}$$

is included in a selected compact subspace $K_q \subset R_q$ (K_q is the *distingueshed* subspace of *acceptable* values, $\mathcal{N}(X)$ is the set of *normal* states for X, and N_q^X is the *normal q-trajectory* of X). Hence we postulate

Axiom M.1. *For any $q \in Morph$ there is an operation η_q taking on values in a metric space R_q (η_q is the q-measurement or q-evaluation).*
1. For any living object X there is a set of states $\mathcal{N}(X) \subset \mathcal{S}$ such that $N_q^X =$

[m]Quando per una causa qualsiasi naturale o artificiale l'influsso anche lievissimo di un'azione perturbatrice persiste in un determinato senso per una lunga serie di generazioni, gli effetti da questi prodotti si sommano in conseguenza del principio di eredità, in modo da produrre in fine radicali modificazioni rispetto al tipo primitivo. Le altre perturbazioni di carattere accidentale possono bensí ritardare temporaneamente un tale processo, ma non possono impedirlo ([11], p. 323).

[n]... ciascuno dei tipi verso cui la trasformazione (cioè la generazione) può esser diretta è assolutamente determinato a priori in tutti i suoi caratteri; cosí che l'ufficio dei fattori evolutivi non è quello di produrre un tipo nuovo, ad arbitrio delle circostanze, ma di scegliere tra tipi possibili, quello che è piú confacente al caso, e che nelle date circostanze il piú utile e il piú conveniente al dato organismo ([11], pp. 347-348).

[o] Assuming a *metric* structure greatly simplifies the following developements. Fixing the real numbers as *constant* space of values would be excessively restrictive, while we can imagine our spaces as subspaces of *euclidean* spaces of suitable dimension. However it is reasonable to expect that weakening the hypothesis to *uniform* spaces could provide a useful generalization. For instance it would enable us to consider qualities whose values are *nonarchimedian*, or even *incomparable*. On the other hand a richer structure of *differential manifold* would facilitate the study of the *dynamic* of evolution, on the basis of the Axiom M.3 below.

$\{\eta_q(X_s) \mid s \in \mathcal{N}(X)\}$ is a compact subspace of R_q for all $q \in Morph$ ($\mathcal{N}(X)$ is the set of normal states for X and N_q^X is the normal q-trajectory of X).
2. For any $q \in Morph$ there is a compact subspace $K_q \subset R_q$ such that a variable object X has the quality q if and only if $N_q^X \subseteq K_q$ (K_q is the distingueshed subspace of R_q).

Now the morphological characterization of species outlined in Section 3 can be made measurable by combining Schiaparelli's *normal type* with our set of *characteristic qualities*. In this way we succeed in reducing the enormous number of dimensions of the entire product space $\mathcal{R} = \prod_{q \in Morph} R_q$, which is the correspective in this setting of the euclidean space $\mathbf{R}^n$ of Schiaparelli's model. In fact the region of $\mathcal{R}$ corresponding to any species S can be taken to be completely determined by its projection onto the more tractable reduced product $\mathcal{R}_S = \prod_{q \in M_S} R_q$. In order to enhance readability, we restrict ourselves to the simplest case $Morphs_S = \{M_S\}$: the general case is quite similar (simply take a polymorphic species as a *cluster* of monomorphic ones). More precisely, we fix a *normal type* for each species and postulate that any member of that species, at its normal states, takes on values *sufficiently close* to the normal type. Moreover, the density of the points representing members of the species in their normal states decreases when the distance from the normal type increases, and tends to zero in a neghborhood of the boundary.

Axiom M.2. *For any species S there exist a subset $M_S \subset Morph$, whose elements are the characteristic qualities of S, and a point $\pi_S \in \mathcal{R}_S = \prod_{q \in M_S} R_q$ (the normal type of S), such that:*
1. there is a real number r_S such that a living object X belongs to the species S if and only if $\prod_{q \in M_S} N_q^X \subseteq B(\pi_S, r_S)$ (r_S is the normal radius for S);
2. for any state $s \in \mathcal{S}$, and any positive $d < r_S$ put

$$L_S(s,d) = \{\, X \mid s \in \mathcal{N}(X),\ (\eta_q(X_s))_{q \in M_S} \in B(\pi_S, r_S) \setminus B(\pi_S, d)\} :$$

then the function $\lambda_S(s,d) = |L_S(s,d)|$ tends decreasingly to 0 as d tends to r_S, uniformly in $s \in \mathcal{S}$.

One could have been more precise about the size of the function $\lambda_S(s,d)$, but any reasonable estimate should depend on a *quantitative* formulation of the Axiom M.3 below, which is far from being available up to now. In fact, Axioms M.1-2 give only a *geometric* (or better *cinematic*) representation of the evolution of the morphological characters. We can obtain a *dynamic* treatment of evolution by introducing two contrasting forces: an attractive one, which is always active and tends to maintain the trajectories close to the normal type, and another force, which is sensible only during the speciation periods, and tends to direct the trajectories towards a new, previously empty confining region. These forces act at each generation, each one trying to move the offspring in the preferred direction. For sake of simplicity, we formulate the axiom for a *simple* operation of generation; however it becomes even more interesting in case of *binary* (sexual) generations, since then one has to take into account the reciprocal positions of both parents in the space $\mathcal{R}_S$ (and possibly also the states). Similarly, if the operation of speciation considered in the axiom preserves the original species, or produces two new species, then the perturbing force is dependent on the states.

Axiom M.3. *Let X belong to the species S, and let G be an operation of generation such that $GX = \{Z^1, ..., Z^n\}$. Put*

$$\xi = \text{mean value}\ \{\, (\eta_q(X_s))_{q \in M_S} \mid s \in \mathcal{N}(X)\,\}$$

and, for $i = 1, \ldots, n$

$$\zeta^i = \text{mean value}\ \{\ (\eta_q(Z^i_s))_{q \in M_S}\ |\ s \in \mathcal{N}(Z^i)\ \}.$$

Then the legth of the vector $\xi\zeta^i$ *is bounded by a constant* $k \ll r_S$*, and its direction depends on a force pointing to the normal value* π_S*, whose intensity is inversely proportional to the distance betweem* ξ *and* π_S*.*

Moreover, if F *is an operation of speciation such that* $FS = S'$*, and* G *is performed during the interval of speciation (i.e. at states included in those of* $\mathcal{E}_S$*), then there is also a force pointing in the direction of* π'_S*, whose intensity is directly proportional to the distance betweem* ξ *and* π_S*.*

The axioms above are only a first, limited step towards the elaboration of a general theory of the evolution of species which should combine the flexibility of the qualitative approach *à lao* De Giorgi with the analytical-geometrical approach *à la* Volterra-Schiaparelli. They should allow for limited previsions (or "postvisions"), so as to permit a better evaluation of the correspondence between the axiomatic framework and the experimental and historical data. Obviously, at the moment these axioms pose much more problems than they can solve:

- which collections are *Morph* and M_S, and which spaces R_q have to be choosen?
- how can one identify the normal types and the normal radii?
- what kind of forces are those claimed for in the Axiom M.3?

These, and other similar questions are the challenge we leave to all interested researchers of the third Millennium.

REFERENCES

1. L. Galleni, M. Forti - An axiomatization of biological concepts within the foundational theory of Ennio De Giorgi. *Biology Forum* **92** (1999), 77-104.
2. L. Galleni, M. Forti - Un'assiomatizzazione di alcuni concetti della Biologia nell'ambito delle teorie fondazionali di E. De Giorgi, in *La Matematizzazione della Biologia: Storia e Problematiche Attuali - Atti del I workshop di Arcidosso, 31/VIII-3/IX 1997* (P. Cerrai and P. Freguglia, eds.). QuattroVenti, Urbino 1999, 119-132.
3. L. Galleni, M. Forti - *Un'assiomatizzazione della genetica all'interno della teoria dei fondamenti di Ennio De Giorgi.* Quad. C.I.S.S.C. **3**, Pisa 2000.
4. E. De Giorgi, M. Forti, G. Lenzi - Una proposta di teorie base dei Fondamenti della Matematica. *Rend. Mat. Acc. Lincei* (9) **5** (1994), 11-22.
5. E. De Giorgi, M. Forti, G. Lenzi - Verso i sistemi assiomatici del 2000 in Matematica, Logica e Informatica, in *Nuova Civiltà delle Macchine* **57-60** (1997), 248-259.
6. M. Forti, F. Honsell, M. Lenisa - Operations, ollections and sets within a general axiomatic framework, in *Logic and Foundations of Mathematics* (A. Cantini, E. Casari and P. Minari, eds.). *Synthèse Library* **280**, Kluwer, Dordrecht 1999, 1-24.
7. M. Forti, G. Lenzi - A General Axiomatic Framework for the Foundations of Mathematics, Logic and Computer Science. *Rend. Acc. Naz. Sci. XL, Mem. Mat. Appl.* **21** (1997), 171-207.
8. M. Forti - The Foundational Theories of Ennio De Giorgi, *Aquinas* **43** (2000), 355-367.
9. J. H. Woodger - *The Axiomatic Method in Biology*, Cambridge Univ. Press, Cambridge 1937.
10. J. H. Woodger - Studies in foundations of Genetics, in *The Axiomatic Method* (L. Henkin, P. Suppes, A. Tarski, eds.), North Holland, Amsterdam 1959, vol.4, 8-428.
11. V. Volterra - Sui tentativi di applicazione delle matematiche alle scienze biologiche e sociali, *Giornale degli Economisti* **23** (1901), 3-28.
12. P. Freguglia - Considerazioni sul modello di Giovanni V. Schiaparelli per una interpretazione geometrica delle concezioni darwiniane, *Atti Mem. Acc. Naz. Sci. Lett. Arti Modena* (7) **14** (1996-97).
13. P. Freguglia - Per una teoria assiomatica delle concezioni darwiniane, I, in *La Matematizzazione della Biologia: Storia e Problematiche Attuali - Atti del I workshop di Arcidosso, 31/VIII-3/IX 1997* (P. Cerrai and P. Freguglia, eds.). QuattroVenti, Urbino 1999, 107-118.

14. M. Forti, F. Honsell, M. Lenisa - An axiomatization of partial n-place operations. *Math. Struct. Computer Sci.* **7** (1997), 283-302.
15. E. De Giorgi, M. Forti, G. Lenzi - Introduzione delle variabili nel quadro delle Teorie Base dei Fondamenti della Matematica, *Rend. Mat. Acc. Lincei* (9) **5** (1994), 117-128.
16. M. Forti, F. Honsell - Positive qualities and the ontological argument. *Ricerche di Matematica* **49** (2000),*Supplemento*, 61-78.
17. B. E. Wilson - A (not-so-radical) solution to the species problem, *Biology and Philosophy* **10** (1995), 339-356.
18. B. Continenza - La specie: definire l'indefinibile?, in *Giochi Aperti in Biologia* (B. Continenza, E. Gagliasso, eds.), Angeli, Milano, 1996, 93-191.
19. M. T. Ghiselin - *Metaphysics and the Origin of Species*, S.U.N.Y. Press, New York 1997.
20. G. V. Schiaparelli - *Studio Comparativo tra le Forme Organiche Naturali e le Forme Geometriche Pure*, Hoepli, Milano 1898.

FOR AN AXIOMATIC THEORY OF THE EVOLUTIONARY DARWINIAN IDEAS: A PROPOSAL

Paolo Freguglia

Dipartimento di Matematica Pura e Applicata
Università dell'Aquila
L'Aquila, Italia
and Domus Galilaeana
Pisa, Italia
fregugli@univaq.it

SUMMARY

In this paper I examine the axiomatic theory of evolutionary Darwinian ideas according to Schiaparelli's work, published in 1898. In particular I propose a formal theory using first-order mathematical logic.

INTRODUCTION

This work is related with the contribution by M.Forti, P.Freguglia, L.Galleni, which appears in this volume. Schiaparelli's work, published in 1898[1], about the evolutionary Darwinian ideas presents a *geometrical model* where a finite, yet very high number of points on a space $\mathbf{R}^n$ represents temporary states on the life of individuals. Every state is characterised by *n* biological parameters, which we assume to be measurable.

THE GEOMETRICAL MODEL

We propose [2, 3], a characterisation of these points belonging to $\mathbf{R}^n$ (which represents temporary states in the life of individuals) with the following axioms:

Ax.1: (*Law of discontinuity*): *There exists a finite set* A *of points of* $\mathbf{R}^n$ *.(These points represent temporary states in the life of individuals).*

Hence we can use on A the properties of metric space and of vector space of $\mathbf{R}^n$, and a vector xy interprets the generation from the individual x to the individual y .

The Application of Mathematics to the Sciences of Nature, edited by Cerrai *et al.*
Kluwer Academic/Plenum Publishers, New York, 2002.

Ax.2: (*Law of determination of species*): *It is always possible to establish a criterion for dividing (co-ordinate partition of)* A *according to the accumulation of its points. We denote by* $P(A)$ *this partition.* (*If* $A' \in P(A)$ *then* A' *represents a species).* See Fig.1.

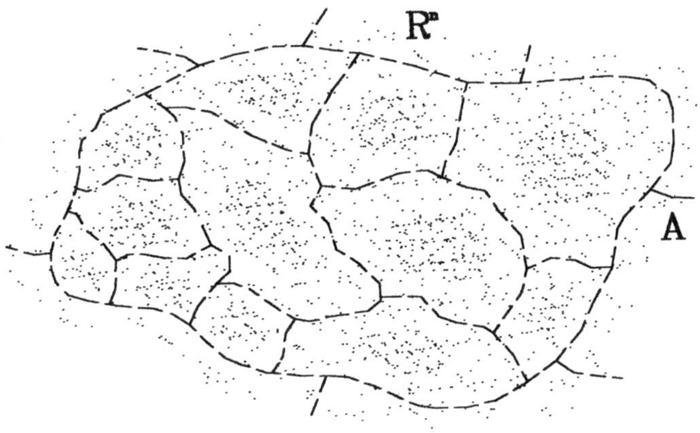

Figure 1.

Ax.3: (*Law of determination of the standard type*): *There are several elements* x' *in* A' *such that other elements* x *of* A' *are the more numerous the nearer* x'. *The set of these elements is denoted by* I(x') *which determines a hypersphere of a subspace of dimensions* n - s *in* $\mathbf{R}^n$. See Fig.2.

It is possible that A' coincides with one I(x') only.

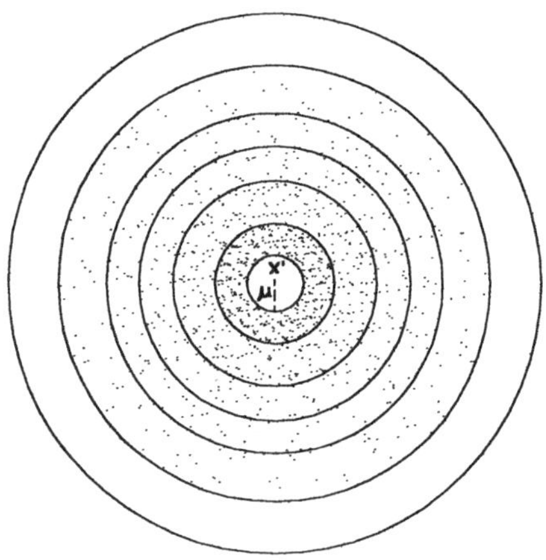

Figure 2.

Ax.4: (*Law of immanence of the standard type*): *There are some points* y° *among those far from* x' , *but belonging to* I(x'), *and some points* y' *very close to* x' *such that the length of a vector* y°y' *is proportional to the distance of* y° *from* x'. See Fig.3.

Ax.5: (*Fundamental law of speciation*): *Let* k_1 *be the measure of the radius of* I(x') *and* xz *a constant vector which represents a persistent perturbing action on the successive* *generations from a* x_1 *very close to* x', *so for a* $t' \in \mathbf{N}$, *a* $h' \in \mathbf{R}^+$, *connected with* t', *exists and* $x_{t'} \in I(x')$ *(and far from* x'*) and the vector* $x_{t'}y_1$ *are such that:*

$$|x_{t'}\, y_1| = h'\, |xz| > k_1$$

where $y_1 \notin I(x')$. See Fig.4.

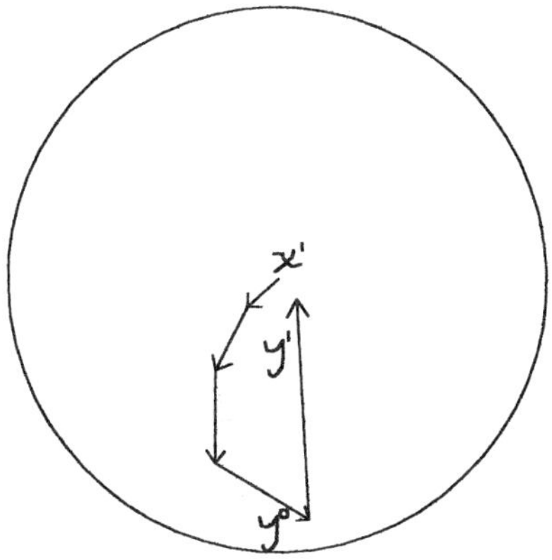

Figure 3.

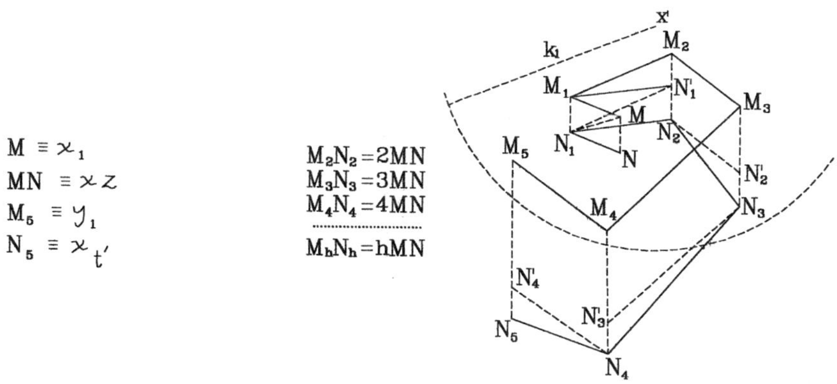

Figure 4.

A PROPOSAL OF FORMALIZATION

Now we give a proposal of formalisation using the logic of predicates of the first order, by modifying partially the previous axioms. Let $(c_1, c_2, \ldots.., c_n)$ be the set of measurable parameters, the number of which determines the dimension of $\mathbf{R}^n$, where a point represents a temporary state on the life of an individual. Hence we propose the following axiom:

Ax.1': *Let* x *be an individual, that is* u(x). *Also let a degree* ε_i *(which can be equal to* 0 *or* 1*) of verifying the parameters* $(c_1, c_2, \ldots.., c_n)$, *associated to any instant* t *of the period* T_x *of the life of* x. *Moreover the parameters for which* $\varepsilon_i = 1$ *are measurable, while the parameters for which* $\varepsilon_i = 0$ *are not measurable* . Symbolically:

$$\forall x \forall t \forall \varepsilon_i (u(x) \wedge t \in T_x \wedge \varepsilon_i \in \{0,1\} \rightarrow f(x, t) = (\varepsilon_1, \ldots, \varepsilon_n)) \wedge \forall \varepsilon_i (\varepsilon_i = 1 \rightarrow \exists \mu_i (\mu_i \in [k_1, k_2] \subset \mathbf{R}^+)$$

where u is the constant monadic predicate "x is an individual" and f is an appropriate functional.

At this point it is necessary to define when an individual x belongs to a species s_j, that is we shall define the constant monadic predicate $s_j(x)$, which depends on *s*-tuples $\varepsilon_1, \ldots, \varepsilon_s$ and $\mu_1, \ldots, \mu_s$. So, for instance, for $s = 4$, by considering a finite interval of five instants, we can have for an individual x the following evolution of the measure of the parameters:

	$\varepsilon_1, \ldots, \varepsilon_4$	$\mu_1, \ldots\ldots, \mu_4$
t_0	1, 0, 1, 1	0.4, 0, 1.3, 1.5
t_1	1, 0, 1, 1	0.3, 0, 0.9, 1.6
t_2	0, 1, 0, 1	0, 0.2, 0, 1.4
t_3	1, 1, 1, 1	0.2, 0.5, 0.3, 1.5
t_4	1, 1, 1, 1	0.5, 0.7, 1.2, 1.3
t_5	1, 1, 1, 1	0.6, 0.7, 1.1, 1.4

Hence the definition of $s_j(x)$ is:

$$s_j(x) \leftrightarrow (u(x) \wedge \varepsilon_i \in \{0,1\} \rightarrow \forall \varepsilon_i \forall t \exists \mu_i (t \in T_x \wedge \tau(\varepsilon_i) = 1 \wedge s \leq n \rightarrow f(x, t) = (\varepsilon_1, \ldots, \varepsilon_s) \wedge (\mu_i \in [k_1, k_2] \subset \mathbf{R}^+)$$

where τ is a functional which allows us to read the dyadic predicate $\tau(\varepsilon_i) = 1$ so: "the value of ε_i will be sooner or later, in a certain instant, equal to 1, and this must happen, even if sometimes ε_i can be equal to 0". Analogously we can define when an individual x belongs to a type $\mathbf{t}_j$. The sets of indices, that we find in the following axioms, are ordered.

Ax.2': *For any individual* x, *a species* s_j, *to which one he belongs, exists.* Symbolically:

$$\forall x (u(x)) \rightarrow \exists! i (i \in I \wedge s_j(x))$$

where I is an appropriate set of indices.

Ax.3': *Every individual belongs to a type* $\mathbf{t}_j$ *of a species* s_j *and a species can have one type only, which coincides with it.* Symbolically:

$$\forall x \forall j \exists i (i \in I_1 \wedge j \in I_2 \wedge u(x) \rightarrow ((\mathbf{t}_i(x) \rightarrow s_j(x)) \vee (\mathbf{t}_i(x) \leftrightarrow s_j(x)))$$

where I_1 and I_2 are two appropriate sets of indices.

Ax.4': *(Law of regular generation) If an individual, which belongs to a particular type breeds another individual, then in general the two individuals are very similar, that is, the measures of their respective parameters are very close to each other, except for particular cases (see Ax.6').* Symbolically:

$$\forall y^\circ \forall y' \forall i (i \in I \wedge u(y^\circ) \wedge u(y') \wedge \mathbf{t}_i(y^\circ) \wedge \mathbf{t}_i(y') \rightarrow (g(y^\circ y') \rightarrow v^\wedge(y^\circ y'))$$

where g is the constant dyadic predicate which we read "y° breeds y'" and v^ is another constant dyadic predicate which we read "y° is very similar to y' ".

Ax.5': (*Law of immanence of the standard type*): *Given an average individual* x' *of a type* $\mathbf{t}_j$ *, let* x *be a individual similar to* x'. *If a succession of generations from* x *exists as far as the individual* y_n*, which is not very similar to* x*, then (if* y_n *can breed) the progeny by* y_n *individual* y *will be very similar to* x.

Using the language of the first-order mathematical logic:

$$\forall x'\forall x\ \forall j(j\in\ I_1 \wedge u(x') \wedge u(x) \wedge \mathbf{t}_j(x') \wedge \mathbf{t}_j(x) \wedge m(x') \wedge v(xx')) \wedge$$

$$\exists y_i\ \forall i\ (i\in\ I_2\ \wedge u(y_i\ ) \wedge \bigwedge\nolimits_{i=1,n} g(xy_i) \wedge l(xy_n)) \rightarrow \exists y\ (g^{\wedge}(y_n y) \rightarrow v^{\wedge}(yx'))$$

where m is the constant monadic predicate "x is an average element or individual of the type $\mathbf{t}_j$", l is the constant dyadic predicate "x is not very similar to y_n", g^ is the constant dyadic predicate "x may occasionaly breed y", v is the constant dyadic predicate "x is similar to y"

and $\bigwedge_{i=1,n} g(xy_i)$ gives the succession of the conjunctions of $g(xy_i)$. So for instance when $n = 3$, considering also $\exists y_i$, explicitly we have:

$$\exists y_1\ \exists y_2\ \exists y_3\ (g(xy_1) \wedge g(y_1y_2) \wedge g(y_2y_3)).$$

Ax.6': *Let* x_1 *be very similar to the average individual* x' *belonging to type* $\mathbf{t}_j$ *of a species* s_i. *If from* x_1 *, by a persistent perturbing action, a succession of individuals could be bred and without this action an other different succession from the earlier succession of individuals could be bred, then really a further different succession of individuals will be bred, which depends on the earlier two. At a moment of this real succession of generations an individual* y' *is bred which does not belong (by his features) to type* $\mathbf{t}_j$ *of a species* s_i .

Symbolically:

$$\forall x'\forall x_1\forall j(j\in\ \ I_1\ \wedge\ u(x')\ \wedge\ m(x')\ \wedge\ u(x_1)\ \wedge\ \mathbf{t}_j(x_1)\ \wedge\ \mathbf{t}_j(x')\ \wedge\ v(x'x_1))\ \wedge$$

$$\exists x_{i+1}\exists y_i\exists x_{i+2}\exists y_i\ \forall i\ (i\in\ I_2 \wedge u(x_{i+1}) \wedge u(y_i) \wedge u(x_{i+2}) \wedge u(y_{i+1}) \wedge \bigwedge\nolimits_{i=1,n} [(g^{\wedge}(x_iy_i) \vee g^{\wedge}(x_ix_{i+1}) \rightarrow g(x_iy_{i+1})]) \rightarrow \exists y'\exists r(r \in\ I_2 \wedge g(x_ry') \wedge \neg\ \mathbf{t}_j(y'))$$

where g^ is read "x_i could breed y_i" ; the scheme $\bigwedge_{i=1,n} ((g^{\wedge}(x_iy_i) \vee g^{\wedge}(x_ix_{i+1}) \rightarrow g(x_iy_{i+1}))$ is iterative. So for $n = 3$, considering also $\exists y_i$ and $\exists x_i$, we have:

$$\exists x_1\ \exists x_2\ \exists x_3\ \exists y_1\ \exists y_2\ \exists y_3\ \exists y_4\ ((g^{\wedge}(x_1y_1) \vee g^{\wedge}(x_1x_2) \rightarrow g(x_1y_2)) \wedge (g^{\wedge}(x_2y_2) \vee g^{\wedge}(x_2x_3) \rightarrow g(x_2y_3)) \wedge (g^{\wedge}(x_3y_3) \vee g^{\wedge}(x_3x_4) \rightarrow g(x_3y_4))).$$

CONCLUSION

It is our intention that this work serves as a basis for a mathematically rigorous description of Darwin's work

REFERENCES

1. G.V.Schiaparelli, *Studio Comparativo tra le Forme Organiche Naturali e le Forme Geometriche Pure*, Hoepli, Milano (1898).

2. P.Freguglia, Considerazioni sul modello di Giovanni V.Schiaparelli per una interpretazione geometrica delle concezioni darwiniane, *Atti e Memorie dell'Accademia Nazionale di Scienze, Lettere e Arti di Modena*, 14:135-158 (serie VII) (1998).
3. P.Freguglia, Considerazioni sul modello di Giovanni V.Schiaparelli per una interpretazione geometrica delle concezioni darwiniane, in *Modelli Matematici nelle Scienze Biologiche*, P.Freguglia (ed.) Quattroventi, Urbino (1998).

FRACTAL COMPLEXITY OF MEMBRANE STRUCTURES IN NORMAL AND NEOPLASTIC CELLS DURING GROWTH AND APOPTOTIC CELL DEATH

Gabriele A. Losa

Laboratory of Cellular Pathology
Institute for Interdisciplinary Scientific Studies (ISSI)
Center for Research in Mathematics and Physics
CH-6601
Locarno
Faculty of Natural Sciences
University of Lausanne
CH-1000
Lausanne
Switzerland
glosa@guest.cscs.ch

ABSTRACT

In contrast to the traditional vision based on concepts of homeostasis, such as linearity and reversibility, the more recent epistemological approach to Nature is based on a set of coherent principles provided by chaos theory and by the development of fractal geometry in this century. Both fields have provided a new mathematical language running counter to the conventional euclidean one, which makes it possible to describe irregular morphologic structures and assess their functional and qualitative properties in complex nonlinear phenomena either physiological, abnormal and tumoral occurring in living organisms. It has been argued that the design principle for living organisms follows fractal rules. In practice, in biological tissues fractal patterns of self-similar structures can be observed only within a scaling window of length and time measurements which has to be established experimentally. The fractal approach has been used to characterize cell membranes in hematopoietic cells as well as in different types of human leukemia, to measure the morpho-ultrastructural complexity which governs plasma membranes and nuclear organelles of normal, abnormal and human cancer cells submitted to a variety of exogenous stimuli, during the hormone-driven growth and/ or the apoptotic process of cell death. More generally, it makes it possible the quantitative link between morphologic-ultrastructural features with functional molecular pathways in line with the structure / function binomial, essential for a real integrative understanding of cell biology.

INTRODUCTION

The living world encompasses an immense diversity of forms and structures, most of which are generated by irregular morphogenetic processes. The functional and metabolic phenomena which take place in living organisms follow non-linear dynamics, rather than the linear law of cause and effect, are irreversible and occur far from thermodynamic equilibrium[1]. Living organisms dissipate energy to ensure an alternation between self-perpetuation far from the chaotic state, where entropy is greatest, and their functional and morphological configuration [2]. Structural features and metabolic processes appear to be highly organized (i.e. far beyond the order) and correlated through a complex system of interactions. This enables organisms to tolerate a wide spectrum of functional variations and structural modifications pertaining to physiological events and hence to sustain their capacity for the adaptive response. The words of Johann W. Goethe (1787) are noteworthy in this context: " A tree is a living being which develops a series of organs consecutively in time which form links with each other and with the entire organism, from a unique and identical morphogenetic idea " [3]. In contrast to the traditional vision based on concepts of homeostasis, linearity and reversibility [4], the more recent epistemological approach to nature is based on a set of coherent principles provided by chaos theory [1] and by the development of fractal geometry in this century [5]. Both fields have provided a new mathematical language which makes it possible to assess and describe morphological structures with complex qualitative properties, such as nonlinear physiological and tumoral phenomena in living organisms, without the recourse to mathematical models which approximate and simplify the reality [6,7,8]. This runs counter to the conventional Euclidean approach, which is mainly suitable for describing ideal geometric bodies and linear phenomena, and is a break with Aristotelian assumptions. It has been argued that the design principle for living organisms follows fractal rules [9]. The fractal approach has already been widely used to characterize cellular ultrastructures and morphologic components at the electron microscopic level, but the number of applications of fractal geometric analysis to resolve optical microscopic images is growing, and interest in the potential use of fractals in immunology, morphological oncology, pathology and physiology is steadily increasing[10,11,12,13]. The need to go beyond subjective descriptions in pathology has led to the adoption of the fractal approach to describe morpho-histological features for diagnostic purposes. The fractal characterization of the appearance of heterochromatin and other nuclear components has already been used for the diagnosis of breast cytology [14], skin diseases such as mycosis fungoides and chronic dermatitis [15] and several other neoplastic conditions, such as squamous cervical epithelioma [16], malignant mesothelioma, metastatic peritoneal carcinoma [17] and prostatic carcinoma [18]. Previous records of physiologic and pathologic states have been used to illustrate the structural and functional complexity of a variety of cells and tissues [19]. The fractal analysis could be adopted to characterize complex ultrastructural features of cell surface and nuclear components during the active phases of apoptosis, a physiological form of cell death, occurring in breast cancer cell cultures. Although cells committed to die undergo a series of functional and morphological changes [20,21,22] it may be difficult to recognize and quantify them in solid tissues for several reasons. First, apoptosis is a relatively rapid phenomenon in vivo and second, apoptotic cells are efficiently removed from a tissue by phagocytic elements [23,24,25,26,27]. On the other hand, the methods currently available do not adequately identify the early phases of apoptosis while the cells continue to appear morphologically normal. Finally, the frequency of measurable apoptosis may be low in living tissues because this is a natural residual process [28,29]. The present paper describes the fractal complexity which governs outer cell membranes and nuclear components of normal, abnormal and cancer cells during growth and apoptotic cell death.

MATERIAL AND METHODS

Isolation of Mononuclear Cell from Peripheral Blood

Samples (20 mL) of heparinized blood obtained from healthy donors and from patients with various forms of lymphocytic leukemia were diluted 1/5 with 9 g/l NaCl solution and layered on Ficoll-Hypaque gradients [30].The samples were centrifuged (30 min/800 x g at 20° C) and the cells at the interface were harvested and washed three times with 0.05 M Tris-K-Mg buffer, pH 7.4. Aliquots of isolated cells were stained with fluorescent conjugated monoclonal antibodies directed against T, B and myeloid surface antigens CD3, CD4, Cd5, CD8, CD14, CD19, CD20, CD34, CD45, and analyzed by flow cytometry with a FACSscan (Becton Dickinson) to identify the immune phenotype.

Cell Cultures

MCF-7 and SKBR-3 human breast epithelial cells were cultured in Dulbecco's Minimal Essential Medium F-12 supplemented with 5% FCS, 2 mM L-glutamine, 100 U/ml penicillin, 100 U/ml streptomycin and 1 μM insulin. Cultures were carried out in a humid atmosphere at 37° C and 5% CO_2. Five days prior to the experiments, cells were replated in medium without phenol red and 5% FCS and stripped of endogenous steroid hormones by incubating with dextran-coated charcoal. The cells were then cultured for 5 min in the presence of 17β-estradiol or dexamethasone, and then harvested using 0.9% NaCl (pH 7.4) containing 1 mM EDTA and processed for electron microscopy. The concentration of both hormones [10^{-9} M] was chosen within the concentration range found under physiological conditions. The apoptotic effect of the calcium ionophore A23187 on SKBR-3 cells was investigated within a concentration interval ranging from 0.01 μM to 20 μM during 24h period of treatment.

Transmission Electron Microscopy

Cell samples (1-5 x 10^6) were fixed for 5-6 h at 4°C with 30 g/l glutaraldehyde, 70 mg/l calcium chloride in 0.1 M sodium cacodylate at pH 7.4. The cell samples were washed and then postfixed for 1-2 h at room temperature with a 2g/100 ml osmium tetraoxide in 0.1 M sodium cacodylate, adjusted with NaCl to 340 mosM in order to minimize artifacts. The samples were then dehydrated with ethanol and propylene oxide before being embedded in Epon. Thin sections with an interference color of gray to silver were cut with an LKB 2128 microtome, mounted on 200-mesh grids coated with a formvar film and finally stained by exposure to 5g/100 ml uranyl acetate for 1 h and then to lead citrate for 30 min. The true section thickness was calculated by using the fold method [31]. Negative micrographs were recorded on plastic plates using a Philips 400-ET microscope.

Immunoelectron Microscopy

The cell phenotype of lymphocytes were identified after staining mononuclear cell populations with the corresponding unconjugated monoclonal antibodies to the surface antigens CD3, CD4, CD8 and CD19. Cell samples were stained for 30 min at 4°C with monoclonal antibodies. After washing with phosphate buffered NaCl solution, the cells were fixed for 30 min at 4°C in a 0.1g/100 ml glutaraldehyde solution, and then incubated at room temperature for 30 min with rabbit anti-mouse IgG-antibody (dilution 1/100). They were then rinsed and finally reacted for 30 min with a mouse peroxidase-antiperoxidase complex. The enzymatic activity was activated by adding diaminobenzidine-H_2O_2 as substrate which was transformed in an electron dense complex. EM thin sections of cell samples were performed as described above. ALL and Hairy-cell leukemia cells were not decorated with monoclonal antibodies for electron microscopy because more than 95% of the cell populations consisted of blasts of monoclonal origin

Profile Extraction

Mononuclear cells from the peripheral blood of healthy donors and of patients with acute lymphocytic and hairy leukemia, MCF-7 and SKBR-3 human breast cancer cells either untreated or receiving 17β-estradiol, dexamethasone or calcimycin were used. EM negatives were taken at 8000 x, 8400 x and 4600 x for lymphocytes and leukemic cells, MCF-7 cells and SKBR-3 cells respectively and then turned into positive pictures with a final magnification of 18400 x, 18100 x and 10580 x. Outlines of plasma membrane and nuclei in which both the external perinuclear membrane (ENM) and domains of condensed heterochromatin bound to the internal nuclear membrane (NMBHC) could be unambiguously identified on the positive prints were selected. Heterochromatin scattered inside the nucleoplasm with the exclusion of the nucleolus region was taken into consideration for the characterization of the apoptotic process in SKBR-3 cells. Hand-drawn or computer acquired contours of the plasma membrane, the ENM and NMBHC were scanned at a resolution of 300 dpi. The data obtained for each contour were stored in binary map patch format (bmp) in a computer. This binary picture was used for the fractal analysis.

The Fractal Dimension [D]

Fractals are objects which exhibit consistent roughness or irregularity at every scale [19]. The outstanding property of fractals is their shape invariance or self-similarity, regardless of scale. An object is called "self-similar" if every part of it looksthe same as the original object after being scaled by an arbitrary factor. The Richardson-Mandelbrot equation [1] provides the mathematical basis for understanding geometric and spatial fractal structures, and for measuring and interpreting them:

$$L(\varepsilon) = N(\varepsilon).(\varepsilon) \qquad (1)$$

where $L(\varepsilon)$ is the length (perimeter) of the closed cell contour or cell component in the image plane, and (ε) is the unit length (yardstick) used to measure the perimeter $L(\varepsilon)$, and

$$N(\varepsilon) = [\, lo / \varepsilon \,]^{D} = lo^{D} \varepsilon^{-D} \qquad (2)$$

where $N(\varepsilon)$ is the number of yardsticks or unit lengths (ε) needed to cover the perimeter $L(\varepsilon)$. $[\, lo / \varepsilon \,]$ is a dimensionless quantity and *lo* a reference scale which does not influence the determination of [D]. [D] is the fractal contour dimension which increases from 1 to 2 with increasing " wiggliness". The dimensional exponent [D] is the fractal dimension which defines the nature of the curve and can be derived by determining the slope (1-D) of the straight line evaluated from a log-log plot of the equation (2) obtained by logarithmic transformation, namely

$$\log [\, L(\varepsilon) / lo \,] = (1-D) \log [\varepsilon / lo \,] \qquad (3)$$

Fractal Dimension Determination

The fractal capacity dimension [D] of the contours of cell membrane and nuclear components was established using the box counting method described elsewhere [32, 33, 34].

The grid lengths in the original pictures corresponded to 20 nm and 20 μm, respectively. The estimation of the fractal capacity dimension [D] is based on counting the non empty boxes N at a certain grid length ε. The fractal dimension [D] was calculated by using the relation:

$$D = \log N(\varepsilon) / \log (1/\varepsilon)$$

The above equation was used for the {log N(ε), log ε} plot and the slope [D] of the linear range was estimated by using a least square fit for the data set. Experimental data were statistically analyzed using the Wilcoxon-Mann-Whitney U-test.

RESULTS

The Fractal Dimension of the Lymphoid Membrane Outline

The values of the fractal dimension D for the plasmalemma outline were recorded for circulating lymphocytes from healthy donors and for lymphoblasts obtained from patients with proliferative disorders, such as acute lymphoid leukemia (ALL of thymic (T) and bone marrow (B) origin) and hairy cell leukemia. These values were significantly lower in T-ALL cells (1.11) than in B-ALL cells at both early (1.15) and late (1.19) stages of differentiation, indicating that the contour of the plasma membrane of these proliferating ALL cells was smoother or less irregular (rough) than that of acute lymphoid leukemia cells of B origin. CD4-T helper lymphocytes, CD8-T suppressor lymphocytes and CD19 B-lymphocytes are differentiated mature cells with a higher fractal dimensions [D] of 1.16, 1.23 and 1.20 respectively (Figure 1).

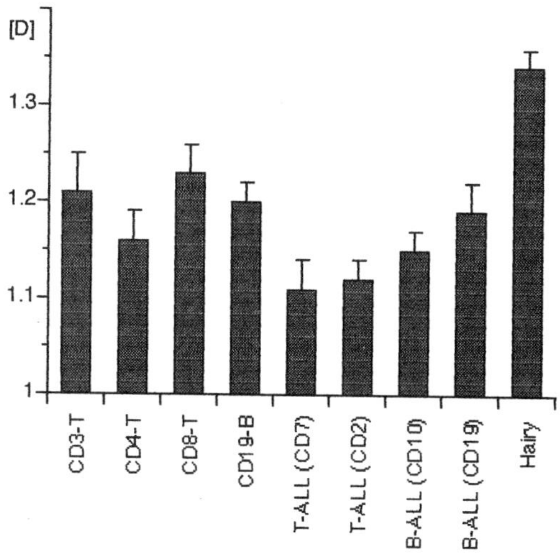

Figure 1. The fractal dimension [D] of the plasma membranes of lymphoblasts from various leukemias and of normal, immunocompetent T or B-lymphocytes. The immune phenotype (CD) of lymphocytes and leukemic cells was determined by flow cytometry after staining with fluorescent conjugated monoclonal antibodies directed against specific membrane surface antigens. The membrane profile was measured for twenty cells in each population. [D] values are given as the mean ± one standard deviation [SD]. CD: cluster of differentiation; ALL: acute lymphoblastic leukemia; Hairy: Hairy cell leukemia.

This shows that these immunocompetent cells have membranes with a more structured (complex) surface than immature leukemic blasts, thus confirming previous reports [35]. The cells from patients with hairy cell leukemia, a chronic form of leukemia, usually displayed a cell surface with a convoluted outline due to numerous membrane protrusions, and this was reflected by the high value of the estimated D (1.34); these leukemia cells probably originate from myeloid-phagocytic precursors, which are already characterized by their rough cell surface.

Chromatin Structure and Nuclear Membranes of Breast Cancer Cells during Hormonal Treatment.

The fractal capacity dimension D of the external nuclear membrane (ENM) and nuclear membrane-bound heterochromatin (NMBHC) was established on nuclear profiles derived from distinct sets of mammary carcinoma MCF-7 cell lines, namely from untreated MCF-7 cells and from cells treated for 5 minutes with a steroid hormone, either 17 β-estradiol [10-9 M] or dexamethasone [10-9 M]. The smallest yardstick or pixel length, that could be applied to scans of electron micrographs taken at the highest magnification (18,100 x) reported in this study was 0.005 μm. The fractal dimension of both nuclear features was therefore calculated from the slope of the straight line established within a scaling window of three orders of magnitude (ε_1-ε_2), i.e. ranging from 0.025 μm to 25 μm (Figure 2).

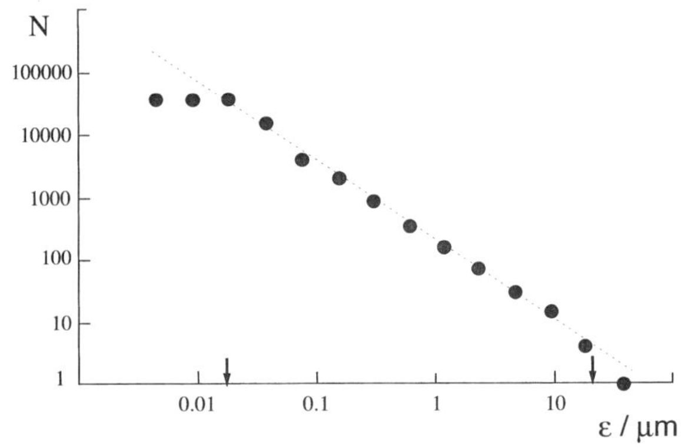

Figure 2. A typical N versus ε plot (log-log scale) of heterochromatin bound to the internal nuclear membrane (NMBHC) from MCF-7 breast cancer cells treated with 10^{-9} M estradiol and determined by the box-counting method. The scaling window is clearly indicated ( ), ε_1 (0.025μm): ε_2 (25μm). The slope of the straight line, which could be plotted within this window, was used to calculate the fractal dimension [D]. Note how the ordinate N (ε) (number of boxes) increases as the pixel length (ε) decreases, showing that the outline length of the NMBHC also increased within the scaling interval ε_1-ε_2.

The mean fractal dimension [D] of the nuclear membrane-bound heterochromatin (NMBHC) changed when MCF-7 cells were treated with both types of hormones. In contrast, the fractal dimension [D] of the external nuclear membrane of steroid-treated cells was unaffected with respect to the untreated MCF-7. However, the change in the outline complexity of the membrane-bound heterochromatin was not the same for both hormones. The fractal dimension [D] of NMBHC was found to be increased (D= 1.42 ± 0.01) in MCF-7 cells treated with 17β-estradiol, but significantly reduced in the presence of the same concentration of dexamethasone (D= 1.34 ± 0.02), when it was even lower than in untreated control cells (D=1.39 ± 0.02). These findings indicate that in cells responsive to the estrogen estradiol, the heterochromatin bound to the inner nuclear membrane showed a more irregular structural pattern than in control cells or in cells cultured in presence of the glucocorticoid dexamethasone [36].

Fractal Analysis of Membrane and Nuclear Components During Apoptotic Cell Death

SKBR-3 cells cultured for 24h in presence of the ionophore calcimycin were likely in the initial phase of the apoptotic death process. This implies that morpho-ultrastructural changes of nuclear and plasma membranes and of intranuclear scattered heterochromatin might still have occurred, although undetectable by conventional inspection (figure 3 a,b).

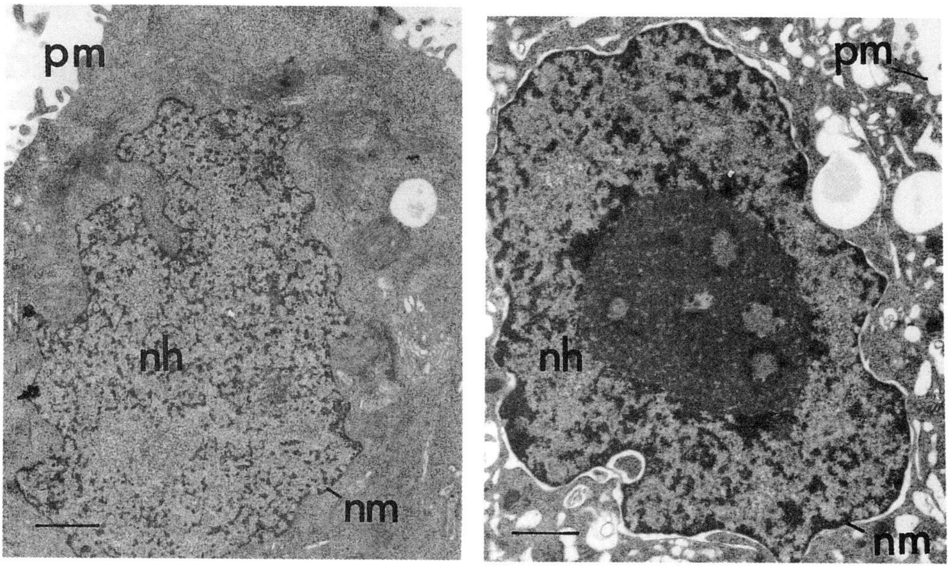

(a) (b)

Figure 3. Electron micrograph of SKBR-3 breast cancer cells: a) untreated and b) cultured in presence of 1 μM calcimycin. Changes in the ultrastructural organization of the nuclear scattered heterochromatin (nh), nuclear (nm) and plasma (pm) membranes evidenced in both cells by fractal analysis cannot be identified by visual inspection. Final magnification: 10580 x (1 cm = 0.94 μm).

Therefore, membrane contours irregularity and gray intensity redistribution of the nuclear stroma have been investigated on electron microscopy pictures of SKBR-3 cells by measuring the fractal dimension [D]. On the whole, untreated control cells had a plasma membrane with an outline significantly more irregular ($p < 0.01$) than that of 1 μM calcymicin-treated cells (24 h), as documented by the higher value of the estimated median D, namely 1.143 versus 1.119. In control cells, the perinuclear membrane showed a somewhat higher irregularity than that of treated cells as documented by the median fractal dimension D of 1.027 versus 1.012. A more relevant finding concerned the reduction of the frequency distribution of grayscale values associated to the scattered nuclear heterochromatin of treated cells and revealed by a median D of 1.571, in comparison to untreated cells with a median D of 1.712. The latter changes of the fractal dimension might be suggestive of a rearranged ultrastructural organization of the DNA-proteins complex occurring during the early apoptotic phases [37].

DISCUSSION

Fractal geometry offers an effective method of determining the optimum conditions for an objective evaluation of the morphologic and structural features of cells and tissues which have no characteristic size and which exhibit self-similar irregular details within scaling windows of measurement [19]. It also makes it possible to make a true comparison of cellular changes occurring in the course of time, provided that morphometric data are collected under appropriate conditions when such morphologic and metabolic changes are measurable and scale invariant. The applications reported in the present study demonstrate the real advantages of the fractal analysis: numerical values could be attributed to peculiar qualitative features of spatial structures, such as irregularity, morphologic complexity and roughness, which can be used to characterize cells of different origins or in different functional and / or pathologic states. Blasts from leukemia patients are strongly proliferating cells with a plasma membrane which has an irregular but smooth profile, with a fractal dimension value [D] significantly smaller than that found for the plasma membrane of normal, mature and immunocompetent T-and B blood lymphocytes. The physiopathological significance and biochemical mechanisms inherent in the observed differences between plasma membrane structure in lymphocytes and in leukemia cells have not yet been established. They may be linked to changes in the configuration and composition of membrane lipids induced by mitogenic or regulatory factors and / or due to the process of neoplastic transformation. In the case of leukemia cells, surface smoothness could result from enhanced shedding of vesicular fragments from the plasma membrane. A cell surface renewal process of this type could enable these cancer cells to evade defense mechanisms and speedup their rate of growth [38]. Membrane shedding is a recognized widespread protective phenomenon used by almost all cancer cells and tissues, but which is also involved in several physiological processes [39]. Recently it has been reported that cells alter the repertoire of their membrane proteins many times as in the course of development and other dynamic functional states [40]. The release of transmembrane substrates and of proteins anchored in distinct ectodomains of the plasma membrane through the intervention of specific metalloproteinases, such as ADAMs, plays an essential role in mammalian development [41]. Bleb formation and the consecutive shedding of membrane-bounded fragments from the cell periphery often occur in cells undergoing apoptotic cell death [42]. For the sake of clarity it should be pointed out that surface smoothness may not be a trait peculiar to leukemia cells. Normal peripheral blood lymphocytes triggered to proliferate by the mitogen phytohaemaglutinin undergo a dramatic loss of roughness of their plasma membrane [35]. Fractal analysis is appropriate for nuclei and nuclear components, which are irregular, self-similar elements characterized by a variable degree of pleomorphism. At the ultrastructural level, even the finest changes in heterochromatin organization (either by unfolding or condensation) or in the nuclear membrane outline of breast cancer cells in response to various steroid hormones, which could not be detected by visual inspection, were revealed by fractal morphometry and numerically assessed using a peculiar fractal dimension. Recent unpublished findings have shown that the nuclear envelope of estrogen-sensitive human breast cancer MCF-7 cells was coarser and more irregular than that of non-estrogen sensitive breast cancer SKBR-2 cells (fractal dimension D, 1.26 versus 1.14). This confirms that the MCF-7 cells possess an higher degree of differentiation which is probably due to the presence of estrogen receptors. Moreower, the fractal approach has proved useful to evaluate subtle morpho-structural modifications of membranes and nuclear components in the early phase of the apoptotic process, which could not be identified by conventional parameters [37]. Recently it has been reported that the contour irregularity of lymphocytic nuclear envelope estimated by fractal analysis can be used to distinguish between lymphocytes from early-stage mycosis fungoides (high D), a malignant cutaneous lymphoma, and lymphocytes from chronic dermatitis (low D), a benign skin lesion [15]. According to these authors, this ability to distinguish between benign lesions and early phases of malignancies underscores the predictive value of the fractal dimension D. These findings combine to illustrate the potential use of fractal analysis to measure structural features, perceive changes reflecting characteristic functional and / or pathological states and carry out diagnoses. Heterochromatin organization, perinuclear membrane appearance, nuclear texture, the outline irregularity of the cell surface and peculiar morphologic features of other organelles can reveal the patho-physiologic state of cells and the biologic characteristics of neoplastic and transformed tissues, but only if they can be handled as reliable quantitative descriptors. This can be achieved by using fractal geometry to quantify irregular cell structures which cannot be evaluated by any single scale of measurement. In this way, cytologic, morphologic and ultrastructural features could be examined at the same time as dynamic functional events arising at the cell surface, in the nucleus or other cellular organelles, and so lead to a deeper comprehension of complex living systems, from birth to death.

ACKNOWLEDGMENTS

The fruitful discussion with Prof. T. F. Nonnenmacher and Dr. R. Graber is acknowledged. This study was supported by grants 31-25702.88 and 31-45756-95/1 from the Swiss National Science Foundation.

REFERENCES

1. I. Prigogine. *Les lois du chaos*, Flammarion, Paris (1997).
2. N. Dioguardi. La medicina, l'incertezza e la complessità. in: *Scienza, Tecnologia, Società alle Soglie del XX secolo*., 47-53. Giuffrè ed, Milano (1997).
3. J.W. Goethe. *Gli scritti scientifici: Morfologia I: Botanica I*. E. Ferrario ed., Il Capitello del Sole srl Publisher, Padova (1996).
4. W.B. Cannon. Organization for physiological homeostasis. Physiol. Rev. 9:399 (1929).
5. B.B. Mandelbrot. *TheFractal Geometry of the Nature*. Freeman, San Francisco (1983).
6. D. Paumgartner, G.A. Losa., E.R. Weibel. Resolution effect on the stereological estimation of surface and volume and its interpretation in terms of fractal dimensions. J.Micr. 121:51 (1981).
7. B.J. West, W. Deering. Fractal physiology for physicists: Lévy statistics. Physics Reports 246:1(1994).
8. T. Mattfeld. Spatial pattern analysis using chaos theory: a nonlinear deterministic approach to the histological texture of tumours. in: *Fractals in Biology and Medicine II*, G.A.Losa, D.Merlini, T.F. Nonnenmacher, E.R.Weibel eds, Birkhäuser Verlag, Basel. (1998).
9. E.R. Weibel. Fractal Geometry: a design principle for living organisms. Am. J. Physiol. 261: 361 (1991).
10. T. Schall. Fractalkine: a strange attractor in the chemokine landscape. Immunology Today 18 (4):147 (1997).
11. S.S. Cross. Fractals in pathology. J. Pathol. 182:1 (1997).
12. G. Landini, J.W. Rippin. How important is tumour shape ? quantification of the epithelial-connective tissue interface in the oral lesions using local connected fractal dimension analysis. J. Pathol. 179:210 (1996).
13. G.A. Losa, R. Graber, G. Baumann, T.F. Nonnenmacher. Steroid hormones modify nuclear heterochromatin structure and plasma membrane enzyme of MCF-7 cells. A combined fractal, electron microscopical and enzymatic analysis. Eur.J.Histochem. 42: 21 (1998).
14. A.J. Einstein, Wu Hai-Shan, M.Sanchez, J. Gil. Fractal characterization of chromatin appearance for diagnosis in breast cytology. J Pathol. 185: 366 (1998).

15. C. Branciardi., C. Miracco, M. Margherita De Santi, A. Perrone, A. Bruni, P. Luzi. Fractal dimension of lymphocytic nuclear membrane in Mycosis Fungoides and Chronic Dermatitis. in: *Fractals in Biology and Medicine II*, G.A.Losa, D.Merlini, T.F.Nonnenmacher, E.R.Weibel eds. Birkhäuser Verlag, Basel (1998).
16. C. MacAulay, B. Palcic . Fractal texture features based on optical density surface area. Use in image analysis of cervical cells. Anal. Quant. Cytol. Histol. 12 (6):394 (1990).
17. F. Sepulcre, J. Grau, M. Garcia-Bonafe, A. Moragas. Fractal approach to the study of chromatin texture. in: *Fractals in Biology and Medicine II,* G.A. Losa, D.Merlini, T.F. Nonnenmacher, E.R.Weibel eds, Birkhäuser Verlag, Basel (1998).
18. T. Irinopoulou, J.P. Rigaut, M.C. Benson. Toward objective prognostic grading of prostatic carcinoma using image analysis. Anal.Quant.Cytol.Histol. 15 (5):341 (1993).
19. G.A. Losa, T.F. Nonnenmacher. Self-similarity and fractal irregularity in pathologic tissues. Modern Pathol. 9(3):174 (1996).
20. M. Raff. Cell suicide for beginners. Nature 396:119 (1998).
21. G.M. Cohen. Caspases: the executioners of apoptosis. Biochem. J, 326:16 (1997).
22. A.H. Wyllie, J.F.R. Kerr, A.R. Currie. Cell death: the significance of apoptosis. Int. Rev. Cytol. 68:251(1980).
23. C. Drachenbery, JC. Papadimitriou. Significance of apoptosis in prostatic neoplasias. Arch.Pathol.Lab.Med. 121: 54 (1997).
24. E.F. Gaffney. The extent of apoptosis in different types of high grade prostatic carcinomas. Histopathology 25:269 (1994).
25. D.J. Allan, A. Howell, S.A. Roberts, T.G. Williams *et al.* Reduction in apoptosis relative to mitosis in histologically normal epithelium accompanies fibrocystic change and carcinoma of the premenopausal human breast. J Pathol. 167:25 (1992).
26. S. Battersby, T.J. Anderson. Proliferative and secretory activity in the pregnant and lactating human breast. Virch. Arch. A Pathol. 413:189 (1988).
27. P. Lipponen *et al.* Apoptosis in breast cancers as related to histopathological characteristics and prognosis. Eur J. Cancer 30: 2068 (1994).
28. H.A.M. Van de Shepop *et al.* Counting of apoptotic cells: a methodological study in invasive breast cancers. J.Clin. Pathol. Mol. 49:214 (1996).
29. G.A. Losa, R. Graber. Apoptotic cell death and the proliferative capacity of human breast cancers. Anal. Cell. Pathol. 16:1(1998).
30. R. Graber, G.A. Losa. Changes in the activities of signal transduction and transport membrane enzymes in CEM lymphoblastoid cells by glucocorticoid-induced apoptosis. Anal.Cell.Pathol. 8:159 (1995).
31. J.V. Small. Measurement of the section thickness. Proc.4th Europ.Conf. Electron Microscopy 1: 609 (1968).
32. G. Baumann, A. Barth, T.F. Nonnenmacher. Measuring fractal dimensions of cell contours: practical approaches and their limitations. in: *Fractals in Biology and Medicine* T.F. Nonnenmacher, G.A. Losa, E.R. Weibel eds, Birkhäuser Verlag, Basel 1994.
33. T.F. Nonnenmacher, G. Baumann, A. Barth, G.A Losa. Digital image analysis of self-similar cell profiles. J. Bio-Medical Comp. 37:131 (1994).
34. J.W. Döllinger, R. Metzler, T.F. Nonnenmacher. Bi-asymptotic fractals: fractals between lower and upper bounds. J. Math. Phys. a: Math. Gen. 31:3839 (1998).
35. G.A. Losa, G. Baumann, T.F. Nonnenmacher. Fractal dimension of pericellular membranes in human lymphocytes and lymphoblastic leukemia cells. Path. Res. Pract. 188:680 (1992).
36. G.A. Losa, R. Graber, G. Baumann, T.F. Nonnenmacher. Steroid hormones modify nuclear heterochromatin structure and plasma membrane enzyme of MCF-7 cells. A combined fractal, electron microscopical and enzymatic analysis. European J. Hist. 42: 1(1998).
37. C. Castelli, G.A. Losa. Morphofractal reorganization of plasma membrane and nuclear components during the apoptotic process of human breast cancer cells. in: *Fractals 2000 in Biology and Medicine III,* G.A. Losa, D.Merlini, T.F. Nonnenmacher, E.R.Weibel eds, Birkhäuser Verlag, Basel (2001).
38. G.A. Losa, D. Heumann, S. Carrel, V. von Fliedner, J.P. Mach. Characterization of membrane vesicles circulating in the serum of patients with common acute lymphoblastic leukemia. Lab. Invest. 55:573 (1986).
39. S. Ladish, Wu Zi, S. Feig, L. Ulsh, E. Schwartz, G. Floutsis, F. Wiley, K. Lenarky, R. Seeger. Shedding of gd2 ganglioside by human neuroblastoma. Int. J.Cancer 39:73 (1987).
40. Z. Werb, Y. Yan .A cellular striptease act. Science 282:127 (1998).
41. J.J. Peschon, J.L. Slack, P. Reddy, K.L. Stocking *et al.* An essential role for ectodomain shedding in mammalian development. Science 282: 1281 (1998).
42. A.H. Wyllie. Apoptosis and regulation of cell numbers in normal and neoplastic tissues; an overview. Cancer Met. Rev. 11:95 (1992).

THE ARC, AN UNEXPECTED AND STILL NOT EXPLAINED ELEMENT OF THE TRACKS OF CREEPING CILIATES

Nicola Ricci, Rosalba Banchetti, Fabrizio Erra

Dipartimento di Etologia, Ecologia ed Evoluzione
Università di PISA
Via A. Volta, 6
56126 PISA
ITALIA

INTRODUCTION TO THE THEME

By its nature itself, biology deals with complex phenomena [1], which cannot be reduced to simple elements: this represents a basic handicap to measure reliably important aspects of the biological phenomena. If, on the one hand, the most elementary biological processes lend themselves to satisfactory quantization (the results obtained in the fields of the molecular biology, the biochemistry, the electro physiology, etc. support this hypothesis), on the other hand a whole series of phenomena exists, which tend to offer striking difficulties to whichever effort of describing them quantitatively: how to measure the mimetism? How to quantify the fitness of a certain adaptation?

The quantitative description of the biological phenomena, on the other hand, has an undoubtful and precise double role: beyond allowing the comparative approach to the study and to the possible comprehension of many biological processes, it enables us to individuate and to penetrate (or, at least, to try to penetrate) the basic phenomena underlying the measured properties. The classic Mendel's breakthrough is one of the clearest examples of this kind of scientific attitude, such a powerful one, that it led to the recognition of the existence of the genes, far before even the meiosis itself was known and understood!

Bonner [2] discussed the behavior as a "pinnacle of complexity" in the widest context of his study about the "evolution of complexity": studies of the animal behavior quantitatively exhaustive are indeed extremely rare, due to the complex nature of the phenomenon investigated. At this point, it must be clearly recalled that the behavior is an adaptive trait of extreme importance for the survival of the single organism, as stated by Heinroth [3] since 1930: the behavior must be considered as the expression of the adaptive answer the organism gives to face the environment (monitored by the stimuli perceived by the sensors and receptors) and its challenges (Fig. 1). In other words the behavior must be

The Application of Mathematics to the Sciences of Nature, edited by Cerrai *et al.*
Kluwer Academic/Plenum Publishers, New York, 2002.

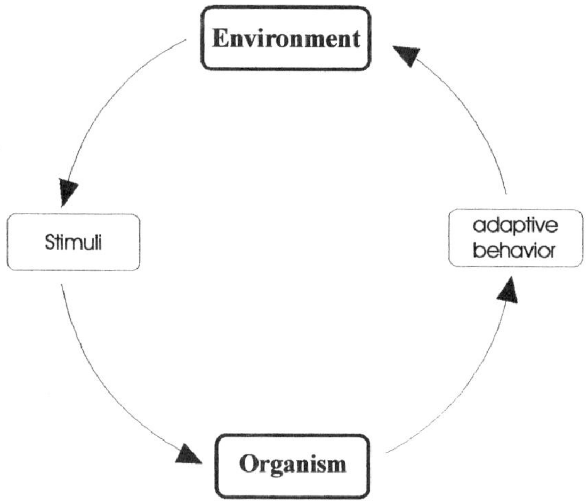

Figure 1. A simplified scheme of the adaptive role of the behavior, between the organism's needs and the environment's challenges/opportunities.

regarded at as a very important expression of whichever behaving species because it mirrors the most basic survival instances of the species itself. In conclusion, the study of the behavior of an animal group must be considered very important, in as much as it investigates its most basic adaptive strategies.

In this context, our model (the behavior of the ciliated protozoa) is very convenient. The protozoa are unicellular eukaryotes (i.e. they share with the animals, their basic eukaryotic nature): this means that they are perfect eukaryotic cells capable of facing (and solving!) the daily problems they meet in their environment, namely that the protozoa can behave as complete organisms. Their average size ranges around values as small as 50-200 μm, so that their biology is mainly determined by small Reynolds numbers [4] and by the consequent lack of any inertial character: the world of the protozoa is the world of the viscoelastic conditions and the behavior of the protozoa is the faithful expression of the resultant of all of the propelling forces acting in each protozoon, instant by instant and point by point, no coasting occurring under any condition.

On the basis of this simple consideration, we can state that the tracks of creeping/swimming ciliates we study since 20 years [5] are the direct expressions of the internal physiology of these organisms [6]. The "simple" organization of the body of the protozoa (considered as an "one cell-one body" organism) significantly conditions their locomotory behavior, which proved [7] to be formed by a succession of discrete, simple elements following each other in much the same way as those (the "runs" and the "tumbles") studied in the bacteria by Berg [8] and Berg and Brown [9]. These elements (Fig. 2) belong to two basic types: the Long Lasting Elements (LLE: they mediate the spatial displacing of the ciliate), i.e. the leftward arcs, A^-, the rightward arcs, A^+ (measured by the radius, r, the central angle, β, the length, l, the duration in time, Δt, and the velocity, v, of the ciliate travelling along them) and the linear segments (S, measured by l, Δt and v), and the Short Lasting Elements (SLE: they mediate the changes of the displacing direction of the track), i.e. the Continuous Trajectory Change (CTC), the Smooth Trajectory Change (STC), the Rough Trajectory Change (RTC) and the Side Stepping Reaction (SSR), characterized by the correction angle, namely the angle (α°) formed by the creeping (or swimming) direction before and after the SLE itself. The complex of these elements has been thoroughly discussed in the context of the concept of ethogram (i.e.: the qualitative and quantitative description, as complete as possible, of all the behavioral patterns of a species) [10] and it was demonstrated to be ever more useful and appropriate for the protozoa than for the metazoa [11,12]. The program written to guide a computer to simulate the CREEping of the CILIates ("CREECILI") [13] and the perfect species-specific tracks it generates represented a strong, although indirect, confirmation of the correctness of our technique for a qualitative and quantitative approach.

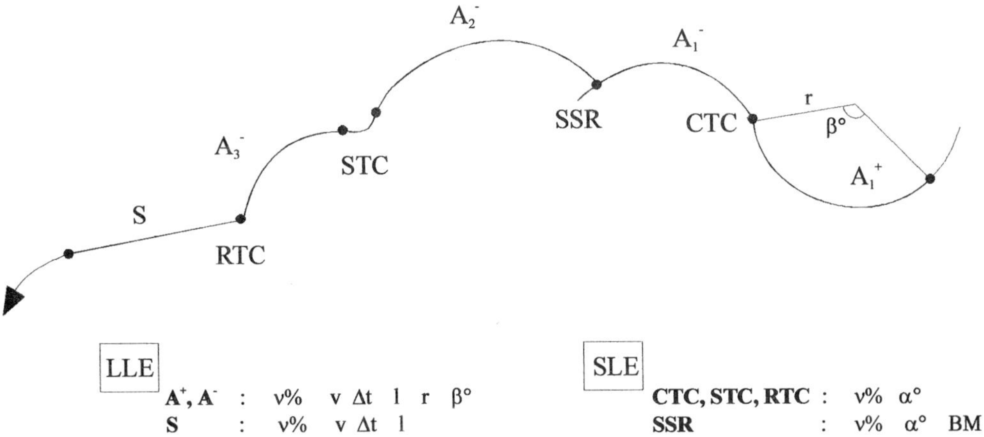

Figure 2. A track of a creeping ciliate with the three LLE and the four SLE: more details may be found in Ricci (1990).

THE ARC

Many facts of different nature demonstrate that the ciliates commonly creep along geometrically perfect arcs: (a) the endless series of the pictures and TV recordings we studied; (b) their mathematical analysis [5]; (c) the perfect tracks simulated at the computer by CREECILI, a program essentially based on the data of the ethograms, in turn largely consisting of the measures of the arcs [13]; (d) the finding that the mathematical simulation of the displacing of a theoretical bi-dimensional ciliate, propelled by the envelope surface of its cilia [14] moves along perfect arcs (Cermelli P., personal communication). A very short and simple consideration will bring into focus the importance of the arc among the ethographic elements of a ciliate. The classic model proposed by Okubo [15] to evaluate the dispersal of an organism took into account three basic parameters of its tracks: the average length of a "run" (*sensu lato*, a generic term referring to a straight displacing element), the average velocity of the organism along the run itself, the average width of the correction angle performed between two successive runs. The occurrence of the arcs performed spontaneously by creeping ciliates somehow weakens the Okubo's model, because the arc represents a geometrically ambiguous element: it represents at the same time both a run-element (Fig. 3, the ciliate displaces from A to B) and a directional correcting element (Fig. 3, the ciliate rotates from its initial orientation D_1 in A, to that in B, D_2, of an angle $\rho° =$ the central angle of the arc itself, $\beta°$). The arcs performed by the ciliates introduce a dramatic and significant perturbation into the Okubo's model, which must be completely redrawn: this new kind of element, on the one hand, cannot be reduced to a simple segment (unless a too large and very significant approximation is made), while, on the other, introduces a new source of directional change (leftwards or rightwards) very difficult to be taken into account together with the "already known" clockwise/counterclockwise correction reactions. Moreover the nature itself of the arc produces a completely new noncorrespond-

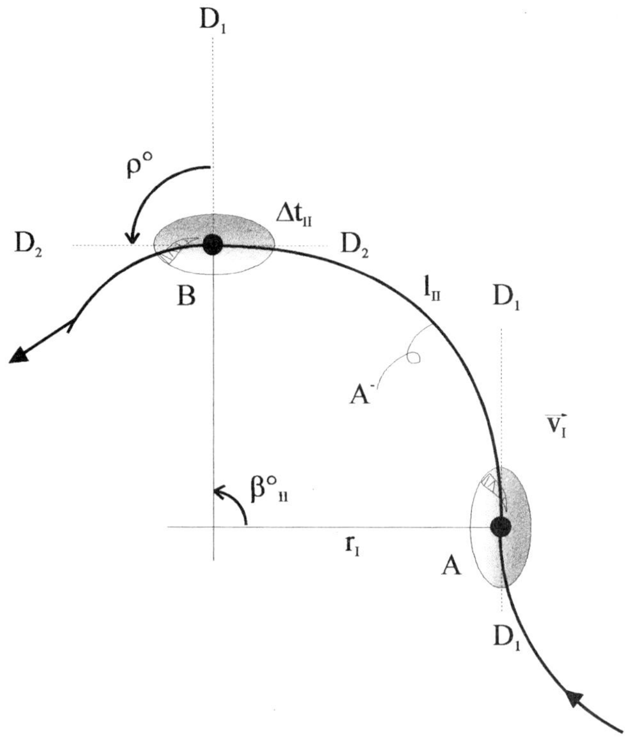

Figure 3. The leftward arc (A^-) followed by a ciliate from A to B: the ciliate rotates of an angle ρ° formed by its direction in A (= D_I) and in B (= D_2); the A^- is defined by five parameters: two primary parameters (the radius = r_I and the velocity = v_I) are set in A, namely at the beginning of the A^- and kept constant throughout the A^- itself; three secondary parameters (the length, l_{II}; the central angle, $\beta°_{II}$; the duration in time, Δt_{II}) go increasing constantly to the end of the A^- itself.

ence between different combinations of the same elements. The overall displacing produced by a track formed by the successive performance of A^+ A^+ A^- A^- is completely different from that of a track formed by the same four elements simply performed in a different temporal succession: A^+ A^- A^+ A^- (Fig. 4B).

A ciliate creeping forward has a rather large number of engines at work: the finding that the engines acting on the left and the right parts of the body work for a while in a way either constantly balanced (if the ciliate is performing a segment) or constantly unbalanced on the left or on the right (when the ciliate travels along a leftward/rightward arc, respectively) was unexpected *a priori* and still is unexplained *a posteriori*. The arc, indeed, can be considered as the LLE performed by a ciliate creeping forward and expressing a constant torque with its engines: the mechanics of the creeping of a ciliate along an A^+/A^- are still to be accounted for.

As shown in Fig. 3, an arc is performed by a ciliate creeping from A to B at an approximately constant velocity with a constant r: in A both of these elements are defined in their value, by the internal state and physiology of the ciliate. For this reason we consider v and r as the "primary parameters" of the arc: they are kept constant throughout the LLE itself. As the time passes, the "secondary parameters" of the arc go increasing steadily (Fig. 3): the central angle, β°, the duration, Δt, the length, l. It has been found [12] that a strong multiplicative correlation occurs between r and β and between v and Δt, respectively (Fig. 5A): no clear explanation of this kind of extremely constant and highly significant correlation has been found so far. When the same data are plotted in a three-dimensional Cartesian space (Fig. 5B, the third axis being the length of LLE considered) they distribute on a saddle surface, clearly describing their occurrence. In the example of

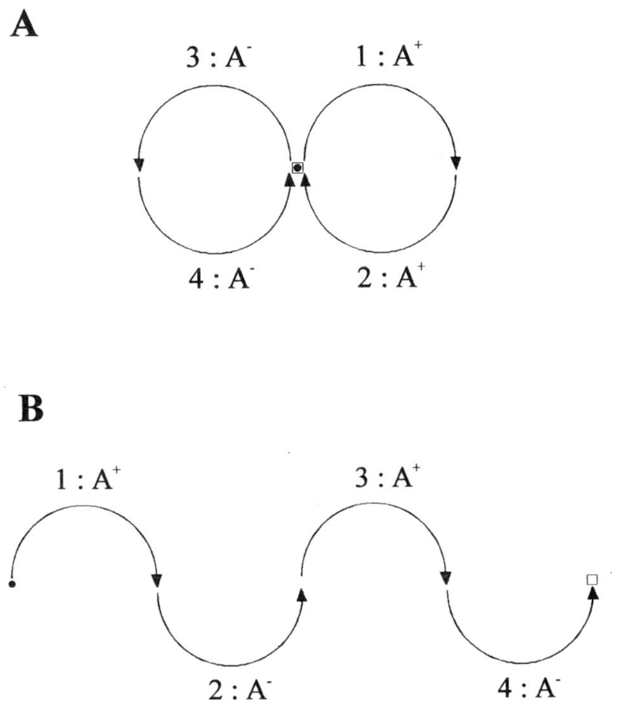

Figure 4. The role of the arcs in the tracks of a creeping ciliate; A: four arcs form a track with a spatial displacement = 0; the black dot = starting point; the empty square = end point: they coincide; B: the same four arcs form a completely different track only because they are performed according to a different temporal trend; the displacement is large. Measured at 24°C; empty circles: 19°C; black triangles: 14°C; small crosses = 9°C. (C): the saddle surface in a different perspective.

Fig. 5B the values of r, β and l, measured in different arcs, performed at four temperatures (24°, 19°, 14° and 9°C), are reported in such a way that it is very easy to visualize the effects of the variation of the temperature in the geometry of the arcs: the lower the temperature, the smaller the radius, the larger the central angle, the shorter the lengths. Although no final conclusion can yet be drawn to explain the correlation occurring between v and Δt above recalled. The fact that not only r and β, but also v and Δt have been found to be significantly correlated is very interesting: while r and β, indeed can apply only to the case of the arcs, v and Δt describe both the arcs and the segments. In other words, we can say that the study, the mathematical handling and the possible solution of the problem posed by the v/Δt correlation should apply to all of the LLE, namely to all of the behavioral patterns mediating the spatial displacing of a ciliate, each LLE expressing the corresponding internal states of the protozoa responsible for the LLE themselves.

Presently, our way of interpreting tentatively the phenomenon (a basic, constant quantum of energy is spent for each LLE by the ciliate) is compared and discussed with the hypothesis proposed by Fronzoni (personal communication) about a transfer function possibly accounting for the observed correlation.

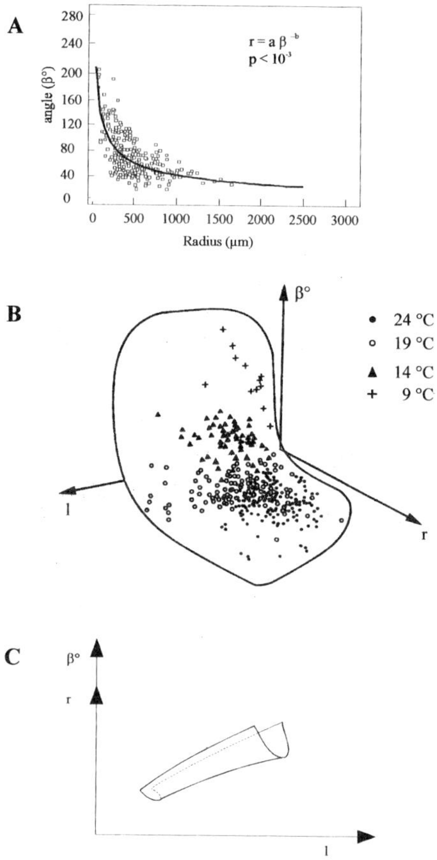

Figure 5. The correlation occurring between the radii and the central angles (A). (B): the saddle surface where the points distribute when plotted in a three-dimensional Cartesian space; black dots refer to values

REFERENCES

1. E. Morin, *Science avec conscience,* Fayard, Paris, 237 pp. (1982).
2. J.T. Bonner, *The evolution of complexity,* Princeton Univ. Press, 259 pp. (1988).
3. O. Heinroth, Ueber bestimmte Bewegubgsweisen der Wirbeltiere, *Sitzungsberichte der Gesellschaft naturforschende Freunde* (1930).
4. E.M. Purcell, Life at low Reynolds numbers, *Am. J. Phys.,* 45: 3-11 (1977).
5. N. Ricci, The ethogram of *Oxytricha bifaria* (Ciliata, Hypotrichida). I. The motile behaviour, *Acta Protozool.,* 20: 393-410 (1981).
6. N. Ricci, Locomotion as a criterion to read the adaptive biology of Protozoa and their evolution toward Metazoa, *Boll. Zool.,* 56: 245-263 (1989).
7. N. Ricci, Qualitative study and quantitative analysis of behavior of ciliated protozoa: principles, techniques, tricks. *In*: *Protocols in Protozoology,* Lee J.J. and Soldo A.T. Eds.Allen Press, Lawrence, Kansas, pp. B-14.1-14.16 (1992).
8. H.C. Berg, The tracking microscope, *Adv. in Optical and Electron Microsc.,* 7: 1-13 (1978).
9. H.C. Berg and D.A. Brown, Chemotaxis in *Escherichia coli* analysed by three dimensional tracking, *Nature,* 239: 500-504 (1972).
10. T. Eibl-Eibesfeldt, *Grundriss der vergleichenden Verhaltensforschung Ethologie.* R. Piper, Verlag, Muenchen, 530 pp. (1967).
11. N. Ricci, The behaviour of ciliated protozoa, *Anim. Behav.,* 40: 1048-1069 (1990).
12. N. Ricci, Ethology of Ciliates. *In*: *Ciliates: Cells as Organisms,* Hausmann K. and Bradbury P.C. Eds., Gustav Fischer, Stuttgart, pp. 403-416 (1996).

13. F. Oliveira-Pinto, N. Ricci and L. Nieri, Simulation of creeping movements of ciliated protozoa, *J. theor. Biol.*, 163: 365-372 (1993).
14. A. Shapere and F. Wilczek, Geometry of self-propulsion at low Reynolds number, *J. Fluid Mech.*, 198: 557-585 (1989).
15. A.Okubo, *Diffusion and Ecological Problems: Mathematical Models*. Springer Verlag, Berlin, 253 pp. (1980).

NATURAL POPULATION AND COMMUNITY STRUCTURE AND DYNAMICS: THE "SUPPLY – SIDE ECOLOGY", THEORY AND THE FIELD DATA

Giovanni Santangelo

Centro Interdisciplinare per lo Studio dei Sistemi Complessi (CISSC),
Dep. Ethology, Ecology and Evolution, Pisa University,
Via Volta 6 I-56126,
Pisa
ITALIA
sant@discat.unipi.it

SUMMARY

A peculiar field of ecology is the study of natural population age structure and dynamics (demography). This field is highly relevant for conservation and management of overexploited or endangered species. In the last years a new approach to population dynamics was developed, the "supply-side ecology", i.e. the study of the process supplying new individuals to natural populations (recruitment). This branch of ecology was carried out mainly by marine ecologists. Several models, based on the clues of density-dependence (the density of adults or new settled individuals limits recruitment) or recruitment dependence (recruitment fluctuates according to larval supply) were developed. Some of these models and experimental field data are here examined and discussed. In several cases density-dependent models, so familiar to mathematicians modeling ecological systems, do not represent the complex reality of marine population dynamics.

INTRODUCTION

All sciences apply models to describe and analyze natural phenomena. All models present an analogy to the real phenomena and such analogies have limitation [1].

Modeling ecological phenomena led to a wide variety of mathematical models, some of which were meaningful and successful [1]. The main conditions to build such meaningful models are: 1) the biological parameters should be accurately chosen; 2) the limitation of the model must be recognized; 3) if the null hypothesis derived from the model is confirmed by experimental data the model must be modified or changed.

The Application of Mathematics to the Sciences of Nature, edited by Cerrai *et al.*
Kluwer Academic/Plenum Publishers, New York, 2002.

Ecologists set out several dynamic models, with the hope of understanding and predicting the trends of population and communities. It should be of the main relevance both for endangered and economically exploited species and environments. In the last ten years they developed a new approach to the study of natural population and community structure and dynamics. This new approach based on the clue that supply of new individuals to populations and communities (recruitment) must deeply affect their structure and dynamics, was called "Supply side ecology" [2, 3, 4]. Mainly marine ecologists developed this peculiar path of population demography.

THE SUPPLY OF JUVENILES TO MARINE NATURAL POPULATIONS

All natural populations are largely dependent both for their age structure and dynamics, on the input of new individuals. This input may depend both on immigration and on reproduction. The input depending on reproduction is cold "recruitment". In other words recruitment was defined as the supply of new individuals to a natural population by means of reproduction [5]. In a population structured into several age classes recruitment may be considered as the first age class (e.g. in a population having a life span of several years, in which therefore several age classes are present, 0-1 year old individuals are considered as recruits). Some authors consider all the young, not reproductive age classes within such a population as recruitment.

Populations may be distinguished into demographically open (having some exchange with other populations) or demographically closed (isolated); it depends both on their nature and on the spatial scale at which they are considered. Marine populations, depending largely on the input of larvae produced elsewhere, are often open. Local populations may be considered as demographically open while wider populations may often be considered as demographically closed. Obviously all the intermediate conditions between these two extremes may be found. Recruitment, as the supply of fry to commercial fish populations was largely studied [5, 6, 7, 8.] The limits of modeling on marine fish population dynamics are due to the completely openness and mobility of the main part of their populations, in which recruits should came from other reproductive areas or populations and new individuals should add or escape via immigration or emigration. Furthermore the boundaries of these populations are often indefinite. In spite of this, commercial fish stocks are generally considered as closed systems by definition [5]. Only some territorial fish population of isolated reefs, producing benthic (that stay on the marine bottom) eggs and sedentary fry may be considered, more reasonably, as nearly demographically closed.

Similarly to fish populations, local reproduction has little or no effect on the most of benthic invertebrate populations because larval recruitment from elsewhere provides the main imput of juveniles [5].

The assessment of recruitment of marine sediment bottom invertebrates (infauna) is difficult, due to the tricky extraction and identification of mobile recruits and adults and to the instability of sediment bottoms. These difficulties are not critical in the study of rocky bottom, sessile (fixed to the bottom) invertebrates: adults and recruits cannot move, new settlers can be easy identified and distinguished from adults and only larval stages are migrant. Furthermore mortality may be easily recorded [9].

Several benthic invertebrates are spawners, releasing floating eggs that should disperse at high distances while other invertebrates brood floating eggs, which shall generate larvae into the seawater [10]. Two different kinds of marine larvae among marine benthic invertebrates can be distinguished: planctotrophic larvae (larvae that feed on plankton) and lecitotrophic larvae (they do not feed during their life but assimilate egg yolk, like planulae larvae of many corals). Of these two kinds of larvae the former are "broadcaster", high distance dispersed, while the latter show a limited dispersion capability, and, in some cases, settle in the proximity of adults (philopatric larvae). In this last case populations may be considered as (nearly) demographically closed. New settled individuals (metamorphosed larvae) are named settlers (and the process is cold settlement).

FACTORS AFFECTING RECRUITMENT

On the basis of the factors determining their recruitment, two different kinds of populations may be distinguished: 1) populations in which recruitment should widely oscillate with time, according with reproduction intensity or larval supply; 2) populations in which density of adults (or settlers) affects recruitment rate. While the former were defined as recruitment-dependent; the latter were defined as density-dependent [5]. In the first case recruitment-limited population density oscillate with time according with oscillations of larval supply. These populations are defined as limited by recruitment. In the second case recruitment is regulated by post-settlement processes causing mortality in new settled individuals [5,9]. These processes are generally considered as dependent on density (of settlers, recruits or adults). When recruits saturate a population, no further recruitment shall be incorporated in the population, even if larval supply shall increase or lasts at high rates for a long time. These populations, defined as "regulated" by density-dependence, also if recruitment signal is not completely cancelled (compensed) by density-dependent postrecruitment mortality, are more numerically constant over time [5]. Regulation occurs when one demographic parameter at list, is density – dependent. Generally, populations characterized by low adult densities are recruitment-dependent, while high density populations are density –dependent.

In some fish population effects of different reproduction successes in consecutive years (pulse recruitment) may be distinguished after many years on the basis of differences in size between the different age classes, as in Yablokov 1986 [11]. These populations are considered therefore as recruitment-dependent. In the majority of fish populations recruitment is regulated not only by reproduction intensity but also by predation on eggs and fry and only seldom by adult fish density. Density-dependence regulation was found only in some benthic, territorial fish population.

Density-dependence in marine hard bottom, sessile communities, was linked usually to the clue of competition for free space [12]. Density-dependence is therefore generally considered as a negative effect of adult (or settler) density on recruitment rate while positive recruitment dependence is considered as unlike [13]. This leads to the development of population models in which density - dependence was included [14]. Since Lotka and Volterra classical papers the idea of density-dependence has deeply affected the ecological models developed by mathematicians. Density-dependence was related to inhibitory effects that population density should have on reproduction and/or recruitment rates and, also, on mortality and growth. Density-dependence may occur between adults, between recruits and adults or between settlers and recruits. In demographic marine models such dependence was included as a function controlling reproduction rate at high population density levels. It was assumed that most existing population models include density-dependence at some level [5, 14]. According to Stephens and coll. and Sutherland [13, 15] other models, based on a positive relationship between adult density and individual fitness (Allee effect, rather common in nature) were rather ignored. Furthermore, the possibility of that natural densities do not always reach levels that lead to competition was seldom considered.

RELATIONSHIPS BETWEEN SETTLEMENT AND RECRUITMENT

Some author defines a primary recruitment limitation and a secondary recruitment limitation. While the former depends on larval supply, the latter depends on post- settlement mortality (i.e. the mortality between settlement and recruitment). Post settlement mortality was considered in many cases as density-dependent. As it is difficult to quantify larval supply and subsequent settlement in marine benthic organisms, these parameters are gathered indirectly by recruitment estimation. According to Connell [16] this estimation is unreliable in all cases in which settler mortality is density-dependent. It is not possible to estimate settlement by recruitment rates if the latter parameter is affected by the age structure of the population, and therefore, the relationship settlement-recruitment should be decoupled. This was found in many marine benthic invertebrates. Connell stated larval supply should affect settler survival at high settlement rates only, but a further careful analysis of the same data by McGuinness and Davis[17] showed a positive relationship exists between settlers and survival at all densities. Other similar data, examined by the same authors gave the same results of "positive density-dependence" between settlers and recruits. In other words the more abundant were the new settled individuals, the higher was their survivor probability. Finally Holm [18] stated that recruitment could still reflect settlement, despite density-dependence, when settlement density was low. Many further studies were carried out on recruitment of barnacle intertidal populations. Carroll [19], in a three-year study on barnacles, found a high variation both in recruitment and in mortality induced by competition and predation. Sutherland (J.P.), in a wide study on recruitment of tropical barnacles, found mortality was independent by adult and recruitment densities but largely dependent on events in the water column [20]. In some sampling areas, however, predation and competition only affect barnacle density.

FURTHER FINDINGS ON DENSITY AND RECRUITMENT DEPENDENCE

According to Caley [5], Hughes [21], and Rougharden [2] were the first to model local open populations of corals and barnacles by dynamic models based on an external input of recruits more than on intrinsic birth rate. Further modelling studies were developed on coral-reef fishes and bryozoans [5]. These studies demonstrated that recruitment should regulate, not only limit local populations. Recruitment-dependent populations too may oscillate around an equilibrium value. Recruitment limitation should lead to a positive correlation between recruitment and adult population density when both are affected by the same factors such as competition or/and predation in a similar way. In such cases fluctuations in recruitment are directly translated in fluctuations of adult population densities. Adult density fluctuations, correlated with pulse recruitment are more evident in populations of species having a limited life span or composed by few age classes. Fluctuations in adult mortality rates should decouple this correlation [5]. Furthermore, when recruitment is inhibited by competition with adults, (density-dependence) this effect may be expressed within the population with a delay, this leads to fluctuations in population size in density - dependent populations. These fluctuations are not linked to recruitment fluctuations. Connell [16] found moreover, that local density of barnacle settlers was determined, in some sites, by larval plankton supply while in others space limitation (density-dependent) operates. A wide variability in barnacle recruitment rates together with density - or predation - dependent mortality, in two close sites, was also found [19].

All this demonstrates, according to Caley [5], that many of the criteria followed in the study of recruitment in the field should be inadequate. Other factors than recruitment – dependence and density-dependence, such as interspecific interacting processes (competition or cooperation and predation) should effect population dynamics. These processes should operate at a spatial scale smaller than larval supply, affecting mainly local survival of settlers[5].

Chesson [22] denied a clear- cut dichotomy between recruitment limitation and density-dependence. According to this author density-dependence and recruitment –dependence together with intra and interspecific interactions should determine marine population densities. He stated, moreover, that fecundity should increase with density in closed populations without density-dependence mortality.

The settlement choice plays a pilot role in determining mortality rate of coral settlers, which, on the other hand, was independent by density [23].

Interspecific interactions such as predation and competition (often mediated by allelochemicals) acting within hard bottom subtidal communities, may reduce greatly larval and settler survival [24, 25].

Finally, all catastrophic events, reducing recruitment by reducing larval supply or spawned amount (in close populations) or favoring the increase of settlement and recruits by eliminating competitors (in open populations) should affect deeply population age structure and dynamics.

All theoretical considerations on recruitment negative density- dependence are in contrast with the Allee effect theory [13] according to which all species that have some kind of advantage by group living show a positive relationship between individual fitness and conspecific density. The same species should reduce their intrinsic rate of natural increase up to zero when their density falls below a certain threshold value. Allee effect was found to really operate in nature, determining local or global extinction of marine (e.g. school fishes and sirenids) and terrestrial (e.g. passenger pigeon) animals.

POPULATION DEMOGRAPHY AND TRANSITION MATRICES

About a decade ago several researches, dealing with hard and soft coral population demography were developed [21, 26, 27, 28, 29]. These studies revealed a tendency of coral population to be mainly self - seeders and to be affected by a strong post recruitment mortality. Babcock[29], Abbiati and coll. [30] and Santangelo [31] developed life tables on these modular animals. Mathematical discrete dynamic models, based on transition probability matrices [32, 33] were developed and applied to coral populations with several discrete, overlapping, age or size classes [28, 29, 30]. In one of these papers [30], dealing red coral (*Corallium rubrum* L 1758), a precious Mediterranean octocoral) the authors developed a demographic model, based on transition Leslie matrix, in which a negative density-dependence of recruit VS adult number or biomass was included. Red coral is a brooder species releasing lecitotrophic, "philopatric" larvae (settling within a relatively short distance from the parent colony). Demographic data were gathered from a coastal, crowded population, composed by small sized individuals, reproductively isolated from other similar populations and considered therefore as at least partly closed [34]. The model developed allowed several realistic simulations of the population dynamics [30]. Successively, the same authors modified the model [35] and two different control modalities of recruitment were considered: low adult biomass (or densities) facilitates recruitment, which, on the contrary, was inhibited by high adult biomass. Experimental field results denied the first version of the model since, as reported in figure 1, a clear-cut positive relationship was found between recruitment and adult colony densities. In this second model, described in figure 2, the ratio recruits/number of planulae larvae produced in each sample was put against adult biomass. The second, modified model on the contrary, seems to be more suitable to represent the field data. As can be seen in figure 3, in which the ratio recruits/number of planulae larvae produced in each sample VS adult biomass in the same sample are reported, experimental (fig. 3) and theoretical curves (fig.2) are similar. The same authors, at present, are trying to verify, by manipulating adult density "in situ", if this new model of interaction between recruits and adults really operates in nature. Moreover, the effects of other biotic or abiotic factors, such as interspecific competition, or different depth or area on recruitment rates are similarly examined. All these experiments are carried out following complex, multifactorial ANOVA experimental models, in which null hypotheses are verified, according to Underwood[36].

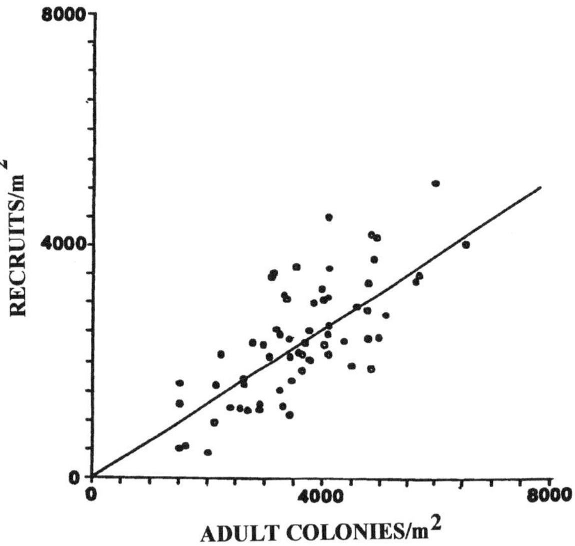

Figure 1. Relationship between adult colony and recruit density: b = 0.64 +- 0.01; n = 65

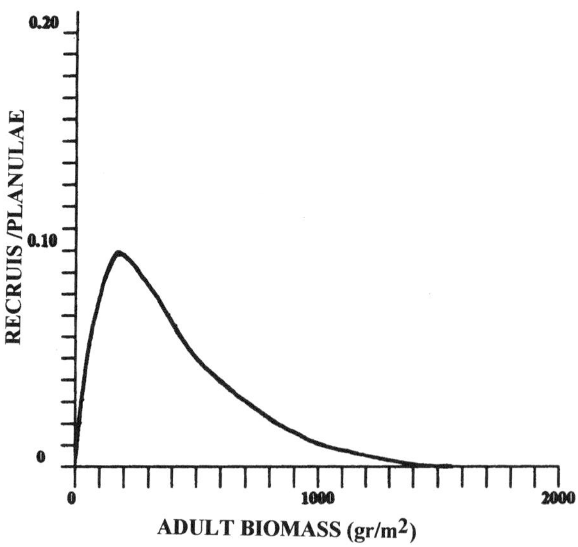

Figure 2. Relationship between the ratio adult density / number of planulae (larval coefficient) and adult colony biomass. The curve is theoretical

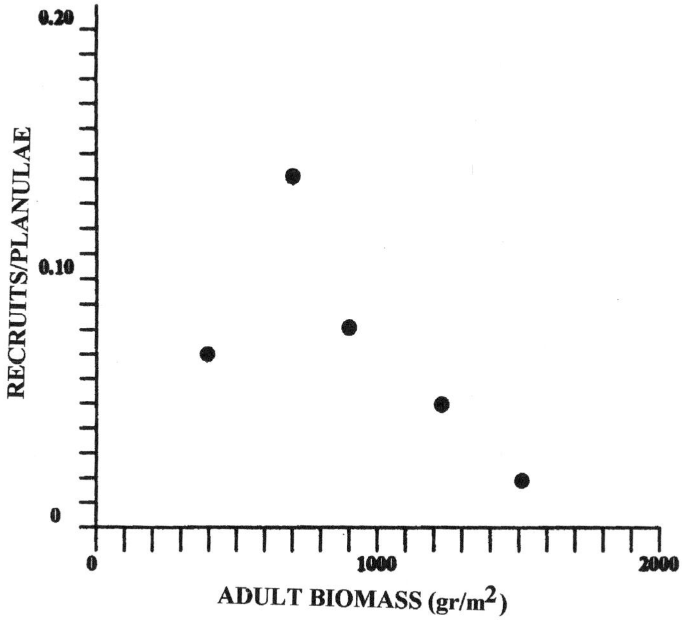

Figure 3. relationship between the ratio adult density / number of planulae (larval coefficient) and adult red coral colony biomass. The curve is based on experimental, field data.

SOME CONCLUSIVE REMARKS

Structure and dynamics of all natural populations are deeply affected by recruitment. Many efforts were made by marine ecologists to identify and describe factors determining recruitment rates. Classic density - dependence models, in which recruitment is affected by density of settlers or adults, are not enough representative of the complex reality of natural dynamics of marine benthic populations, which are mainly open. At the other hand recruitment-dependent models (in which the population density depends on larval supply) are similarly inadequate. Recruitment limitation and density-dependence should act together and/or in turn to determine population structure and dynamics.

The complex interactions, which determine recruitment in natural populations, cannot be reduced to simple density-dependence or recruitment-dependence. Other factors, such as interspecific interactions and environmental variability operate in natural communities. Population dynamic models, based on field data, must include these factors. A multifactorial approach is therefore required to evaluate the contribution of recruitment and other limiting and regulating processes in determining population density and population age structure. Realistic field experiments, based on multifactorial experimental designs, in which natural densities are manipulated and abiotic factors are included, should provide new knowledge on effects of selected factors on recruitment rates.

REFERENCES

1. T. Fenchel *Ecology: Potential and Limitations*. Excellence in Ecology. Ecology Institute Oldendorf/Luhe D. (1987).

2. J. Roughgarden, Y. Iwasa, C. Baxter Demographic theory for an open marine population with space-limited recruitment. *Ecology* 66: 54-67. (1985).
3. J. Roughgarden, S. Gaines, H. Possingham. Recruitment dynamics in complex life cycles. *Science* 241: 1460-1466 (1988).
4. R. Levin Supply-side ecology. Existing model of population structure and dynamics of ecological communities have tended to ignore the effect of the influx of new members into the communities. *Science* 234: 25-27.(1986).
5. M.J. Caley, H. M. Caley, Carr, M.A. Hixon, T.P. Hughes, G.P. Jones and B. A. Menge Recruitment and the local dynamics of open marine populations. *Ann. Rev. Ecol. Syst.* 27: 477-500. (1996).
6. P.L. Chesson Spatial scale in the study of reef fishes: a theoretical perspective. *Austral. J. Ecol.* 23: 209-215.(1998).
7. G.P. Jones The importance of recruitment to the dynamics of a coral fish population *Ecology* 1:1691-1698. (1990).
8. D.H. Cushing *Population production and regulation in the sea.* Cambridge University Press, Cambridge. (1995).
9. C. Todd Larval supply and recruitment of benthic invertebrates: do larvae always disperse as much as we believe? *Hydrobiologia:* 375/376: 1-21 (1998).
10. R. Babcock Fine-scale spatial and temporal patterns in coral recruitment. *Proceedings of the 6th International Coral Reef Symposium.* Vl. 2: 635-639. (1988).
11. A.V. Yablokov *Population Bio*logy. MIR Pub. Moscow 1986).
12. G.H. Branch Competition between marine organisms: ecological and evolutionary implications. *Oceanogr. Mar. Biol. Ann. Rev.* 22: 429-593. (1984).
13. P.H. Stephens, W.J. Sutherland. Consequences of the Allee effect for behaviour ecology and conservation. *Trends Ecol. Evol.* 14: 401-405. (1999).
14. S. Tuljapurkar, H. Caswell *Structured-population Models in Marine, Terrestrial, and Freshwater Systems.* Pop. Comm. Biol. Ser. Chapman & Hall N.Y. (1997).
15. P.A. Stephens, W.J. Sutherland and R.P. Freckleton.. What is the Allee effect? *Oikos 87*: 185-190. (1999)
16. J.H. Connell. The consequence of variation in initial settlement vs. post-settlement mortality in rocky, intertidal communities. *J. Exp. Mar. Biol. Ecol.* 93: 11-45. (1985).
17. K. A. McGuinness and A. R. Davis. Analysis and interpretation of the recruit-settler relationship. *J. Exp. Mar. Biol. Ecol.* 134: 197-202. (1989).
18. E.R. Holm Effects of density-dependent mortality on the relationship betweenrecruitment and larval settlement. *Mar. Ecol. Prog. Ser.* 60: 141-146. (1990).
19. M. L. Carroll Barnacle population dynamics and recruitment regulation in South-Central Alaska. *J. Exp. Mar. Biol. Ecol.* 199: 285-302. (1997).
20. J.P. Sutherland *Recruitment regulates demographic variation in a tropical intertidal barnacle.* Ecology 7(3): 995-972. (1990).
21. T.P. Hughes. Recruitment limitation, mortality, and population regulation in opened systems: a case study. *Ecology* 71: 12-20. (1990).
22. P. Chesson Recruitment limitation: a theoretical perspective. *Aust. J. Ecol.* 23: 234-240. (1998).
23. R. Babcock and C. Mundy Coral recruitment: consequences of settlement choice for early growth and survivorship in two scleractinians. *J. Exp. Mar. Biol. Ecol.* 206- 179-201 (1996).
24. A.R. Davis, J.A. Butler and I. Van Altena. Settlement and behaviour of ascidian larvae: preliminary evidence for inhibition by sponge allelochemicals. *Mar. Ecol. Prog. Ser.* 72: 117-123. (1991).
25. M. Maida, P.W. Sammarco and Coll.. Effects of soft corals on scleractinian directional recruitment directionall allelopathy and inhibition of settlement. *Mar. Ecol. Prog. Ser.*121: 191: 191-202. (1995).
26. T.P. Hughes Population dynamics based on individuals size rather than age: a general model with a reef coral example. *Am. Nat.* 123-:778-795. (1984).
27. P. W. Sammarco and J.C. Endrews Localizatized dispersal and recruitment in Great Barrier Reef corals: the helix experiment. *Science* 239: 1422-1424. (1988).
28. N. Gotelli Demographic models for Leptogorgia virgulata, a shallow water gorgonian. *Ecology* 72(2): 457-467. (1991).
29. R. Babcock Competitive demography of three species of scleractinian corals using age and size dependent classifications. *Ecol. Monogr.* 61(3): 225-244. (1991).
30. M. Abbiati, G. Buffoni, G. DiCola, G. Santangelo. Red coral population dynamics: stability analysis and numerical simulation of time evolution of perturbed states. In: *Perturbation and Recovery of Terrestral and Acquatic Ecosystems: 219-228.* Ellis Horwood Ldt. ed, London, (1991).
31. G. Santangelo A focal point in Corallium rubrum rersearch: the determination of population age structure. In: *Les Sspèces Marines à Protegere en Mediterranee*, C.F. Boudouresque, Avon M., Gravez V. ed., Gis Publ., Fr.; 61-63. (1991).

32. M. B. Usher Developments in the matrix model. In: *Mathematical Models in Ecology*. Jeffers ed. Blackwell, London. (1972).
33. M. Pascual, H. Caswell. The dynamics of size-classified benthic population with reproductive subsidy. *Theor. Pop.Biol.*: 39: 129-147. (1991).
34. M. Abbiati, G. Santangelo, S. Novelli. Genetic variation within and between two Tyrrhenian population of the Mediterranean alcyonarian Corallium rubrum. *Mar.Ecol. Prog. Ser.* 95: 245-250. (1992).
35. M. Abbiati, B. Buffoni, G. Caforio, G. DiCola, G. Santangelo. Harvesting, predation and competition effects on a red coral population. *Neth. J. Sea Res.* 30: 219-228 (1992).
36. A. J. Underwood. *Experiments in Ecology*. Cambridge Univ. Press. (1997).

HISTORY OF SCIENCE

THE HISTORY OF THEORETICAL POPULATION ECOLOGY: WHICH ROLE FOR MATHEMATICAL MODELING?

Luciano Andreozzi

Università degli Studi di Bari
Piazza Umberto I, 1
70121 BARI
Italia

OVERVIEW

The use of mathematics in biology has always been controversial. At variance with economics, which became a formalized discipline quite early in its development, biologists had always been suspicious about the use of formal models in their field. During the now long history of bio-mathematics, supporters of this approach have devised quite a large number of arguments to persuade skeptics that the marriage between mathematics and biology was a good deal for both disciplines. Beside the obvious reason that formal models lead, hopefully, to quantitative predictions that are easier to falsify, mathematical models have been frequently defended because of their (alleged) heuristic value. According to this view, mathematical models' most important contribution to biology consists in their ability to highlight some phenomena, or some relationships among phenomena, that could not be understood without formalization. A very clear defense of this point of view is to be found in one of the early classics of mathematical population ecology, Vito Volterra's "Variazioni e fluttuazioni del numero d'individui in specie animali conviventi", first published in 1926.

> We first try to express with words how the phenomenon develops; then we translate these words into mathematical terms. This induces us to write differential equations. Then, if we follow the methods of calculus, *we are lead much further than we would arrive through ordinary language and reasoning*, and we can formulate precise mathematical laws.[1]

The same point of view is repeated in the short monograph *Les Associations Biologiques*, Vito Volterra wrote with Umberto D'Ancona in 1935.[2] After repeating that "calculus can lead us much further than ordinary language and reasoning", they add that the heuristic value of mathematics had been demonstrated by the mathematical models elaborated by Volterra and others in the ten years elapsed since the publication of "Variazioni e fluttuazioni". As an example, they cite the oscillations in predator-prey

The Application of Mathematics to the Sciences of Nature, edited by Cerrai *et al.*
Kluwer Academic/Plenum Publishers, New York, 2002.

systems, which in their view were first shown by Lotka and Volterra with their models (see below).

A similar defense of the applications of mathematical models based on their heuristic value is today a commonplace among mathematical biologists.* In the introduction to his classical *Evolution and the Theory of Games*, John Maynard Smith claims that mathematical models are useful just because they allow the description of phenomena that could not be discovered through ordinary language (p. v).[5] Remarkably enough, he supports his thesis mentioning just the oscillations in predator-prey systems.

However, not all biologists are willing to concede that mathematical models can play an heuristic role. The following passage taken form Ernst Mayr's *The Growth of Biological Thought* is particularly clear on this point.[6] According to Mayr,

> several instances are known in the history of biology, perhaps many more if one would really look for them, where a law, principle or generalization was ignored when first stated because it was phrased in words rather than in the form of mathematical equation. When finally stated mathematically, it was hailed and generally accepted. (p. 851)

The thesis Mayr presents in this passage clearly clashes with the one Volterra (and Maynard Smith) defended, since he claims that, as long as biological sciences are concerned, mathematics has always played a role one could term "rhetoric" rather than "heuristic" (although this isn't Mayr's terminology).* On this view, mathematical models have seldom (if ever) produced new propositions that could be of any use for biologists; they only helped in giving a particularly clear form to propositions scholars had already discovered. As a matter of fact, mathematical models have been particularly effective in this rhetoric role: Mayr himself recognizes that in many cases the mere fact that a "law, principle or generalization" received a mathematical formulation sufficed for it to gain a definitive acceptance within the scientific community.† However, Mayr's granting mathematical models just a rhetoric function (no matter how effective it might have been) was obviously meant to diminish their relevance for biology.

This is one of those controversy that can only be resolved through an historical analysis, because any "abstract" discussion concerning the heuristic role mathematics could

---

* See for example Karl Sigmund's very readable discussion in his *Games of Life*[3]. I have discussed the growing emphasis on the heuristic value of mathematical models in theoretical population ecology elsewere[4].

* Mayr states that the use of mathematics in the works of many biologists can be considered as "window dressing".

† The historical example Mayr cites to support his thesis is the Hardy-Weinberg principle, whose content had been anticipated (in a non-mathematical form) by W. E. Castle in 1903. However, Mayr had a much more personal reason to resent the success of mathematical models in biology. In 1939 he studied the bird fauna in New Caledonia in terms of a dynamical equilibrium between colonization and extinction, which he presented in his famous *Systematic and the Origin of Species*, published in 1942. However, Mayr laments that "this was ignored for twenty-five years, until MacArthur and Wilson phrased it in mathematical terms in their theory of island bio-geography." (p. 851. Mayr stresses this point in several other occasions[7,8]) It is also worth mentioning that this position is all but uncommon among biologists. During a Symposium of the Society for Experimental Biology held at Oxford in 1952, C. H. Waddington expressed his perplexities about the use of mathematics in biology in terms which resemble closely Mayr's view on the same topic. He wrote that mathematical genetics "did not achieve either of the two results which one normally expects from a mathematical theory. It has not, in the first place, lead any noteworthy quantitative statement about evolution. The formulae involve parameters of selective advantage, effective population size, migration and mutation rates etc. most of which are still too inaccurately known to enable quantitative predictions to be made or verified. But even when this is not possible, a mathematical treatment may reveal new types of relation and of process, and thus provide a more flexible theory, capable of explaining phenomena which were previously obscure. It is doubtful how far the mathematical theory of evolution can be said to have done this. Very few qualitatively new ideas have emerged from it."[9] (p. 186, this passage is quoted in William Provines' detailed reconstruction of the influence of mathematical population genetics on the evolutionary syntesis[10].)

play in biology is utterly pointless. Quite obviously, the only way to decide which side of this controversy to take is to examine a set of "discoveries" usually attributed to the use of formal models to see whether, and to which extent, they had been anticipated by scholars not involved in mathematical studies. In other words, one should take Mayr's invitation seriously to actively search for examples of "laws" and "principles" that had been originally ignored when presented without formalization, and that were universally accepted (or widely discussed) once presented as consequences of some mathematical model.

The birth of mathematical approaches to ecology during the 20.s and the 30.s (the so called *golde age of theoretical ecology*[11]) represents a particularly fertile field for this sort of analysis, because in that period Volterra, Lotka and few other pioneers created almost from scratch the basis of a large part of what is now called theoretical population ecology. The study of their works allows us to spot the novelties (if any) introduced by mathematical models with respect to earlier, non formalized treatments. A similar research would be more difficult or misleading when applied to works published after the Second World War, because at that time mathematical models were already part of the theoretical ecologists' toolkit (albeit a contested one).

In this paper I shall sketch such an investigation, taking into consideration some of the most relevant results usually attributed to the use of mathematics during the *golden age*. In the first and in the third section I shall show that it's all but difficult to find non mathematical anticipations of almost all the results usually attributed to Lotka and Volterra, either in the works of celebrated authors as Darwin, Spencer and Malthus, or in more obscure publications due to less known Italian naturalists as Lorenzo Camerano and Antonio Berlese. Considering that the years following the publication of Lotka and Volterra's classical works witnessed a huge proliferation of the literature concerning the topics they addressed,* we are apparently forced to side Mayr's thesis that mathematical models, far from being a source of fresh ideas in biology, played almost exclusively an highly effective rhetoric role. As long as the applications to theoretical ecology are concerned (these results probably cannot be extended to other fields like the history of population genetics), while mathematical models didn't contribute to the discovery of previously unknown phenomena, they attracted researchers' attention upon some biological truths that had been overlooked until then. Section two is devoted to a brief description of some recent developments in the history of theoretical population ecology which suggest that the effectiveness of mathematics during the *golden age* could have induced scholars to embrace some explanation of observed phenomena that turned out to be less satisfying than competing ones not based on mathematical reasoning.

However, this is too a pessimistic representation of the applications of mathematics to ecology. In the conclusions I shall try to show that things are slightly more complicated, and that there are ways of defending a weaker version of Volterra and Maynard Smith's claim concerning the relevance of mathematics for theoretical ecology, which is based on more complex examples than the simple oscillations of predators and preys in two population models.

FLUCTUATIONS IN PREDATOR-PREY MODELS

Volterra's studies in mathematical biology started in the last months of 1925, when his perspective son-in-law Umberto D'Ancona asked him to find some mathematical way

* Phenomena like oscillation in predator-prey systems, competition for ecological "niches" and logistic growth became the center of a wide program of research which involved both the empirical and the theoretical side. There are quite a few surveys of this literature[12,13,14,15,16].

of explaining the surprising increase of predators that took place in the Upper Adriatic Sea during the interruption of fishing due to the First World War.[17] Volterra's solution, published in his 1926 "Variazioni e fluttuazioni", was based on mathematical predator-prey models, that had already been developed by Alfred Lotka.[18,19] To model the interaction between a predator and its prey Volterra used the following couple of differential equations:

$$\frac{dx_1}{dt} = x_1(\varepsilon_1 - a_1 x_2)$$

$$\frac{dx_2}{dt} = x_2(-\varepsilon_2 + a_2 x_1)$$

where x_1 and x_2 represent the densities of preys and predators respectively, and ε_i and a_i are constant coefficients. Volterra showed quite easily that x_1 and x_2 oscillated periodically around their equilibrium values $x_1 = \varepsilon_2/a_2$ and $x_2 = \varepsilon_1/a_1$. These oscillations are represented in picture 1, which is taken from Volterra's paper.

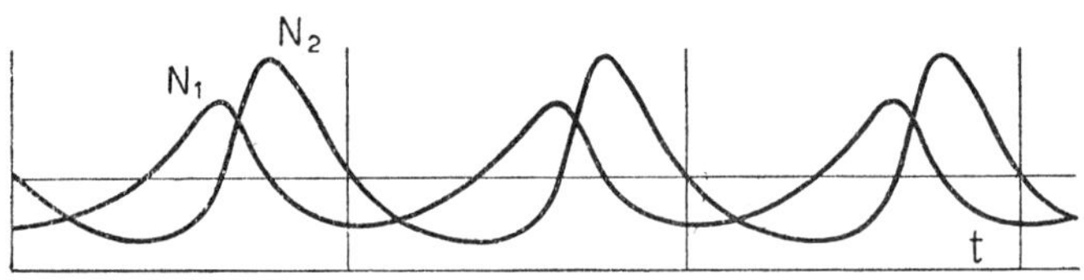

Figure 1

In their 1935 monograph *Les associations biologiques*,[2] Volterra and D'Ancona did not claim that oscillations in animal population where first noticed by Volterra. They admit that "the existence of fluctuations has been observed and studied more than once, both from a qualitative and from a quantitative point of view" (p. 78) although the "vast majority of the authors [...] tried to explain them in terms of meteorological or cosmic phenomena", while it was only with the introduction of Volterra's (and Lotka's) models that these oscillations had been attributed to the "biological phenomena generated by the coexistence" of predators and preys.* (p. 80) In other words, Volterra and D'Ancona do recognize that oscillations of animal populations where all but a novelty when Volterra started his mathematical research in 1925, but they deny that "non mathematical" authors had been able to see their origin in the "internal" predatory relationships among species.

The most famous example Volterra and D'Ancona cite to support their thesis are the researches the British biologists Charles Elton started in the early 20.s concerning the fluctuations of populations of small mammals living in the Hudson Bay (hares, lynxes and lemmings).† His studies where based on the data the Hudson Bay Company gathered for about two centuries concerning the number of furs traded, which revealed the existence of

* D'Ancona repeated this point in all his later writings on the same topic[20,21].

† There are some good historical surveys of Elton's contribution to contemporary ecology[12,22].

wide oscillations of all the populations, with a constant amplitude and a period of approximately ten years.

In 1924, few months before Lotka published his *Elements of Physical Biology* and Volterra started his collaboration with D'Ancona, Elton published his first paper on this topic, in which he explained the observed oscillations as a consequence of periodic climatic variations due to the occurrence of sunspots.[23] Picture 2, which reproduces picture 6 in Elton's paper, shows the curve of sunspots. Black dots represent the years of the maxima in the population of hares, and are meant to illustrate the link between these two phenomena.

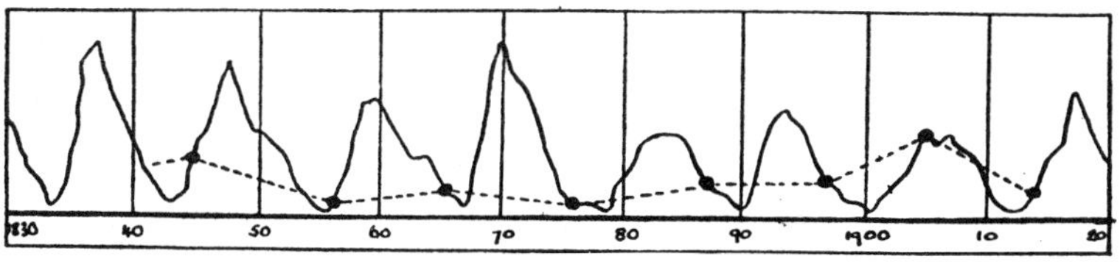

Figure 2

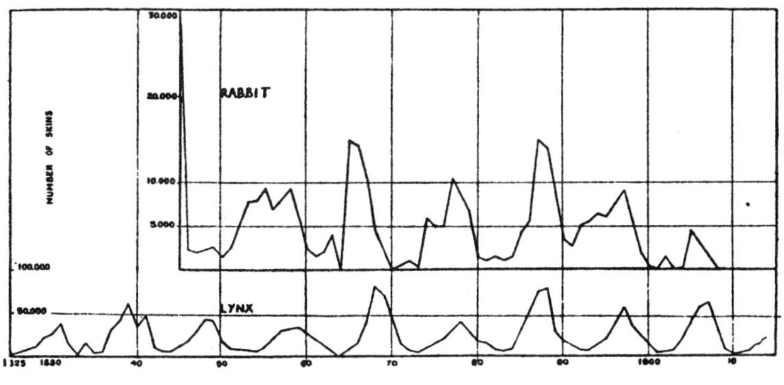

Figure 3

Even if Elton recognized that hare and lynxes' oscillations were somewhat linked, he failed to see that they could occur without any external influence, as the one exerted by sunspots. He presented this idea only in following papers,[24] and there is no doubt that the mathematical treatment of similar issues proposed by Lotka and Volterra played a major role in changing his mind.* In a letter to Volterra written in 1931, Elton himself provides the following description of his "conversion".

> I agree completely with your main thesis, i.e. that external factors are not in themselves necessary to cause fluctuations of wild animals. In fact I have been led to postulate both the existence of an external climatic cycle, and also a process similar to the ones you have described from mathematical data, in explaining the rabbit-lynx 10-year cycle in Canada.[26]

Contemporary scholars in ecology are accustomed to consider the fluctuations of hares as linked to the similar fluctuations of lynxes. In his 1924 paper[23] Elton conceived such a relationship in the following terms: sunspots were directly responsible only for the oscillations of hares, while the population of lynxes oscillated because of the periodic dearth of hares, which are their chief food. To highlight the link between these two phenomena, Elton plotted in the same graph (which is here shown as picture 3) hares and lynxes' oscillations, which is the direct ancestor of the "best-known graph in ecology"[13], represented in figure 4.

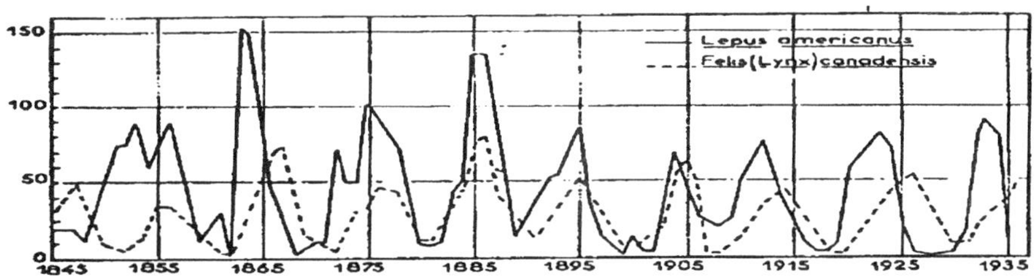

Figure 4

As a matter of fact, Elton's conversion to the "mathematical" explanations provided by Lotka and Volterra had never been complete. Although during the 30.s he abandoned completely his original explanation based on sunspots, Elton never endorsed without reservation the "endogenous" nature of oscillations derived from mathematical models. What made him suspicious was that these oscillations occurred *in phase* on the two sides of the Rocky Mountains, which apparently could only be explained by a large scale climatic oscillation. Beside the occurrence of sunspots, whose influence proved to be hard to

* On the other hand, it's hardly surprising that the models with one predator and one prey attracted his attention. These models seems to be tailored to the hare-lynx association, since, as Elton himself wrote in a later paper, the Canadian lynx "lives with Rabbits, follows the Rabbits, thinks Rabbits, tastes like Rabbits, increases with them, and on their failure dies of starvation in the unrabbited woods."[25] (p. 241)

demonstrate, Elton failed to produce other factors that could be held responsible for climatic variations large enough to generate such a uniform oscillation. These difficulties induced him to assume an extremely cautious position on this point.

> The long run of figures now available proves conclusively that the hypothesis put forward by Elton [1924][23] of control of this cycle by sunspots, acting through climatic cycles, is not true. [...] We have at present no clue at all to the nature of the factors controlling this enormous wild-life rhythm in the northern forests, except that it seems almost certain that climatic fluctuation must play a controlling part. The cycles operates exactly parallel on both sides of the Rockies; we have unpublished notes suggesting that the introduced snowshoe rabbits on Anticosti have developed a cycle corresponding to the mainland one; the whole Canada keeps in step without ever getting right out of phase[25] (p. 242)

Elton's skepticism, however, must not diminish the importance that mathematical models had in shaping his view. Even if he was never entirely convinced that these models could be the *only* explanations of the phenomena he observed, he doubtless become aware of the possibility of an "internal" nature of the oscillations in animal populations only after reading Lotka and Volterra's papers.[*]

We can thus grant Volterra and D'Ancona the right of citing Elton's personal development in support of their view concerning the heuristic importance of mathematical models in biology. Yet, there are reasons to believe (and probably Volterra and D'Ancona knew) that Elton's development had been the exception rather than the rule in the history of the *golden age*. The existence of non-mathematical anticipations of the "internal" nature of oscillations had been noticed by D'Ancona since his first paper concerning the effects of fishing in the Upped Adriatic Sea.[27] In that paper D'Ancona quoted the Italian entomologist Antonio Berlese (1863-1924), who believed that coexisting animal species could oscillate independently from any external factor (p. 50). According to Berlese,

> [w]e usually say that living species equilibrate each other regulating reciprocally the numerical development of their individuals. But it would be more precise to say that this is a disequilibrium, since it involves continuous oscillations in which one species overcomes another, some disappear, other substitute them etc., in such a way that these variations can be represented as continuous oscillations, since the same path is done over and over.[28,†] (p. 208)

Another and more important anticipation Volterra and D'Ancona didn't know in 1926 came from the Italian biologist Lorenzo Camerano (1856-1917). In his paper "Dell'equilibrio dei viventi mercé la reciproca distruzione" Camerano clearly states that "the equilibrium of living creatures" due to competition and predation is usually reestablished after a long series of oscillations analogous to the oscillations of a pendulum or a diapason.[29] Camerano's paper, which would deserve a longer presentation than we can provide here,[‡] clearly demonstrates that through "ordinary reasoning" one could have

---

[*] Sharon Kingsland notices that to Elton "the generation of oscillations through internal causes was new and unexpected".[12] However, she also stresses that they were all but unexpected to scholars belonging to other fields such as economic entomology (p. 129). In the next few pages we shall see that, as a matter of fact, most of the non mathematical anticipations of Lotka and Volterra's results came from entomologists.

[†] See also the passage quoted in D'Ancona's paper (p. 50). Quite interestingly, Berlese clearly distinguished between oscillations due to the predator-prey relationships from those due to climatic factors, which he called "fattori meteorici".[28]

[‡] Camerano's view concerning the "equilibrium of nature" has been recently discussed by Joel Cohen[30].

reached most of the results mathematicians reached through their formal models. The following passage should suffice to prove this.

> Suppose that for some reason the vegetation in a given locality happens to grow abnormally. This increase in vegetation will necessarily result in an increase in phytophagous animals. Following the rule already noted which states that animal grow in direct proportion to the available food. This increase in phytophagous animals will, for the same reason, promotes an increase in carnivores who prey upon herbivores. This in turn will promote the development of animals which prey on these predators. The increase of parasites will cause an increase of animals which prey on these parasites. Finally, this will increase the endoparassites of these predators.
> Now, an increase of endoparassites will tend to reduce the number of animals in which they live, that is predators, resulting in the growth of their prey.
> In our case, this will increase the number of parasites, and this will again result in a decline of other predators. Consequently, the number of phytophagous animals will increase, which in turn diminish the quantity of vegetation. […]
> If the same causes continue to exist […] the same cycle will be repeated again.

These are two anticipations Volterra and D'Ancona could have discovered simply reading the Italian scientific literature, but there had been many other anticipations in the rest of the world. For example, the Canadian entomologist William R. Thompson (1887-1972) – another pioneer in the mathematical study of biological associations – started his mathematical investigations in the attempt of explaining the violent oscillations that had been frequently observed by his mentors Paul Marchal (1862-1942) and Leland O. Howard.* According to Howard, for example, "whenever a plant-feeding species […] transcends its normal abundance to any great extent, there is always a great multiplication of its natural enemies, and this multiplication is so great as to reduce the species even below its normal level", which sets the stage for a new cycle. The observation of a parasite-host biological association, for example, usually reveal

> a great fluctuation in numbers, their parasites rapidly increasing immediately after the increase of the host species, overtaking it numerically and reducing it to the bottom of another ascending period of development.[33†]

Both Howard and Marchal were applied scientists, who based their descriptions on empirical data they gathered in their job as entomologists. But other, more theoretically oriented, authors arrived at the same conclusions with more speculative methods. For example, Lotka was originally prompted to study the oscillation in animal species by the following passage taken from Herbert Spencer's *First Principles*.

> Every species of plant and animal is perpetually undergoing a rhythmical variation in number – now from abundance of food and absence of enemies rising above its average, and then by a consequent scarcity of food and abundance of enemies being depressed below its average […] amid these oscillations produced by their conflict, lies that average number of species at which expansive tendency is in equilibrium with surrounding repressive tendencies[19] (pp. 61-2)

---

* Few things are known by historian of biology regarding these two entomologists.[31] One of the best sourceses is still an article written by Thompson in 1939, which contains a detailed history of the early attempts entomologists made of explaining fluctuations in the abundance of populations of insects.[32]

† Similar remarks can be found in contemporary papers wrote by Paul Marchal[34,35].

One might think that it is hardly surprising that these anticipations didn't play considerable role in the development of theoretical population biology, and that applied ecologists were so impressed by Lotka and Volterra's results. For, apart from Herbert Spencer's passage, all these quotations come from rather obscure publications. However, even the great classics of the history of population studies contained clear non-mathematical anticipations of the "internal" character of population oscillations. Probably the most well known example is Thomas Malthus' population theory, exposed in his *Essay on the Principle of Population.*[*]

> We will suppose the means of subsistence in any country just equal to the easy support of its inhabitants. The constant effort towards population, which is found to act even in the most vicious societies, increases the number of people before the means of subsistence are increased. The food, then, which before supported eleven millions, must now be divided among eleven millions and half. The poor consequently must live much worse, and many of them be reduced to severe distress. The number of labourers also being above the proportion of work in the market, the price of labour must tend to fall, while the price of provisions would at the same time tend to rise. During this season of distress, the discouragements to marriage and the difficulty of rearing a family are so great, that population is nearly at a stand. In the meantime, the cheapness of labour, the plenty of labourers, and the necessity of an increased industry among them, encourage cultivators to employ more labour upon their land, to turn up fresh soil, and to manure and improve more completely what is already in tillage, till ultimately the means of subsistence may become in the same proportion to the population as at the period from which we set out. The situation of the labourer being then again tolerably comfortable, the restraint to population are in some degree loosened; and, after a short period, the same retrograde and progressive movements, with respect to happiness, are repeated.
>
> This sort of oscillations will not probably be obvious to common view; and it may be difficult even for the most attentive observer to calculate its periods. Yet [...] in the generality of old states, some such vibration does exist... [37] (p. 20)

It might be safe to conclude that any attentive reader of the works written by demographer and biologists between the XIX and XX century would have found hardly any reason to be surprised by the conclusions Lotka and Volterra arrived at with their models. At variance with Volterra and D'Ancona's claim, the endogenous nature of oscillations in predator-prey relationships had already been described several times before it received a mathematical representation.

FURTHER REFLECTIONS ON PREDATORS, PREYS AND MATHEMATICAL MODELS

Although mathematicians had not been the first to understand the endogenous nature of biological oscillations, their models played a fundamental role in convincing many biologists that predator-prey systems can oscillate for "endogenous" reasons. In a review article published in 1941, the British entomologist Paul de Bach claimed that mathematical models had been one of the chief factors in the spreading of this view among biologists.[38]

---

[*] There has been several attempts of using standard Lotka-Volterra equations to model Malthus theory of pupulation[36].

However, supporters of mathematical approaches had few reasons to be happy. Elton's data, which were the only empirical support not derived from laboratory populations, gave mathematical models just an intuitive confirmation.* As a matter of fact, during the *golden age* nobody did anything more than claim that, by and large, oscillations revealed by these data seemed to agree with mathematical models' predictions. When more accurate studies involving standard statistical tools began, it became rapidly clear that Elton's data were hard to explain by means of Lotka-Volterra equations. In 1973 the American bio-mathematician Michael Gilpin published an article on the *American Naturalist* (whose provocative title was "Do hares eat lynx?") in which he used Elton's data to estimate the parameters of a couple of cubic Lotka-Volterra equations.[41] His conclusions were discouraging: the result of his analysis was "worst that bad". Gilpin had to admit that the only way to reconcile Lotka-Volterra models with the existing data was to assume that hares were predators particularly greedy of lynxes.

The failure of Lotka-Volterra models in explaining Elton's data generated a large amount of alternative explanations. A recent article published on the *Journal of Theoretical Biology*, distinguishes two chief alternatives in the current literature to the classical explanation based on some version of the Lotka-Volterra equations.[43,†] The first hypothesis claims that while hare population oscillates because of a time lag in its response to changes in the abundance of vegetation, lynxes oscillate because their preys do. According to the second hypothesis, oscillations are due to several environmental factors, among which climatic variations are particularly relevant. Some of those who share this view endorse a thesis which is close to Elton's, according to which the controlling factors are due to the periodic occurrence of sunspots.[45] Remarkably enough, the reemergence of the original view which considers environmental factors a key element in the explanation of these oscillations is largely due to the fact that, as Elton noticed more than fifty years before, oscillations are surprisingly synchronized in areas separated by insurmountable barriers like the Rocky Mountains.

The success the Lotka-Volterra hypothesis gained during the *golden age*, which seems to be unjustified *ex-post*, leads us to make some further reflections which seem to confirm another of Mayr's perplexities concerning the use of mathematics in biology. According to Mayr, biologists have frequently suffered from an inferiority complex toward formalized disciplines like physics. For example,

> [w]hen Darwin calculated that the earth must be more than a thousand million years old in order to account for the phenomena of geology and phylogeny, Lord Kelvin pronounced him emphatically wrong on the basis of calculations of the heat loss of a globe of the size of the sun: 24 million years the maximum he could allow. It is rather amusing with what assurance Kelvin assumed the correctness of his own and the error of the naturalists' age determination. Since biology was an inferior science, there could be no question where the error was. (p. 41-2)[6]

Mayr believes that biologists was so ready to accept Kelvin's conclusions (which turned out to be mistaken) because they where presented as a consequence of a complicated mathematical procedure, instead of a non-formalized argumentation. The story of Elton's data apparently confirms this intuition: Lotka-Volterra models initially defeated Elton's non-formalized argumentation because they were based on mathematical formalization. The rhetoric success of formalized arguments is likely to lead scholars astray since, in

---

* On the other hand, it is a well know fact that biologist Georgii Gause failed to obtain clear-cut empirical verifications of the oscillations of predator-prey models even in laboratory conditions.[39,40]

† E. Pielou presents a similar classification and considers the Lotka-Volterra hypothesis a "long discarded" one.[44]

biology as in other sciences, not necessarily their conclusions have a better grounding than those based on ordinary language.

OTHER MATHEMATICAL RESULTS

The previous two sections have been hopefully clear in demonstrating the main point of this paper, so I can quickly discuss two other results obtained by mathematicians of the *golden age*: the "Volterra principle" and the "principle of competitive exclusion".

The "Volterra principle" states that in a predator-prey model, the uniform destruction of the two populations produces an increase in the average number of preys and a decrease in the number of predators. This is the result Volterra used to explain the increase of predators (selchians) in the Upper Adriatic Sea observed by D'Ancona during the suspension of fishing due to the First World War. That this result could be considered a genuine contribution mathematics gave to population studies was hard to accept, since, as Volterra knew since 1926, Darwin described a similar phenomenon in the third chapter of the *Origin.*[46] (p. 51)

> The amount of food for each species of course gives the extreme limit to which each can increase; but very frequently it is not the obtaining food, but the serving as prey to other animals, which determines the average numbers of a species. Thus, there seems to be little doubt that the stock of partridges, grouse, and hares on any large estate depends chiefly on the destruction of vermin. If not one head of game were shot during the next twenty years in England, and, at the same time, if no vermin were destroyed, there would, in all probability, be less game than at present, although hundreds of thousands of game animals are now annually killed.*

On the other hand, it has been frequently noticed that the "Volterra principle" amounts to the rather obvious proposition that sometimes the number of mice is kept under control by cats. It is intuitively obvious that in similar circumstances to kill in equal number both mice *and* their natural predators (cats) can *increase* the number of mice rather than diminish it. One might wonder whether a trivial result like this would have gained such a large audience had not Volterra put it in mathematical terms.

The "principle of competitive exclusion" states that competition among two species that "struggle for the same nutriment" to use Volterra's words (or compete for the same "ecological niche" in Elton's terminology), leads inevitably to the extinction of one of the two species. Even if apparently Volterra wasn't particularly interested in this result, since Second World War it played an increasingly important role in theoretical population ecology. Again, it wasn't a novelty when Volterra first stated it in 1926, since Darwin devoted to the informal presentation of a similar result an entire paragraph of the third chapter of the Origins, which is tellingly titled "Struggle for life most severe between Individuals and Varieties of the same Species".

> As species of the same genus have usually, though by no means invariably, some similarity in habits and constitution, and always in structure, the struggle will generally be more severe between species of the same genus, when they come into competition with each other, than between species of different genera. We see this in the recent extension over parts of the United States on one species of swallow having caused the decrease of another species. The recent increase of the missel-thrush in parts of Scotland has caused the decrease of the song-thrush. [...] We can

* Volterra quotes this passage in "Variazioni e fluttuazioni",[1] p. 19.

> dimly see why the competition should be most severe between allied forms, which fill nearly the same place in the economy of nature; but probably in no one case could we precisely say why one species has been victorious over another in the great battle of life.

It is hardly surprising that both the non-mathematical anticipations of Volterra's results were due to an authority as Charles Darwin. To my knowledge, Mayr' suggestion of looking for non mathematical anticipations of the results we usually attribute to the use of mathematics has not been followed by the community of the historians of biology.* It cannot be excluded, and it is rather likely indeed, that a systematic search through less known texts will reveal the existence of a wide variety of other anticipations, both in the biological literature and in works devoted to somewhat related fields like economics and demography.†

CONCLUSIONS

The story we told in the previous paragraphs apparently confirm Ernest Mayr's thesis: all the most important results we usually attribute to the mathematical models elaborated during the *golden age* (from the oscillations in predator-prey models to the principle of competitive exclusion) had been formulated more than once well before Lotka, Volterra and other started their investigations. Furthermore, once put in mathematical terms all these results became the center of an intense scientific research, both empirical and theoretical. All this can be taken as a proof that in the history of theoretical ecology mathematical models played what I've termed a rhetorical role rather than an heuristic one.

Probably some readers will find it difficult to accept such a pessimistic view of the use of mathematics in biology. After all, in the previous pages I've mentioned (because both Volterra and John Maynard Smith did) the most elementary discoveries of the *golden age*. It is reasonable to believe that if one takes into consideration more complicated models, the search for non mathematical anticipation becomes much harder. For example, Lotka-Volterra predator-prey models predict that inter-specific competition (represented by a coefficient Verhulst-Pearl larger than zero) dampens the oscillations, and thus that the equilibrium between predators and preys (if it exists) is asymptotically stable. To my knowledge, none of the authors who discussed the existence of oscillations in predator-prey models ever mention this fact.‡ Analogously, in two papers published in 1936 Andreï N. Kolmogoroff (1903-1987) and the bio-mathematician Valdimir Kostitzin demonstrated that two populations predator-prey models could have stable limit-cycles.[50,51] Although this result is apparently interesting only from a mathematical point of view, it is somewhat important in the explanation of Elton's data, for the amplitude of the oscillations of hares and lynxes remains roughly constant for more than one century, which can only be explained by means of a stable limit cycle.[52] Similar conclusions might be drawn

* This is probably due to the fact that, Giorgio Israel notices, the history of mathematical modeling in biology and social sciences is still largely unexplored.[47]

† For example, the classical reconstruction of the "principle of competitive exclusion" due to Garrett Hardin quotes a certain number of non-formal anticipations, some of which due to social scientists like Sismondi[48] (Darwin's indebtedness with Sismondi seems to be confirmed by more recent research[49]). Among them, the following passage by J. Grinnel is particularly clear. "[e]very animal tends to increase at a geometric ratio, and is checked only by limit of food supply. It is only by adaptations to different sorts of food, or modes of food getting, that more than one species can occupy the same locality. Two species of approximately the same food habits are not likely to remain long enough evenly balanced in numbers in the same region. One will crowd the other." (This passage was quoted in Hardin's paper[48])

‡ For example, Camerano distinguishes the cases in which oscillations are dumped from those in which they remain constant, without linking this fact to inter-specific competition.

considering more complicated competitive models (like those Lotka presented in a paper published in 1932[53]), which give the "principle of competitive exclusion" a more complicated form, allowing, for example, for the coexistence of several competing species when there are several "ecological niches".

The (provisional) conclusion we can draw from these considerations is that probably both Mayr and Volterra are right in what they say concerning the applications of mathematics to biology. Mayr's thesis that mathematics only played a rhetoric role, because all its result had already been discovered without it, applies chiefly to the most elementary results obtained by the mathematics of the *golden age*. When more complicated models are at stake, usually the search for non-mathematical anticipations becomes much harder, and thus probably Volterra might be right in advocating the use of mathematics because of its heuristic fertility. There is only one risk in this sort of argument. It is a platitude to say that mathematics can easily lead us much further than ordinary language. Any mathematician worth the name can easily construct hundreds of models whose behavior cannot be predicted without the aid provided by calculus. But these models are not necessarily of interest for any scientific discipline, even if their variables are called "population densities", "temperature" and so on. In other words, we are usually not interested in the mere unpredictability of the results obtained via formalization, but in advancing our understanding of natural phenomena. Really interesting mathematical models are those complex enough to generate results hardly predictable without them, but which retain some relationships with the phenomena they are meant to explain. Probably they are neither as numerous as Volterra hoped in the 30.s, nor completely absent as Mayr would us to believe.

REFERENCES

1. Vito Volterra, Variazioni e fluttuazioni del numero d'individui in specie animali conviventi, *Memorie del Regio Comitato Talassografico Italiano*, CXXXI (1927), English trans. in F. M. Scudo and J. R. Ziegler, *The Golden Age of Theoretical Ecology: 1923-1940,* Springer Verlag, Berlin (1978).
2. Vito Volterra and Umberto D'Ancona, *Les associations biologiques au point de vue mathématique*, Hermann, Paris (1935).
3. Karl Sigmund, *Games of Life: Explorations in Ecology, Evolution and Behaviour*, Oxford University Press, Oxford (1993), 2nd edition Penguin Books (1996).
4. Luciano Andreozzi, Vito Volterra and Umberto D'Ancona on the effects of fishing in the Upper Adriatic Sea and their mathematical representation. Reflections on a celebrated episode in the history of population ecology, forthcoming on *Sciences. et Techniques en Perspective.*
5. John Maynard Smith, *Evolution and the Theory of Games,* Cambridge University Press, Cambridge (1982).
6. Ernst Mayr, *The Growth of Biological Thought,* Harvard University Press, Cambridge (Mass.) (1982).
7. Ernst Mayr, Where are We. Genetics and Twentieth Century Darwinism, *Cold Spring Harbor Symposia on Quantitative Biology*, 24: 1-14 (1959).
8. Ernst Mayr, *Evolution and the Diverisy of Life*, Belknap Press, Cambridge (Mass.) and London (1976).
9. C. H. Waddington, Epigenetics and evolution, in *Symposia of the Society of Experimental Biology*, 7, Academic Press, New York (1953).
10. William B. Provine, Role of mathematical population geneticists in the evolutionary synthesis of the 1930's and 40's, in L. Solomon, *Mathematical Models in Biological Discovery*, Lecture notes in Biomathematics, Springer-Verlag, Berlin Heidelberg (1977).
11. Francesco M. Scudo and James R. Ziegler, *The Golden Age of Theoretical Ecology: 1923-1940,* Springer Verlag, Berlin (1978).
12. Sharon E. Kingsland, *Modeling Nature. Episodes in the History of Population Ecology,* The University of Chicago Press, Chicago-London (1985).
13. Sharon E. Kingsland, Mathematical figments, biological facts: population ecology in the thirties, *Journal of the History of Biology*, 19: 235-256 (1985).
14. Ana Millàn Gasca, Volterra's biomathematics and biologists' biological facts, *Hist. St. in the Phys. and Biolog. Sc.*, 26: 347-403 (1996).

15. Francesco M. Scudo, The 'golden age of theoretical ecology': a conceptual appraisal, *Rev. Euro. des Sc. Soc.*, 22, 67 (1986).
16. Herbert Andrewartha, The use of conceptual models in population ecology, *Cold Spr. Harbor Symp. on Quant. Biol.*, 22: 219-236 (1957).
17. Luciano Andreozzi, Vito Volterra organizzatore scientifico e la nascita della biologia matematica in Italia, forthcoming on *Nuncius*.
18. Alfred Lotka, Analytical note on certain rhythmic relations in organic systems, *Proc. of the Nat. Acad. of Sc.* , 6 (1920).
19. Alfred Lotka, *Elements of physical biology*, Williams and Wilkins, Baltimore (1925), republished as *Elements of Mathematical Biology*, Dover, New York (1956).
20. Umberto D'Ancona, *La lotta per l'esistenza,* Einaudi, Torino (1942).
21. Umberto D'Ancona, *The Struggle for Existence*, Engl. transl. by A. Charles and R.F. Withers, Bibliotheca Biotheoretica, Leiden (1954).
22. George Evelyn Hutchinson, *An Introduction to Population Ecology*, Yale University Press, New Haven and London (1978).
23. Charles S. Elton, Periodic fluctuations in the number of animals, their causes and effects, *Brit. J. of Exper. Biol.*, II (1924).
24. Charles S. Elton, *Animal Eecology*, Sidgwik and Jacson, London (1927), 2nd enlarged edition New York, MacMillan (1935).
25. Charles S. Elton and Mary Nicholson, The ten-year cycle in numbers of the lynx in Canada, *J. of Anim. Ecol.*, XI (1942), pp. 215-244.
26. *Volterra Archive*, letters from Elton to Volterra, letter number 1.
27. Umberto D'Ancona, Dell'influenza della stasi peschereccia del periodo 1914-18 sul patrimonio ittico dell'Alto Adriatico, *Memorie del Regio Comitato Talassografico Italiano*, CXXVI (1926).
28. Antonio Berlese, Considerazioni sui rapporti tra piante, loro insetti nemici e cause nemiche di questi, *Redia*, IV (1906).
29. Lorenzo Camerano, Dell'equilibrio dei viventi mercè la reciproca distruzione, *Atti della Regia Accademia delle Scienze di Torino*, XV (1880).
30. Joel E. Cohen, Lorenzo Camerano's contribution to early food web theory, in *Frontiers of Theoretical Biology. Lecture Notes in Biomathematics*, 100, Simon A. Levin (ed.), Springer-Verlag, New York (1994): 351-359.
31. Pascal Acot and Jean-Marc Drouin, L'introduction en France des idées de l'écologie scientifique américaine dans l'entre-deux-guerres, *Rev. d'Histoire des Sci.*, 50, 4: 461-479 (1997).
32. William R. Thompson, Biological control and the theories of the interactions of populations, *Parasitology*, XXXI: 301-388 (1939).
33. Leland O. Howard, A study of insect parasitism, *U.S. Department of Agricolture, Division of Entomology, Techincal Series* no. 5 (1897): pp. 5-57.
34. Paul Marchal, L'équilibre numérique des espéces et ses relations avec les parasites chez les insects, *Comp. Rend. Soc. Biol.,* XLIX, 49: p. 129 (1897).
35. Paul Marchal, The utilization of auxiliary entomophagous insects in the struggle against insects injurious to agricolture, *Pop. Sc. Montly*, 72: 352-70, 406-419 (1908).
36. Richard M. Goodwin, *Essays in Economic Dynamics*, The Macmillan Press, London (1982).
37. Thomas R. Malthus, *An Essay on the Principle of Population*, edited by Patricia James, Cambridge University Press, Cambridge (1989), first published in 1803.
38. Paul De Bach and Harry S. Smith, Are Population Oscillations Inherent in the Host-Parasite Relation?, *Ecology,* XXII (1941): 363-369.
39. Georgii Frantsevich Gause, *The Struggle for Existence*, Williams and Wilkins, Baltimore (1934), reprint New York, Dover (1971).
40. Georgii Frantsevich Gause, *Vérifications Expérimentales de la Théorie Mathématique de la Lutte Pour la Vie*, Hermann, Paris (1935).
41. Michael E. Gilpin, Do hares eat lynx?, *Amer. Naturist*, 107 (1973).
42. E. Leigh, The ecological role of Volterra's equations, in M. Gerstenhaber , (ed.) *Some Mathematical Problems in Biology,* The American Mathematical Society, Providence (1968).
43. Jordi Bascompte, Ricard V. Solé, Norbert Martinez, Population cycles and spatial patterns in snowshoe hares: an individual-oriented simulation, *J. of Theor. Biol.*, 187: 213-222 (1987).
44. E.C. Pielou, The usefulness of ecological models: a sotck-taking, *The Quart. Rev. of Biol.*, 56 (1981).
45. A.R. Sinclair, J.M. Gosline, J.M. Krebs, C.J. Boutin, S. Smith, R. Boonstra, Can the solar cycle and climate syncronize the snowshoe hare cycle in Canada?, *Amer. Naturalist*, 141: 173-198 (1993).
46. Charles Darwin, *On the Origin of the Species*, in *The Works of Charles Darwin*, edited by Paul H. Barrett and R. B. Freeman, William Pichering, London (1988), vol. 15 (first published in 1859).
47. Giorgio Israel, *La Mathématization du Réel. Essai sur la Modélisation Mathématique*, Èditions du Seuil, Paris (1996).

48. Garrett Hardin, The competitive exclusion principle, *Science,* 131 (1960).
49. P. A. Riley, The origin of the principle of competitive exclusion: was Darwin influenced by Sismondi?, *Linnean*, 2 (3): 20-22 (1986).
50. Andreï N. Kolmogoroff, Sulla teoria di Volterra della lotta per l'esistenza, *Giornale Istit. Ital. Attuari,* 7 (1936).
51. Vladimir A. Kostitzin, Sur les solutions asymptotiques d'équations différentielles biologiques, *Compt. Rendus de l'Académ. des Sc.*, 203: 1124-1126 (1936).
52. Robert May, On relationships among various types of population models, *Amer. Natur.*, 107: 46-57 (1973).
53. Alfred Lotka, The growth of mixed populations: two species competing for a common food supply, *J. of the Wash. Ac. of Sc.*, 22 (1932).

THE MATHEMATICS IMPLIED IN THE LAWS OF NATURE AND REALISM, OR THE ROLE OF FUNCTIONS AROUND 1750

Jean Dhombres
EHESS
54, bd Raspail
75006, Paris
FRANCE

To do mathematics in order to perform an investigation of Nature has always been a critical moment in the history of thought. The two sciences at stake, one being Physics and the other Mathematics, claim with good and sound reasons to have their source of inspiration in two completely different worlds. One is the human mind, or the internal world and intellect; and the other one is Nature, or the external world. Because they saw no experiments working along the process of mathematics, many historians of science, at least in modem times, have even excluded mathematics from the scene of science. They are not historically right, but they are not completely wrong epistemologically. More than once has physics been looked at as an experimental ground to verify mathematics. For many years during the nineteenth century it was thought that non-Euclidean geometry had to be checked by geodesic measurements [1]. But from a mathematical point of view this geometry was understood only when one understood that such verifications were irrelevant [2]. This intellectual experience led to a completely new role assigned to mathematics and models slowly became a must in the sciences, while a purist like Hardy described the mathematician as a creator of patterns [3]. But patterns of what?

There have been many attempts in order to solve the basic opposition between Physics and Mathematics, and they cannot be looked at indifferently in the course of history as if it were the eternal opposition between realistic truth and formal truth. Certainly, one should take care of the almost perennial struggle against mathematics. A struggle as old as science itself: mathematics is regularly perceived as a formal but tyrannical imposition upon knowledge. And this knowledge should be set free in order to grasp the various forces of the world. Sometimes claimed to be unreachable by just the realm of numbers. Chateaubriand, in a splendid rhetorical move, even said that only merchant people could reduce civilization to exchange of mathematical quantities [4], which is a romantic way to say that mathematical realism is not to govern the mankind. But there is as well the old opposition to any explanation of Nature which is more than a classification of accidental qualities, and this also leads to a clash as classification is not the usual quality attributed to mathematics. Except by Descartes, who is so reluctantly seen as a mathematician by classical historiography. He however set forth an explanation for the role of mathematics in the philosophy of nature: not just a tool, but a model proving that

The Application of Mathematics to the Sciences of Nature, edited by Cerrai *et al.*
Kluwer Academic/Plenum Publishers, New York, 2002.

truth is accessible*. Is there any other way than mathematics to provide a logical foundation to the ambition of science? The old-age accusation against mathematics about its lack of reality has to be put in its local perspectives.

One serious difficulty in writing history of science, and particularly when it concerns what has been uniformly called mathematical physics, is to really put some history in the process. And particularly to say what does one mean when claiming a theory explains something in the world. We know for sure that there are strong epistemological differences between the time of Aristotle - by the way not so Euclidean as has been claimed -, the time of Oresme - by the way not so Aristotelian as usually said, of Galileo - officially antiaristotelian, of Newton - who was desperate not to be able to write according to the Greek canon, of Einstein - who achieved Newton's dream and opened completely new fields for cosmology which Newton had closed, and our time - which it would be stupid to define in few words. But we have difficulties in explaining in each case a different interplay between physics and mathematics; we often claim that the difference just lies in the different levels of mathematical sophistication all those times experimented. We then give to the sentence *eadem sed aliter* the only sense of time running, which some interpret as progress. And common sense claims that there is no progress at all in mathematics, which is supposed to be born perfect. Historians of science generally accept to say that a physicist has to adapt the available tools from mathematics for his own purposes, or has to create new tools. In this last case, he then has to take the part of a mathematician. And this is what gives its rhythm to history of science. It is not frequent to see a different kind of explanation†, namely the urge from a mathematician having coined new concepts, created new objects, or new methods, to try seeing their use for an explanation of the physical world. To say it in a different way, most explanations we provide for the history of mathematical physics, refuse to take care of the platonic mood of most mathematicians [5]. They consider, perhaps with no deeper philosophy in mind, but certainly with a will to kill history, that mathematical objects always lie somewhere in front of them, thus in the external world, and not after they were invented or discovered‡. Could physics be the only way to put some history in the development of mathematics?

MATHEMATICS AND HISTORY OF SCIENCE

Because the present workshop aims at undertaking a dialogue between active scientists and historians of science, it immediately meets the question of the role of mathematics, a vocabulary which earlier would have been the role of geometry, or the role of analysis, or the role of algebra. I may better describe in French the question of specialization by using the vocabulary of "mathématisation", "géométrisation", or "algébrisation". But there is no substantive for the process of Analysis, and just "Analyse" is used. Which is certainly due to the role of the analytical thought in the *Bildungsprozess* of Science. Do I say that to avoid the dialogue with scientists? Has it ever been cut by historians of science? Perhaps yes, when one considers as nowadays that a scientist of the past cannot be read or understood from what he wrote and said, but from all the hidden

* Pertinent to my question is the fact that so many scholars raise the question of th lack of Cartesian geometry in Descartes' physics, for example in Le Monde, ou Traite de la Lumiere (premiere edition francaise en 1664; see the recent edition by A. Bitbol-Hesperies et J. P. Verdet, le Seuil, Paris, 1966).

† This explanation is the one generally given to describe Newton's Principia, Fourier's Analytic Theory of Heat, Schroedinger's equation. In spite of the fact that one is confronted by the question of Galileo's new uses of the theory of proportions, of Maxwell's transformation of quaternions, and of von Neumann's use of Hilbert spaces.

‡ There recently has been plenty of books on epistemology, in order to reevaluate what realism means for mathematics; very few take this problem on an historical basis, as if there was only a unique realism, almost independent of the mathematical theories it concerns, The present paper will show one of such cases of historical misinterpretation of realism.

facts upon which he kept silent. In this trend of thought, mathematics has often been described as a curtain, used in order to hide all what may be irrational in the making of science. This is what Feyerabend jokingly assigned as a role to the analytical style upon which I immediately turned while trying to investigate the relative position of mathematics and physics. If a historian has to become a policeman in order to state what was scientifically incorrect in the whole life of a scientist, he then cuts himself from the whole process of the production of science. Because it is a process of objectivity which involves more than one mind.

What has mathematics to do with the construction of objectivity? One may argue that there is still something "scientific" which could be kept unknown under what science claims to be, and that under the cover by mathematics lies the true process of science. Historians of science always wished to scientifically understand how scientific ideas germinate, but in a very unscientific way some claim that everything plays its part in the process, except what mathematics provides. This is the very old and very obnoxious answer given by obscurantism.

Who cares about statements which claim that there is no science, or no scientific method as well, and that everything which has been said concerning ways of reasoning in the sciences is a pure construction of the mind of perverse scientists? Perverse in the sense that scientists attempt to convince everybody science possesses a universal way to truth or to efficiency (is there any difference?), as in fact it works just as in the magic world, or in the religious world, or in the social world. And the use of mathematics is a clever disguise to avoid showing the similarities and deep analogies between all these words. Social sciences are in the trend of enlarging their scope and, in a paradoxical way, pretend that all the sciences behave alike. I find much more interesting the fact that there existed a constant claim for exclusion of ideas and methods in the sciences: exclusion of some kinds of reasoning, of some kinds of arguments, of some kinds of objects, etc. And therefore exclusions of methods automatically coming from one science to another [6], which explains the process of separation of the various sciences. And in this long-term process of exclusion, I consider that mathematics was particularly instrumental. Even on a social point of view; a refusal of any knowledge in mathematics has for a long time been the surest way to stay outside science in the making[*].

But this kind of refusal has drastically changed in our age: in the human and social sciences, a very superficial exhibition of mathematics has more than once been used as a proof for the validity of other kinds of arguments[†]. Mathematics certainly did not play for all the sciences just the part of an aristocratic reference. And even if it is thought so, what does this primacy given to mathematics mean on an epistemological point of view? As scientists are everywhere acclaimed to be efficient, could it be possible that they maintained such a role to mathematics just by following tradition? Yes, if one reduces science to a social process of conquest from the formal sciences over other ways of knowledge.

I wish to give some historical flesh to the bare bones of general consideration by looking at the way mathematics and physics combined during the Enlightenment in order to

---

[*] Is there another interpretation to the famous sentence: Nobody enters here who does not know Geometry? I do not wish to reject as only cultural and political an order of the Empress dowager who urged Chinese intellectuals to just learn mathematics from Westerners.

[†] This attitude led to the intellectual and pleasant reaction by J. Bricmont and A. Sokal. *Les impostures intellectuelles*, O. Jacob, Paris (1997). After some ridiculous French answers to this pamphlet. and perhaps due to an interesting rewriting in the French usual terminology by J. Bouveresse *Prodiges et vertiges de l'analogie. De l'abus des belles-lettres dans la pensée*, Raisons d'agir Editions, Paris (1999), one begins to recognize there is first to think anew the place attributed to history of science in the history of ideas. See also J. Starobinski, *Action et réaction, Vie et aventure d'un couple*, Librairie du XXe siècle, Ed. du Seuil, Paris (1999) a book which proved useful for the debate Buffon/Clairaut upon which I'll be focusing.

prepare for a new concept, the concept of function. I will have to show why it is relevant to the question of the sense of reality in mathematics.

The interest of focusing on this concept is that we have now such an ordinary notion of what a function is that we cannot easily conceive it had not always been present to the mind of scientists. To say it in another way, our epistemology of a function is purely a pedagogical one, delivered from any historical, or even genetic point of view. History of science has this interesting quality to show that ideas do not come at once from Heaven, and even common places have their history. Can it show how ideas grow? Every historical case, and also is the function case, has to be established for itself and, at least logically, this history does not prove more than the accidental circumstances of the case used. The historical proof, one is inclined to believe in our thinking of the 20th century, is more suitable when one is able to provide all the circumstances, of all kinds, from the personality of the discoverer to the *Zeitgeist* when discovery took place. And there the concept of function is well adapted to avoid this requirement of exhaustive description which has always been a good excuse not to think in history of science. In the sense that there was no perfectly defined discovery for the concept of function. And one may even say that the slow process of establishment of the concept did not belong to mathematics only, but could be looked at as the mathematical answer, that is a quantitative answer, to the very old metaphysical question about causes and effects.

I am not trying to say that science solved this very old question just by eventually adopting mathematics; a solution was provided which, due to mathematics, excluded many other ideas concerning the linkage between causes and effects. In the sciences, causality was no longer viewed as a general question, but a rather restricted one. Which effects can exactly be measured from a given cause? It can be agreed that this question is not a mathematical one, but a typical question of physics, of the new physics issued from Galileo, Descartes, and Newton. Twentieth century physics was to change once more causality, and as well the use of functions. I began by saying that the link between mathematics and physics is difficult to tell because definitions of what physics and mathematics were, are historically very sensitive.

A METAPIIYSICAL CONTROVERSY ABOUT FUNCTIONS: BUFFON VERSUS CLAIRAUT IN 1748-1749

If I begin with a quotation from Buffon, the natural historian so famous for his superb descriptions of the animal world, it is because he precisely puts the emphasis on the mathematical meaning of what a law of nature is.

> *Il est utile pour les Sciences d'établir de manière certaine la proposition que j'ai avancée, savoir, que la loi de l'Atraction, et même toute autre loi physique, ne peut jamais être exprimée que par un seul terme, et qu'une nouvelle vérité de cette espèce peut prêvenir un grand nombre d' erreuers et de fausse applications dans la sciences physico-mathêmatique; j'ai cherché plusieurs moyens de la démontrer* [*].

[*] Buffon, Seconde addition au Memoire qui a pour titre : Reflexions sur la loi de L'Attraction, *Hist. et Mém. Acad Sc. Paris* (1745) [1749], p. 580. As usual for texts in French of the 18th century , the spelling has been modernized, but punctuation is maintained. However I kept the original spelling in some of the quotations to follow in order to show that in the fury of the debate, some papers were so hastily written that numerous mistakes occurred. But some mistakes were corrected; a copy of the volume for year 1745 kept at the library of Nantes has many mistakes, which have disappeared in other copies, such as the one kept at the library of Basel for instance.

This sentence occurred within a polemical debate about Newton's law of attraction, and this was a very normal debate for science: how was one to try complicating Newton's use of so simple a law in order to answer to observations in the astronomical world which did not seem to fit with mathematical and physical deductions from Newton's law? How to regularize "irregularities"? This last expression meant the discrepancy between predictions and observations, in both cases measurements being more and more precise. The quest for precision, both practical and theoretical, would have to stay as the basic description of what science achieved in the Enlightenment*. Precision then led to discrepancies and they had to be explained.

By using the new mathematical tool of Calculus, and by going form observations to causes, Newton's acclaimed method had been to produce a general principle expressed in the mathematical form of an attraction between masses. And he too provided the mathematical ways of deducing from this law the orbits of the planets, each considered as alone with the Sun. If one maintained the same attraction law, it was mathematically far more difficult to deduce the movement of the Moon, attracted both by the Earth and by the Sun. Newton launched a technique, which could roughly be described as a technique of approximation: it consists in considering that the Moon indeed had an elliptic orbit, around the Earth (at one of the foci of the ellipse), as if the Sun did not play a part, but to take care of this part by supposing that the ellipse was rotating with time. Now, with such a supposition in mind, and from computations on results from precise observations, one may deduce a period of nine years for the revolution of the main axis of the ellipse. Keeping the model of a revolving ellipse as an approximation, and for sure Newton's law, Clairaut tried to compute the period in a more refined way than Newton: he then found 18 years for the period. The difference, as Clairaut explained, was of the same order as the difference deduced from the far older supposition of a Moon rotating around the Earth on a circular orbit. This was serious. Because the technique used was already an approximation, Clairaut thought that the present form of the mathematical law of attraction was also an approximation. Clairaut maintains Newton's principle of a law for attraction, but rejects the form of the law and, as has been done since, also rejects the mathematical approximation conducted by Newton for the movement of the Moon.

Confronted to similar problems, authors will propose to change the coefficient 2 in the expression of the attraction between two masses as a function of the inverse distance, and take some number close to 2. Clairaut thought he could correct by adding some terms to the law to solve up to the precision now required, and he proposed a term with the power 4 of the inverse distance between two masses. What was more important, he pointed out there was a general problem to be solved, the three body problem, and a movement to approach by approximation, the equation of the Moon. Under a pompous title, he published a preliminary report to the Academy of sciences in Paris: *Du Système du Monde dans les Principes de la Gravitation Universelle,* which was read on the 15th of November 1747, but had been hastily deposited earlier in September, before the holidays regularly taken by the Academy during October. We know the reason why Clairaut was in such a hurry: d'Alembert published a paper on the same subject in the same issue of the academic volume, and precise rumors were that Euler was preparing a paper on the movement of Saturn, another issue on the three body problem.

Then Buffon, who had years before translated into French the English version of Newton's *Method of Fluxions* †, argued in various ways against Clairaut's hypothesis, and

---

* I certainly deny any value here to the traditional view of development of mathematics in the age of Euler and Lagrange as a neglect of principles and rigor, by a sort of embarrassment of riches, a wealth of ideas, methods of computations, etc.

† Newton had written very early, in Latin, what became his *Method of Fluxions and Infinite Series* in an English translation which appeared posthumously in 1736, and was due. to John Colson. Buffon translated this translation into French, *La méthode des fluxions et des suites infinies*, Paris (1740). And this book was considered as a triumph of Newtonian thought in France, even if physics and astronomy were not at all

in particular in a mathematical way. And he did so by giving a metaphysical interpretation to the idea of what a function is. This is what I wish to examine.

If a law of nature, viewed as a function, has at least two terms in its mathematical expression, it cannot but mean that the law expresses the influence of more than one cause, so claimed Buffon in a paper read at the Academy on January, 20th 1748.

> *Dès qu'on voudra employer deux termes, on détruira l'unité de la qualité physique; parce que deux termes représenteront deux variations differentes dans la même qualité, c'est-à-dire deux qualités au lieu d'une: deux termes sont en effet deux mesures, toutes deux variables et inégalement variables, et dés lors elles ne peuvent être appliquées à un sujet simple, a une seule qualité* [7].

This seems to Buffon impossible as precisely a law of nature selects one cause, and one cause only. Why is this a metaphysical reason? Not only because of the vocabulary used by Buffon, which he tries to make as scholastic as possible ("sujet simple", "qualité"), but because there is the requirement of a mathematical expression, or representation of the law, to have a real meaning, and not just a quantitative meaning. When two terms are proposed in the expression of Newton's law, the first requirement for Buffon is to explain on a physical basis all the terms which are used. And some months later he argues that a mathematician like Clairaut, who tentatively attempted to modify Newton's law, has to justify the "numbers" or "coefficients" he introduced on top of those already present in the usual formulation of Newton's law.

> *Il faut donc que M, Clairaut commence par nous prouver que les nombres sont des êtres reels actuellement existans dans la Nature, ou que les coëfficiens sont des qualités physiques* [*]

A word like "qualité" cannot but refer to the Aristotelian vocabulary, and was understood as an opposition to mathematics, or to the view of mathematics as the instrument to deal with quantities. It is particularly interesting to notice that Buffon does not refer to geometry, as a source for evidence and as in the traditional Euclidean way as a resource for providing reality to mathematical arguments. He refers to "numbers". Numbers do not have a reality by themselves and bear no ontology at all. In other words, for Buffon a law which is a relation between numbers is not just a mathematical law; it means something more as it analyses reality. And he justifies his question to Clairaut by a question which has often be the way to refuse mathematics in physics.

> *Existe-t-il dans la nature un coëfficient comme un 4, un 5, un 6, etc. et n 'y a-t-il pas de l'absurdité a supposer qu'un nombre puisse exister réellement, ou qu'un coëfficient puisse être une qualité essentielle à la matière ? Il faudrait pour cela qu'il y eût dans la Nature des phénomènes purement numériques.* [†]

---

concerned by this purely mathematical book. See P. Brunet, *L'introduction de la théorie de Newton en France*, Paris (1931). In fact, Buffon had added a preface to his translation where he glorified Newton to diminish Leibniz. It was a particularly mundane position.

[*] I. Buffon, Seconde addition au Mémoire qui a pour titre: Reflexions sur la loi de L'Attraction, *Mém. Acad.Sc Paris* (1745) [1749], p. 582. On the same page, Buffon writes Clairaut's proposed new law in the form : $a/x^2+1/mx^4$, and he argues about the meaning of m thus placed in an unusual position: "or je demande ce que c'est que ce coefficient m, il est clair qu'il ne depend, ni de la masse, ni de la distance... .

[†] *Idem.* This text is dated 11th June 1749.

This is all the more astonishing for us as, from Cantor and others, we have inherited a complete reorganization of mathematical knowledge, and Arithmetic numbers and Set theory come first. If he adopts a way to refuse mathematics in physics, Buffon does it in the name of a rigor attained in mathematics. At least a historical stage of the rigor. What Buffon retains in his mind is the old interdiction about the Euclidean theory of proportions: it cannot be used between quantities which have not the same quality (the technical language was to say that one cannot compare but homogeneous quantities). Certainly, it was understood by a man like Buffon that a notion like velocity could be unambiguously defined even if it was the quotient of a length by a time, yet one should not add two completely different quotients to measure the force of attraction. Because no valid determination of the value of the coefficient of the second term could be sought for, and Buffon comments: "une quantité quelconque ne peut jamais être mesuré que par une autre quantité de même genre" [*]. Therefore he made no objection to the numerical coefficient of the first term because it was the universal constant of attraction.

The answer provided by Clairaut is a general disqualification of Buffon, and in the wording which was recognized as valid by the Academy.

> *Tous ces articles du nouveau Mémoire de M. de Buffon, ne me paraissant avoir aucun sens assez positif pour entreprendre d'y répondre, je passerai à ceux qui sont plus susceptibles d'examen.* [8]

The word "positif" is not used indifferently here, and already means an exclusion of any argument coming from metaphysics. Precisely because it deals with philosophy of nature, the Academy had not to deal with arguments from another kind of knowledge. But Clairaut had clearly understood that, at a deeper mathematical level, the objection made by Buffon against any complication of Newton's law of attraction, relied on a particular conception of what a function was, and obliged the mathematician to clarify it. I will come to that in a moment. For the time being, I would like to show that Clairaut avoids the problem of homogeneity and the question of meaning:

> *Je demanderai encore a M. de Buffon ce que c'est que des phénomènes numériques du même genre que le coefficient m.* [8]

Clairaut considers that the question about the nature of the coefficient is a "prétention aussi bien-fondée que si on vouloit trouver dans la Nature l'existence des lettres et des mots qu'on emploie à définir les êtres qui existent réellement" [†]. Laplace, sixty years later, when confronted with an expression of two terms for the law of capillary which was not deduced from Newton's gravitation law, will immediately distinguish between chemical action and proper capillary action [9]. And so will do Fourier, once he will have obtained the mathematical law for the propagation of heat: he will go further and will establish the now classical dimension analysis, in order to clearly state the independent physical qualities which are involved through quantitative representations [10]. Coefficients will have thus obtained a status in physics. A second stage in mathematical physics, a stage of the 20th century, will be to understand the physical nature of such coefficients, and to link coefficients intervening in attraction theory, in thermodynamics, and so on. Understanding coefficients could be a possible description of the program of solid state physics at its beginning.

I certainly do not claim that Buffon could have guessed such future to his problem. Clairaut's answer just shows that the meaning of Buffon's question was not perceived in the typically analytical mood mathematical physics was to take. Because Clairaut had already a

---

[*] *Idem.*

[†] *Idem*, pg. 585.

different view on functions, and coefficients were no longer having a meaning related to physics, but to functions. This is what I would like to show. But before attacking this debate on functions, I have to first turn to a completely different approach, by looking at the rhetorical which was being displayed. Was there anything scientific being discussed?

THE MODE OF THE SALONS, AND A DISQUALIFYING DISCOURSE USING MATHEMATICS

Buffon is certainly joking while using a very old form of writing on philosophy of nature; and I do not think he wants to be seriously taken on those words. He is using the mode of a pastiche, in the way it was usual to do in the *salons* (imitation of an author, wit at the expenses of someone, etc.). Here is a good example of a reasoning using syllogism, which certainly was no longer used in any writing in natural history:

> *Une loi en Physique n'est loi que parce que sa mesure est simple, et que l'échelle qui la représente est non seulement toujours la même, mais encore qu'elle est unique, et qu'elle ne peut être représentée par une autre échelle; or toutes les fois que l'échelle d'une loi ne sera pas représentlée par un seul terme, cette simplicité et cette unité d'échelle qui fait l'essence de la loi, ne subsiste plus, et par conséquent il n'y a plus aucune loi physique.* [7]

And Buffon changes his mind.

> *Comme ce dernier raisonnement pourrait paraître n'être que de la Métaphysique, et qu'il y a peu de gens qui la sachent apprécier, je vais tâcher de le rendre sensible en m'expliquant davantage .* [7]

But Clairaut is writing exactly in the same mood, not perhaps for the same salons, adding a kind of sarcasm which will be so common among the authors of the *Encyclopedie* which was to begin two years later. He is using the pastiche of theological arguments. Here are two examples. The first one is to answer the argument of Buffon on the simplicity of Newton's law which could be destroyed by adding two terms:

> *Nous convient-il de vouloir décider si le Créateur a donné la vertu attractive a la matière- par deux décrets différents, ou s'it l'a douée de deux forces à la fois par un seul acte de sa volonté ?* [11]

And the second is an answer to the question of the meaning of the second coefficient.

> *Et si on fait dépendre l'attraction de quelque vertu métaphysique que Dieu aurait donnée a la matière par des raisons qui nous sont inpénétrables, par quel moyen fixerons-nous cette attraction si ce n'est pas des faits ?* [12]

I certainly may go on with various quotations about the debate, which prove pleasant wit from both parts. But we are told that we have no more time to joke in history of science! Thus I am pushed towards a disqualification of Buffon's arguments. Because he was ignorant in mathematics. *Grosso modo*, this is the position most historians adopted, and some were happy to find this division in the time of the Enlightenment. This meant a divorce among the so-called Philosophers, a divorce between those linked to mathematics or abstract reasoning and providing analyticity to this world, and those inclined to natural history and life sciences, providing its naturality*. Buffon himself understood the danger of such an interpretation, and still using the style of pastiche, this time Euclidean pastiche, he tried to reconcile with the mathematical world.

> *Je vais tâcher de rendre mon objection plus intelligible en la traduisant en calcul. Ce sera la seule réplique que je ferai à sa réponse.*
> *La loi de l'Attraction par rapport à la distance, ne peut pas être exprimée par deux termes.*
> *Demonstration I.*
> *Supposons que* $1/x^2 \pm 1/x^4$ *représente I'effet de cette force par rapport à la distance x, ou, ce qui revient au même, supposons que* $1/x^2 \pm 1/x^4$ *qui représente la force accélératrice, soit égale à une quantité donnée A pour une certaine distance; en résolvant cette équation la racine x sera ou imaginaire, ou bien elle aura deux valeurs différentes: donc à différentes distances, I' Attraction serait la même, ce qui est absurde: donc la loi de l' Attraction par rapport à la distance ne peut pas être exprimée par deux termes. CQFD.* [13]

A careful reading of the papers, even in the order they are preserved within the academic volume, which is not the order of reading at the Academy, and still better unpublished papers which can be read in the archives, leave the impression, at least to a mathematician, that Buffon had no well defined line of attack. He certainly was horrified that one could acknowledge Newton's law had to be reorganised †. And he reacted too quickly and in too dogmatic a way as it is rarely so simple to show a contradiction in so general a law of Nature as the one provided temporarily by Clairaut:

> *Je trouve qu'il est plus court d'assurer la loi de l'attraction telle qu'elle est, et de faire voir que la loi que M. Clairaut veut substituer à celle de Newton, n'est qu'une supposition qui implique contradiction.* [14]

Clairaut soundly claims that he has just made an hypothesis, because so is the imaginative power of mathematics. But the future of any hypothesis is to be reasonably attacked ‡ . And the attack to come must bring new phenomenas, not just "logical" and elementary arguments.

> *Servons-nous dans cette épreuve des armes qui doivent toûjours être dans les mains des Mathématiciens, je veux dire le calcul & les observations* [15] .

Buffon brings no argument from observations, no computation at all, and quotes Newton in a very scholastic way. This position is what determined a change in the tone of debate, and certainly a change of style. The first answer of Clairaut to Buffon is really worked out, putting arguments after arguments in a very professional way, and taking

* Is it possible to do it on the basis of the very old reproach to mathematics to lack a physical sense of reality ? At least the split among philosophers deserves a better analysis, and for instance to investigate the differences between Diderot and d' Alembert.

† As the official translator, Buffon was perhaps more horrified by some comment from Clairaut, who accused Newton to « hide » the way along which he had thought by not publishing himself his Method of fluxions. It is interesting to notice that Buffon never reacted to the main reproach of Clairaut to Newton, the serious difficulty of understanding on which principles Newton based his explanations of the movement of the Moon.

‡ Il est vrai que la généralité de ma loi pourroit bien un jour être détruite par quelques nouveaux phénomènes, mais je les attends, & de plus je chercherai à les découvrir avec le même intérêt & la même activité, que ceux qui augmenteroient la vrai-semblance de ma supposition". Clairaut. Réponse aux Réflexions de M. de Buffon, sur la Loi de l' Attraction et sur le mouvement des Apsides. *Mém. Acad Sc Paris* (1745) [1749], p. 542.

seriously care of possible mathematical objections from Buffon. Even if he may end a scientific argument in the typical ironical style of a salon:

> *Si jamais j'apprends qu'on a vû la Lune autrement que ronde, je me rendrai a cette explication.* [15, pg 546]

Buffon gives up the scientific style in his second answer. Perhaps because Clairaut had stated that valid counter arguments should be close to his own, that is extremely specialized in celestial mechanics (with the three body problem for which he claims to be the first one to bring a coherent solution as it led to an equation of the Moon). He claimed that there can be no a priori arguments from mathematics as mathematics is there applied, and there is no "natural" argument as Nature "*n'indique point que la loi du quarré soit la seule*" [15, pg 547]. And for this Clairaut quotes as well Newton's Opticks where other laws are sought for in order to explain capillarity, the spherical form of bubbles, etc. and thus, in a mundane fashion, he poses as a better follower of Newton than Buffon appeared so easily to be by just dogmatically saving Newton's law [*].

Buffon refuses to see the reproach which Clairaut had made to Newton, which is not a reproach about the induction of a law in general, but about the kind of physical principles he used when confronted by more complicated phenomena.

> *Dans la plupart des endroits difficiles il emploie un trop petit nombre de paroles à expliquer ses principes, tandis qu'il parait se livrer avec complaisance aux détails et aux vérités de calculs sur lesquels les lecteurs ne se feraient aucun scrupule de s'en rapporter à lui, lorsqu'ils possedent ce qu'il faut pour en trouver les demonstrations* [16] .

A typically rhetorical act of Buffon is then to claim he has exhausted all recognized means of reasoning, and all contradict Clairaut.

> *Les raisons métaphysiques, mathématiques et physiques, s'accordent donc toutes a prouver que la loi de l'attraction ne peut être exprimée que par un seul terme, et jamais par plusieurs termes* [17] .

On the contrary, Clairaut was precise about his own intellectual debate about modifying Newton's law of attraction. He could see about this law as an approximation of the true law, the true function because he is not losing the value of all previous deductions from Newton's law (still valid because the second term plays there almost no influence), and adds better explained phenomena, such the movement of the Moon.

> *Un résultat aussi contraire aux principes de M. Newton, me porta d'abord à abandoner entièrement l'attraction, mais en faisant attention ensuite... il me parut aussi difficile de le rejeter que de l'admettre* [†] .

The end of the quarrel is a rather theatrical statement of Clairaut to "les Géomètres", by which he says that his new views on the problem of the equation of the Moon leaves him now the possibility to reconcile rather precisely with Newton's theory of attraction. A

---

[*] Capillarity is thus seen as an "irregularity" to Newton's law, but to be explained by Newton's method.

[†] Clairaut, Du systéme du Monde dans les principes de la gravitation universelle,·*Mém. Acad. Sc Paris* (1745) [1749], pp. 336-337. Because he wanted to defend his hero's scientific status, Roger quotes only the fIrst part of Clairaut's sentence, as if it was a definitive condemnation of Newton. See. *Buffon.un philosophe au Jardin du Roi*, Fayard, 1989 ; English translation, Comell University Press (1997), pp.55.

"secret" Mémoire had been deposited at the Academy on the 21st January 1749 and a second sent to London for the Royal Society * on the 26th.

But this proved no triumph at all for Buffon, who is now excluded from the world of Geometers, as Clairaut precisely adds.

> *On verra, lorsque je les donnerai au public, que tout ce qui a été dit sur cette matière, ne m' a pu être d'aucun secours pour le résultat que j'avance, et qu'il n'y sera pas question de raisons vagues, mais de principes sûrs et appliqués suivant les règles que prescrit la Géométrie.* †

Clairaut's objection to Buffon is not only a disqualification of Buffon ‡ : it is a question of scientific style. A law is not a theory, and a theory is not just an analytical sentence summarized by a law. There is a lot more when mathematics and physics interplay, and where innovation also plays its part.

> *Mon but actuel est uniquement d'avertir les Géometres qui s'intéressent à cette question, qu'après l'avoir considérée de nouveau sous un point de vue qui n'avait pas encore été envisagé de personne, je suis parvenu à concilier assez exactement les observations, sans supposer d'autre force attractive que celle que suit la proportion inverse du carré des distances: du moins les différences que j'ai trouvées entre mes résultats et les observations sont-elles assez légères pour pouvoir être attribuées à l' omission de quelques élémens que la théorie ne peut employer que très difficilement, et qui sont heureusement de peu d'importance* [18] .

The technique used by Buffon in the course of the quarrel was not really to defend the same line of thought, but to benefit from any move of his adversary. And thus to learn from him mathematical details which he then interprets in his own way to get a contradiction. Such is the fact that an equation supposed to explain some physical, or real phenomenon, could also have a negative solution, or even imaginary solutions. A century earlier, Descartes had explained how the degree of a problem explained such questions. Buffon knows that his way of explaining cannot be accepted compared to the style of argument of Clairaut; he himself regrets that "analogy" does not provide enough argumentation.

> *Mais comme ce raisonnement préliminaire, quelque bien fondé que je le croie, pourrait être contredit par les gens qui font peu de cas de la force des analogies, et qui ne sont accoutumés à se rendre qu'à des démonstrations mathématiques, M. Newton a cru qu'il valait beaucoup mieux établir la loi de l'attraction par les phénomènes memes, que par toute autre voie* [7, pg 493]

Buffon seems to believe Newton could have proved in an Aristotelian way but due to the domination of mathematicians was then obliged to prove by mathematical induction.

---

* For a thoroughly conducted explanation of the work of Clairaut, there is the remarkable book by J. L. Greenberg, *The Problem of the Earth's Shape from Newton to Clairaut. The Rise of Mathematical Science in Eighteenth-Century Paris and the Fall of "Normal" Science*, Cambridge University Press (1995).

† Avertissement de M. Clairaut au sujet des Mémoires qu'il a donnés en 1747 et 1748. sur le systeme du monde, dans les principes de I' Attraction, *Mém. Acad. Sc Paris* (1745) [1749], p. 577.

‡ The way Clairaut puts Buffon outside the "community of geometers" could be analyzed more carefully: but it comes rather early. Just for his first answer, Clairaut [12 pg 544] says in a rather direct way that Buffon does not know enough, and this can be seen in the following sentence, wher the worst is to come, in *cauda venenum*: "Ceux qui ont un peu examiné ces sortes de problèmes, qui connaissent les méthodes par lesquelles on trouve l'attraction d'un corps dont la figure est donnée, voient ce qu'on peut attendre d'un pareille supposition, avec autant de facilité que l'on juge de la grosseur des corps a la vue simple".

And this leads Buffon to enter on a mathematical ground for which he lacks the most recent expertise.

Does a sort of physical failure in the explanation of the use of mathematics to state a physical law automatically discard what could be a new analytic and contrived theory? Buffon thinks so, as he cannot conceive mathematics as able to provide concepts to physics. The problem was not the use of mathematics in physics, but really the use of sophisticated mathematics to state what should be simple principles. Buffon stayed along a particular line of realism, closer to just conservatism or reaction, and his idea of Descartes' use of simple and clearly distinct fictions had no historical flexibility [*].

FUNCTIONS OF AND IN THE QUARREL

Therefore, the use of arguments on functions to which I now restrict myself, and the answers by Clairaut, give a good photography of the concept around the middle of the century. When the various papers of Clairaut and Buffon appeared in 1749, Euler had already published a year before his *Introductio in Analysin Infinitorum* where he assigned function to be the object to Analysis. And where he explained how to perform the analysis of a function, for instance an exponential for which he provided the famous entire series. Buffon had worked on the *Method of Fluxions* where the main tool was also a theory of infinite series. For him, some functions like monomial have a "simple" form, just because they are not decomposable into a sum of other simple forms. His untold argument certainly relies on Descartes 'method of indeterininates', and this was so well-known that it was not necessary to state it. It says that a polynomial cannot be written in two different ways once ordered according to ascending powers of the variable. Or, that a monomial cannot be written as a sum of monomials using different powers. This can be seen in the way Buffon explains that a writing $1/x^2 \pm 1/x^4$ par $1/x^n$ was not possible, but requires the coefficient n to depend on x. He then goes on, in a rather unclear way, to say that "*une quantité qui doit être nécessairement un nombre compris entre 2 et 4, pourrait cependant devenir infinie, ce qui est absurde*" [19] .Clairaut answers in a dry way, disqualifying Buffon:

> *Pour peu qu'on soit initié dans l' Algèbre, on ne saurait penser que l'on represente* $1/x^2 \pm 1/x^4$ *par* $1/x^n$

If monomials are independent functions, they are not the only possible functions for physics. Buffon had not understood the idea that "algebraically" independent functions became a tool to work with functions, and no longer a measure of simplicity. This is what Euler explained. It thus became clear that a sum of terms in the expression of a function could not mean a synthesis or a sum of causes; it meant that every mathematical function could be reduced in the same way to a sum of terms; that it was its nature to have an infinite number of parameters which are the coefficients of the power series. The explanation from series got rid of the idea that there were "simple" functions, to be distinguished from "transcendental" ones. Or, better said, if the difference between such functions could be expressed by algebra and for algebraic purposes (or purely mathematical purposes), physics only provided "infinite series". And this claim was made because functions in physics occurred from differential equations, for which a good technique of solutions was to compute with infinite series. If this point was not expressed in the book of Euler, it had already been made clear by a paper of d'Alembert "Sur les cordes vibrantes" [†]

---

[*] In other words, Buffon thinks that mathematics makes no progress, and does not obtain new simple ideas.

[†] The vibrating chord problem has already been the focus of a long series of papers; a survey was done by C.A. Truesdell in the *Opera Omnia* of Euler for the literature before 1960. Nothing similar occurred for modern literature, and I feel too lazy to summarize now.

But Euler at this time did not try to provide for a function a definition directly linked with the general idea of causality and of dependence between two quantities; to a function he assigned a form: the infinite power series representing a function was its canonical form. Euler could not think that power series were not the unique form for a ftmction. For Clairaut, what was the concept of function was not considered as simple to understand; this is what he explains in the course of the debate with Buffon, already giving a larger understanding of a function than Euler.

> *Comme ce qu'on appelle en Algèbre fonction, c'est-à-dire quantité composée d'une autre suivant une formation quelconque, est communément difficile à saisir, on ne croit pas que ces quantités plus générales que les simples puissances, puissent servir à représenter une loi suivant laquelle doit agir la nature qui est toujours simple.* [11 pp 539-540] .

Simplicity in mathematics is not a question of point of view; but depends on progress. Later in the quarrel, he precisely names some functions, which could not have been ignored by Buffon, but his main purpose is to state that such functions are as "natural" as the now familiar powers of a variable, and there are far more "natural" functions.

> *Mais pour proposer de possibles fonctions, plus composées encore que celles que M. de Buffon rejette, sera-t-il nécessaire de lui prouver auparavant qu'il existe dans la Nature des quotients de quantités complexes, des radicaux, des logarithmes, des sommes intégrales, etc. toutes expressions dépendantes de l'algorithme des Géomètres* [*] *, et qui n'ont aucun rapport avec l'existence des quantités physiques qu'elles peuvent servir à exprimer.* [20]

Is it possible to make sense of the apparent contradiction in this sentence, where nature (physics) appears to be different from physics! I have refused the easy way of discarding the debate because of its mundane aspects. And I think that just a repetition of the word "nécessaire" after the word "rapport" could provide an answer. The second part of the statement refers to the discussion on the power of algebra: there is no necessity for an analytic formula to be useful to represent reality, outside the power of computation it gives. Algebra is a way of writing in a condensed form a numerical table, and this will become for instance the case in naval architecture, where instead of "exact" integral formulae, one has preferred "empirical" formula, mainly expressed in terms of powers. The first part of the sentence is to claim that the new tool, differential and integral calculus, is not an artificial tool, but lies in Nature itself and therefore brings no artifact.

Empirical law are what Buffon refuses, as he perfectly understands, precisely with the help of power series, that any function may be reached, or at least approximated, by adding one term after another.

> *Au lieu de* $1/xx$ *mettre* $1/xx + 1/mx^4$ *me paraît n'être autre chose que d'ajuster une expression de telle façon qu'elle corresponde à tous les cas; ce n'est plus une loi physique que cette expression représente, car en se permettant une fois de mettre un second, un troisième, un quatrième terme, etc. on pourrait trouver une expression qui dans toutes les lois d'attraction représenterait le cas dont il s'agit en l'ajustant en même temps que mouvements de l'apogée de la Lune et aux autres phénomènes.* [7 pg. 497]

---

[*] If such an expression -algorithme des géomètres -may generally denote integral and differential calculus as a whole, I do not think it might mean here just the expansion of a function into power series (what we call Taylor's formula).

There cannot be a function to denote attraction in order to explain the equation of the Moon, and another one for another phenomenon, so he argues. Clairaut agrees that there should be only one function if attraction is really the only cause, but to prove which function, one has to determine by successive approximation from all the known phenomena linked with attraction. One phenomenon only, for example the roughly elliptic form of the orbit of a planet, cannot provide the exact form of the function. Buffon thought that Newton ended mathematics!

For the use of functions in physics, Buffon, Clairaut and Euler deeply differed. For Euler, a function was a mathematical concept, which could certainly be used to do physics, but which could not be used a priori to do physics: there remained the need to show the validity of this use of a function in physics. As functions had a specific form. Or, which is the same, the use of a function to determine a physical object, and for instance a law of nature, was a reduction, and a decision of the mind. Such a position will bring serious difficulties while using trigonometric series to explain vibrations, and Euler will change his mind on what a function is [21 and references therein]. For Buffon, a function was already a physical object, holding all the power of reality, and not at the disposal of the mathematician like a number, which was seen as an artificial object, and certainly not an object of nature. Clairaut had already the view that functions coming form physics could be general, but simple in the sense that this was just the meaning of a function. Simplicity is a quality of nature, thus function representing a natural phenomenon has to be simple. On the contrary, Clairaut argues that analytical simplicity is not the simplicity of Nature.

> *Serait-il possible de quitter la voie des phénomènes, pour connaître les forces qui agissent dans la Nature, et de les vouloir déterminer par le plus ou le moins de simplicité d'une expression analytique?* [20 pg 585]

He refutes Buffon's nominalism, and his lack of physical realism. Had not Buffon already divided the attraction law into two phenomena, as if the parallelogram of composition of forces, a very useful tool in physics, and a rather realistic tool as well, was to explain different causes.

> *Le mouvement des Planètes dans leurs orbites est un mouvement composé de deux forces: la première est une force de projection dont l'effet s'exercerait dans la tangente de l'orbite, si l'effet continu de la seconde cessait un instant; cette seconde force tend vers le Soleil, et par son effet précipiterait les Planètes vers le Soleil.* [7 pg 493]

In the same volume of the Academy, d'Alembert speaks in a very different way: decomposition of forces is a tool of the analytical mind, but in the physical world there just exists one phenomenon.

> *Toutes les planètes (en y comprenant le Soleil) agissent les unes sur les autres en s'attirant mutuellement, et par conséquent elles décrivent chacune son orbite particulière dans I'espace absolu, de manière que le Soleil n'est pas réellement en repos.* [22]

In his book on Buffon, J. Roger claims that his hero gave up mathematics because he was too much on the realistic side. We may differently say that Buffon maintained the idea of a physical realism, which was outdated. Within thirty years, the realistic point of view in gravitation theory will have seriously changed. There will be the introduction of a function, the potential function invented by Laplace. It will replace the idea of a force, particularly when dealing with the task of explaining discrepancies between observations and computations deduced from Newton's law. Those computations will be completely rethought, and this will lead to perturbation theory. If we restrict to the use of functions in this transformation, potential function will be the mathematical and successful way to transfer Newton's law from discrete masses to all continuous matter. The use of a function like potential could then be described as a transformation of physics by integral calculus in several variables. But one has to explain something further. If in a way something remained realistic with the potential function, even mediated by integral calculus, one could not in general compute it directly and precisely. Just because, in the case of the Earth, there was an ignorance of the repartition of masses, and a lack of knowledge of the form of the Earth, which prevented any real computation of the potential. In fact, to find the form of the Earth became the new question which was raised, in order to see the influence of this form on the computed orbits of planets [*] . Therefore, the potential function V had to be determined from something else; it proved to be a partial differential equation, at least valid outside the attracting body: $\Delta V = 0$. It is this equation which replaced Newton's law and proved to be the new simple idea. This is what Laplace explained in a remarkable paper *Théorie des attractions des sphéroides et de la figure des planètes* [23] I a paper written in 1782, just after Laplace had understood the meaning of a technique of orthogonal polynomials due to Legendre. With this use of functions, celestial mechanics was transferred from the realm of physics to the world of mathematics. And even this world only was then able to say which analogies could be made between phenomenas like hydrodynamics or propagation of the sound: in all such domains, the same partial differential equations worked. Fourier's discovery was to show that heat also followed the same equation. And Fourier pretended that there was no difference between the way Nature works and Analysis: these cannot be a better affirmation for mathematical realism.

CONCLUSION

In this presentation, I avoided mathematical technicalities, and therefore was not able to really do what I had in mind, which was a history of functions explaining the birth of potential theory, a major theory in mathematical physics of the eighteenth century. But, using mathematical allusions, and then a dialogue with those who know about potential theory, I argued that an academic confrontation between two scientists like Clairaut and Buffon, futile as it may appear because apparently nothing serious came out of it, was an excellent observatory. In order to show that science, at least science using mathematics, whether "normal science" or "revolutionary science", is not achieved through solutions of logical contradictions, and is no more achieved through *ad hoc* modifications. There is a need for a transformation of the questions raised, and in the present case it showed there was another question behind the modification of Newton's law in order to have a coherent perturbation theory. There was a need to modify the mathematical tools used: a function was to be understood as a solution of a differential equation, or as an "integral" as was then said, and such an equation as the basic concept of physics. This was to become the new sense of reality for mathematics and thus for physics as well.

REFERENCES

1. J.-C. Pont, *L'aventure des Parallèles. Histoire de la Géométrie Non Euclidienne, Précurseurs et Attardés*, Peter Lang, Berne { 1986).
2. J.-D. Voelke, *La Seconde Naissance de la Géométrie Non Euclidienne*, thèse, Université de Genève {1998).

[*] Already, Clairaut tried to explain to Buffon, at least in his first answer, that the form of the Earth is important in the problem, but still unknown.

3. G. H. Hardy, *A Mathematician's Apology*, Cambridge University Press (1940); trad. fr., Paris, Belin {1985).
4. J. et N. Dhombres, *Naissance d'un Pouvoir: Science et Savnyes en France, 1793-1824*, Payot, Paris (1989), chapter IV.
5. J. Dhombres, Un platonisme spontané: le témoignage contradictoire des textes mathématiques, in M. Panza. J.M. Salanskis. *L'objectivité Mathématique. Platonismes et Structures Formelles*, Masson Paris (1995), pp. 5-22.
6. J. Dhombres, Une histoire de l'objectivité scientifique et le concept de postérité, in Des Sciences et des techniques: un débat, *Cahier des Annales*, t. 45 (1998), pp. 127-148.
7. Buffon, Réflexions sur la loi de l'attraction, *Mém. Acad. Sc Paris* (1745) [1749J, p. 498.
8. Clairaut, Réponse au mémoire de M. de Buffon, *Mem. Acad Sc Paris* (1745) [1749], p. 584.
9. J. Dhombres, Mathématiques superficielles ou étendue de la mathématique ; la théorie de la capillarité selon Laplace, *Revue d'Histoire des Sciences* (1989), ILII, pp. 43- 77.
10. J. Dhombres, J. B. Robert, *Joseph Fourier, Créateur de Physique Mathématique*, Berlin (1998). Paris, chap. 8 et 9.
11. Clairaut, Réponse aux réflexions de M. de Buffon, sur la loi de l' attraction et sur le mouvement des apsides, *Mém. Acad. Sc Paris* (1745) [1749], p. 541. Some editions have "Réfléxions".
12. Clairaut, ref 11, p. 539.
13. Buffon, Addition au mémoire qui a pour titre réflexions sur la loi de l' attraction, *Mém. Acad. Sc Paris*, (1745) [1749], pp. 551-552. Certain editions have the writing "dictance".
14. Buffon, Réflexion sur la loi de l'attraction, *Mém. Acad. Sc Paris* (1745) [1749], p. 496.
15. Clairaut. ref 11, p. 543.
16. Clairaut, Du Systeme du monde dans les principes de la gravitation universelle, lu le 15 novembre 1747, *Mém. Acad Sc Paris* (1745), [1749], p. 329.
17. Buffon, Seconde addition au Mémoire qui a pour titre Réflexions sur la loi de l' attraction, *Memo Acad. Sc. Paris* (1745) [1749], p. 583.
18. Clairaut, ref 12 p. 578. In certain editions, it is written: "de peut d'importance".
19. Buffon, Addition au ref 14, pp.552
20. A. Clairaut, Réponse au nouveau mémoire de M. de Buffon, *Mém. Acad. Sc Paris*, 1745, [1749], p. 583.
21. J. Dhombres, Les présupposés d'Euler dans l'emploi de la méthode fonctionnelle, *Revue d'Histoire des Sciences*, 1987, XL/2, pp. 179-202, and references therein.
22. d' Alembert, Methode Generale Pour déterminer les orbites & les mouvemens de toutes les planètes, en ayant égard à leur action mutuelle, *Mém. A cad. Sc. Paris* (1745) [1749], p. 365.
23. Laplace, *Mém. Acad Sc., Paris* {1782) [1785], pp. 113-196; *Oeuvres de Laplace*, Gauthier-Villars, Paris. 10, pp.339-419.

GEOMETRY, THE CALCULUS AND THE USE OF LIMITS IN NEWTON'S PRINCIPIA

Niccolò Guicciardini

Dipartimento di Filosofia
Università di Bologna
via Zamboni 38
40126 Bologna
Italia

THE DEBATE ON THE PRINCIPIA'S MATHEMATICAL METHODS

Since its publication in 1687 the *Principia* originated much debate. In particular, during the first decades of the eighteenth century the mathematical methods employed by Newton were criticised or defended by the small number of experts who could read the magnum opus with sufficient competence.[1] Under its classic façade the *Principia* hides a panoply of mathematical methods: series, infinitesimals, quadratures, geometric limit procedures, classical theories of conic sections and higher curves, interpolation techniques, and much more. How should the science of motion be mathematized? During Newton's lifetime this question was still unanswered. It is only in the 1730s, mainly thanks to the work of Euler, that the mathematical community became convinced, at first on the Continent, that the calculus, most notably differential equations, was *the* appropriate language for 'dynamics'.[2] Nowadays, a student of 'Newtonian mechanics' will find the language used in the post-Eulerian era somewhat familiar. On the contrary, the language of the *Principia*, burdened by geometrical diagrams, the theory of proportions, almost devoid of symbolical expressions, leaves our student, even a tenacious one, perplexed.

Recently the *Principia* has been the object of renewed interest among mathematicians and physicists. A reappraisal of the fertility of geometrical methods in dynamics, the plethora of celebrations for the tercentenary of the publication of Newton's masterpiece, the belief that technical history of science can be used for didactic purposes, are some of the factors which have induced some working scientists to open and comment the *Principia*. The results have been often profound. Scientists of the calibre of Chandrasekhar, Arnol'd and Feynman have scrutinised some of Newton's demonstrations in

The Application of Mathematics to the Sciences of Nature, edited by Cerrai *et al.*
Kluwer Academic/Plenum Publishers, New York, 2002.

depth. Some American mathematicians and physicists, such as Weinstock, Cushing, Pourciau, Nauenberg, Brackenridge, Erlichson, have turned to technical history, revealing details of several propositions, lemmas and corollaries of the *magnum opus*. This is indeed a good moment for Newtonian scholarship, since working scientists have been able to improve our understanding of Newton's mathematical methods considerably. General historians of science should be grateful for these contributions and avoid to reject them on the basis of academic disciplinary preconceptions.

Did Newton reach his solutions by cogent arguments? Are there in the *Principia* existence and uniqueness demonstrations of such solutions? Did Newton possess a method which he applied in order to reach these solutions? What is the relation of this method with the calculus? These questions have been debated by the mathematicians and the physicists who recently commented the *Principia*. The purpose of this paper is to show that attention to the context can help answering some of the above technical questions.

TEXT: THE DIRECT PROBLEM OF CENTRAL FORCES, PROPOSITIONS 6–13, BOOK 1

By *central force* Newton means a force whose magnitude varies with distance from a given point (the *centre of force*) and which is directed towards or away from the centre of force. In sections 2 and 3, Book 1, Newton considers a 'body' (we would say a 'point mass') accelerated by such a force. By *direct problem of central forces* late-seventeenth-century natural philosophers meant the problem of determining the law of force, i.e. the variation of force magnitude as a function of distance from its centre, when the trajectory (a plane curve) and the centre of force (a point lying on the plane of the trajectory) are given.

In Propositions 6, Book 1, Newton provides a geometrical representation for central forces. (In the second edition (1713) of the *Principia* Proposition 6 was considerably expanded and modified.) Proposition 6 reads as follows (see fig. 1):

> If a body P, revolving about the centre S, describes the curved line APQ, while the straight line ZPR touches the curve at any point P; and QR, parallel to distance SP, is drawn to the tangent from any other point Q of the curve, and QT is drawn perpendicular to the distance SP; then the centripetal force will be inversely as the solid $SPquad \times QTquad/QR$, provided that the magnitude of that solid is always taken as that which it has ultimately when the points P and Q come together.
>
> For in the indefinitely small figure $QRPT$ the nascent line-element QR, if the time is given, is as the centripetal force (by Law 2) and, if the force is given, is as the square of the time (by Lemma 10), and thus, if neither is given, is as the centripetal force and the square of the time jointly, and thus the centripetal force is as the line-element QR directly and the square of the time inversely. But the time is as the area SPQ, or its double $SP \times QT$, that is, as SP and QT jointly, and thus the centripetal force is as QR directly and $SPquad$ into $QTquad$ inversely, that is, as $SPquad \times QTquad/QR$ inversely. Q.E.D.[3[454]]

The idea underlying Newton's demonstration of Proposition 6 is the following. The central force acting on the orbiting 'body' changes in both magnitude and direction. However, the smaller the arc PQ, the smaller are these changes. Newton makes it clear that one has to consider a limiting situation: the 'figure $QRPT$' is to be taken

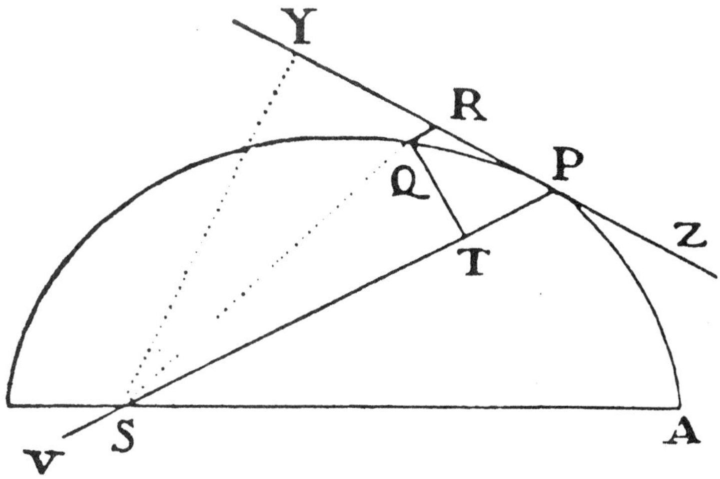

Figure 1. *Principia*, Prop. 6, Book 1.[3[454]]

as 'indefinitely small'. If these limiting conditions are accepted, one is justified in considering the force as constant both in magnitude and in direction: the body traverses the arc PQ under the action of a constant force directed parallel to PS. At P the body has a velocity directed along the tangent ZPR. By inertial motion (if the force did not act) the body would reach R in the same time it reaches Q. The deviation from inertial motion QR, this 'nascent line-element', is from first principles (the second law of motion and Lemma 10, Section 1, Book 1) proportional to the force times the square of time. In fact point Q is reached by composition of inertial motion along PR and uniformly accelerated motion along RQ. The deviation from inertial motion for uniformly accelerated motion is proportional to the square of time (Lemma 10), it is proportional to the force (Law 2) and is in the direction PS (Law 2). One can thus write:

$$QR \propto F \cdot \Delta t^2 \tag{1}$$

Newton makes it clear that this formula, valid for constant forces, can be extended to variable central forces, provided that the limit, when 'P and Q come together', is taken. Note that Newton, as always, considers a 'body' of given mass and, writing proportions, does not make explicit the constant mass term. Force is thus equated to acceleration.

In order to complete the demonstration of Proposition 6 one has just to consider that, since the force is central, Kepler's area law holds. Thus the area SPQ (a triangular area since the limit of the ratio between the vanishing chord PQ and arc PQ is 1) is proportional to time. In the limiting situation considered, the area SPQ is equal to $\frac{1}{2}(SP \cdot QT)$. And at last:

$$F \propto QR/(SP \cdot QT)^2. \tag{2}$$

Formula (2) is employed by Newton in the solution of several instances of the direct problem of central forces. These problems have the following structure: given a plane curve O and a point S in the plane of that curve, determine the central force directed towards, or away from, the force centre S which accelerates a 'body' making it describe the orbit O.

Newton's standard technique for solving the direct problem consists in determining the limit of the ratio of QR/QT^2 as Q tends to P, and in applying formula (2). (Notice that in formula (2) SP remains constant.)

In Propositions 11–13, Book 1, Newton shows that when the orbit is a conic section and S is placed at one focus,

$$QR/QT^2 \propto 1/L, \tag{3}$$

where L is a constant (the *latus rectum*), and the ratio QR/QT^2 is evaluated, as always, as the first or last ratio 'with the points Q and P coming together'. Therefore the force varies inversely with the square of distance.

DEBATE: THE ANALYTICAL VERSUS THE GEOMETRICAL INTERPRETATIONS OF NEWTON'S LIMITS

The mathematics of Newton's solution of the direct problem of central forces has often puzzled modern commentators. Newton's limit procedures are familiar to calculus-trained scientists. On the other hand, Newton's approximations are not regulated by the algebra of $\epsilon - \delta$ inequalities (e.g., when the trajectory is locally approximated by a Galilean parabola). What is the relationship between Newton's limits and Cauchy's limits? Or, in a more historically contextual way: what is their relation with Leibniz's differential calculus, or with Newton's method of fluxions?

These questions have often been raised in the recent technical studies devoted to Newton's *Principia*. The divergency of answers can be best explained with an example. In the 1996 issue of *The Mathematical Intelligencer* the mathematician S. K. Stein, a recipient of the Lester Ford Award for expository writing, published a lucid paper with the title 'Exactly how did Newton deal with his planets?'.[4] This paper is devoted to the above outlined Newtonian treatment of the direct problem of central forces. Stein opens his paper as follows:

> In 1987, the tercentenary of Newton's *Principia*, I decided to mark the occasion by reading his treatment of planetary orbits. Just how did he approach the subject if he didn't use the calculus? It took me a while to adjust to his way of thinking, but eventually his approach, a mix of geometry and limits, became quite clear.[4[6]]

After a delightfully clear presentation of the *Principia*'s Propositions 6-13, Stein reaches the following conclusion:

> There is absolutely no evidence that Newton first used his calculus, developed some 20 years before he wrote the *Principia*, to treat these problems and then recast his work in geometric terms.[4[11]]

But here the editor of *The Mathematical Intelligencer* adds a note which nullifies all the efforts exerted by Stein to convince the reader of the geometrical peculiarity of the *Principia*. The editor's note – the last words of the paper – reads as follows:

> The geometric terms look like calculus to me.[4[11n]]

The exchange of views between Stein and the editor typifies the divide between two tendencies in which one can group the technical commentators of the *Principia*: let us call them the analytical and the geometrical interpretations.

The great champion of the analytical interpretation has been Chandrasekhar with his monumental commentary *Newton's Principia for the Common Reader.*[5] As is well-known, In Section 1, Book 1, Newton presents his geometrical method of limits – the so-called method of first and last ratios – on which Propositions 6–13 depend. According to Chandrasekhar Newton here 'presents the elementary notions of differential calculus (without explicitly so stating)'.[5[43]]

On the other hand Arnol'd in his stimulating booklet on Huygens, Barrow, Newton and Hooke has underlined the geometric character of Newton's demonstrations.[6] According to Arnol'd, Newton's geometric approach has several advantages which have been underestimated. He writes:

> It must be said that Newton made all the discoveries contained in the *Principia* without using analysis, although he had command of it at this time. He proved everything that was required by means of direct elementary geometrical arguments more or less equivalent to analysis (and not by translating analytical calculations into geometrical language).[6[26]]

Arnol'd's provocative essay is aimed at showing that, in the golden age of mathematics, geometrical thinking could yield results which were to be painfully achieved via analysis only centuries later. Results 'obtained by modern mathematicians not without difficulty and even with the help of computers, were already known at that time'[6[9]], namely when geometers of the calibre of Huygens, Hooke, Barrow and Newton thought in terms of diagrams, physical intuition and graphical techniques. So, according to Arnol'd, Barrow knew already the essential theorems of the calculus, Hooke approached inverse-square orbits via graphical techniques, Newton had a topological proof of a theorem on the transcendence of Abelian integrals, Huygens 'by means of elementary geometrical constructions' [6[53]] anticipated results achieved much later in 'symplectic geometry, the calculus of variations, optimal control, catastrophe theory'[6[8]]. Newton's treatment of attraction inside a spherical shell is taken as an example of

> how it is possible to solve problems from potential theory without analysis, without knowing the theory of harmonic functions, or the fundamental solution of Laplace equation, or the single and double layer potentials. Similar arguments, preceding the rise of analysis, often occur in papers of those times and turn out to be very powerful. Here is an example of a problem that people like Barrow, Newton and Huygens would have solved in a few minutes and which present-day mathematicians are not, in my opinion, capable of solving quickly: to calculate
>
> $$\lim_{x\to 0} \frac{\sin\tan x - \tan\sin x}{\arcsin\arctan x - \arctan\arcsin x}$$ [6[27–8]]

When seen from a purely internalistic point of view Newton's geometric limit procedures are open to quite contrasting interpretations. They can be seen as equivalent to calculus, or they can be seen as peculiarly geometric. In the first case the numerous propositions of the *Principia* which depend upon them are seen as based on calculus techniques. In the second case these propositions are appreciated as based on demonstrative techniques which are alternative to the familiar analytical techniques of modern mathematics.

CONTEXT: FROM THE ANALYTICAL TO THE SYNTHETIC METHODS OF FLUXIONS

Against Descartes

When he was 'in the prime of his youth', Newton – starting from Descartes's *Géométrie*, Wallis's *Arithmetica infinitorm* and, most probably, Barrow's direct instructions – began expanding the realm of what he termed the 'new analysis'. The results of his *anni mirabiles* was the 'method of series and fluxions' described at length in two essays which were to remain unpublished for several decades: the *De analysi per aequationes numero terminorum infinitas* (composed in 1669 but published in 1711) and the *De methodis serierum et fluxionum* (finished in 1671 but published in 1736). In later years Newton referred to this outstanding algorithm for tangents and quadratures as the 'analytical method of fluxions', contrasting it to the 'synthetic method of fluxions' which we will consider just below. The 'analytical method of fluxions' is what nowadays we call, somewhat improperly, the 'calculus'. As a matter of fact, the method, revealed to the happy few who had access to the manuscript the *De analysi* and *De methodis*, was a mix of algorithmic techniques and geometrical argumentation: it was quite different from the abstract calculus we are used to study nowadays. Its objects were curves, rather than functions, its algorithmic rules were based on series and cancellations of infinitesimals ('moments') rather than epsilontic limits. Typically Newton began by considering a problem related to the properties of a curve. He proceeded by geometrical argument, often – but not always – referring to classical sources. He then, as Descartes had taught, translated some geometric relations into an equation involving variables and constants. Here the algorithm began: infinitesimals and series expansion were employed. Newton understood that tangent- and area- problems were one the inverse of the other and, accordingly, built the first 'integral tables' ('catalogues of fluents', as he named them) of the history of mathematics. This method was 'analytical' both because it was an extension of the 'new analysis of the moderns' and because it was a problem solving technique rather than an axiomatic theory.

Just after completing the *De methodis*, Newton composed an *addendum* where he developed a 'more natural approach' to the fluxional method – an approach 'based on the genesis of surfaces by their motion of flow' and 'which will come to be still more perspicuous and resplendent if certain foundations are, as is customary with the synthetic method, first laid'.[7[3:282–3,328–31]] Whereas in his early writings Newton represented the fluents with algebraical symbols, in this new approach he referred directly to geometrical figures. These figures, however, are not static, as in classic geometry: they must be conceived to be 'in motion'.

In a successive work entitled *Geometria curvilinea*, dated tentatively by Whiteside around 1680, Newton introduced the idea of 'first and ultimate ratios'. Also the *Geometria curvilinea* is concerned with geometric flowing quantities. But, whereas in the *addendum* infinitesimal quantities generated by motion are referred to as 'moments', here Newton talks about 'nascent or vanishing parts'. In Axiom 6 he states that:

> Fluxions of quantities are in the first ratio of their nascent parts or, what is exactly the same, in the last ratio of those parts as they vanish by defluxion.[7[4:427]]

In the Preface of the *Geometria curvilinea* Newton distances himself from the use of infinitesimals. The demonstrations in the *Geometria curvilinea* are not based on moments, they rather depend on the determination of limits of ratios of 'vanishing parts'.

The geometrical limits of the *Principia* which we have seen at work in the treatment of the direct problem of central forces have their origins in Newton's reworking, carried on in the 1670s, of the analytical method of fluxions into a synthetic method. This new method is synthetic both because it is entirely geometrical and because it has an axiomatic structure.

Newton's motivations for reworking his analytical method of fluxions into a synthetic form are quite complex. Here it will suffice to say that during the 1670s his admiration for ancient geometry grew considerably. This new approach to classic mathematical sources was part of a general interest that Newton developed towards the ancients: in theology, chronology and alchemy. He also became, mainly for theological worries, an ardent anti-Cartesian. He wished to distinguish himself from the 'modern' mechanistic philosophers. They thought to be superior to the ancients, but Newton was convinced that the ancients had possessed the truth about God and Nature. Furthermore, the "Ancients" had also the best mathematics. Most of their knowledge was lost, but the time was ripe for true interpreters who could restore their elegant, concise and powerful methods of discovery and proof. As Newton wrote:

> Indeed their [the Ancients']method is more elegant by far than the Cartesian one. For he [Descartes] achieved the results by an algebraic calculus which, when transposed into words (following the practice of the Ancients in their writings), would prove to be so tedious and entangled as to provoke nausea, nor might it be understood. But they accomplished it by certain simple propositions, judging that nothing written in a different style was worthy to be read, and in consequence concealing the analysis by which they found their constructions.[7[4:277]]

Newton rejected symbolic style not only because of his classicism, but also because of his desire to relate mathematical reasonings to existing entities. A theme which was to become a *leitmotif* during the priority controversy with Leibniz was the ontological content of the method of fluxions. According to Newton, fluents and fluxions are really exhibited in *rerum natura*. In the 1710s, when opposing Leibniz, Newton contrasted the safe referential content of his method of fluxions with the lack of meaning of the differential calculus. In Leibniz's calculus, according to Newton, 'indivisibles' occur, but the use of such quantities is not only a departure from ancient tradition, it also leads to the use of symbols devoid of referential content. In the anonymous account to the *Commercium epistolicum* Newton wrote:

> We have no ideas of infinitely little quantities & therefore Mr Newton introduced fluxions into his method that it might proceed by finite quantities as much as possible. It is more natural & geometrical because founded on *primae quantitatum nascentium rationes* wch have a being in Geometry, whilst *indivisibles* upon which the Differential method is founded have no being either in Geometry or in nature. [...] Nature generates quantities by continual flux or increase, & the ancient Geometers admitted such a generation of areas & solids [...]. But the summing up of indivisibles to compose an area or solid was never yet admitted into Geometry.[7[8:597–8]]

Nature and geometry are the two key concepts: they allow Newton to defend his synthetic method of fluxions because of its continuity with ancient tradition as well as its ontological content.

Newton's admiration for the ancients's *prisca sapientia* and his anti-Cartesianism were not an exception in Restoration England. His suspicion towards the abstraction of algebra and infinitesimal techniques were shared – for different reasons and with different nuances – by many of his English contemporaries: Hobbes and Barrow being two examples of a geometrical reaction against such promoters of the new analysis as Oughtred and Wallis.[8] Also Newton's option for geometric limiting techniques can be found in many seventeenth century mathematicians. Malet has edited a manuscript by James Gregorie in which a theory of limits is proposed that resembles very closely Newton's theory of first and ultimate ratios.[9] The idea that the infinitesimal techniques popular after Cavalieri and Torricelli could be translated in rigourous form as limit arguments, somewhat reminiscent of Archimedean exhaustion, was stated time and again. Even Leibniz, as it appears in a manuscript composed in 1676 and edited recently by Knobloch, was interested in retranslating calculus quadratures in a form based on geometric limiting techniques.[10]

De Gandt defines Newton's synthetic method of limits as a geometry which is neither classic geometry nor calculus – a mathematical style which he aptly terms a 'geometry of the ultimate'.[11] Newton's geometrical limit procedures are not classical: they could be conceived only by a man well acquainted with the analytical method of fluxions. A translation of these limit procedures into the analytical method is of course possible. For example, formula (2) has been given analytical fluxional form by Routh and Brougham.[12] But this translation often spoils the evidence and informative content of Newton's geometric proofs. It happens very frequently that a solution by geometry becomes very complicated when translated into algebraic terms. And *vice versa*, what is solved in a few lines by algebra, becomes impracticable when translated into geometric terms. The issue of the translatability of Newton's geometric limit procedures into analytical form, and of the advantages and disadvantages of such a translation was debated by Newton's contemporaries.[1]

This brief historical excursus can help us in understanding the limitations of both the analytical and the synthetic interpretations adopted by modern interpreters. The historian of mathematics De Gandt has developed a convincing interpretation of Newton's geometric limits and we refer the reader to De Gandt for further information.[11]

Newton's are not Cauchy's Limits

Newton pointed out that the method of first and ultimate ratios rests on the following Lemma 1:

> Quantities, and also ratios of quantities, which in any finite time constantly tend to equality, and which before the end of that time approach so close to one another that their difference is less than any given quantity, become ultimately equal.[3[433]]

Newton's *ad absurdum* proof runs as follows:

> If you deny this, let them become ultimately unequal, and let their ultimate difference be D. Then they cannot approach so close to equality that their difference is less than the given difference D, contrary to the hypothesis.[3[433]]

This principle might be regarded as an anticipation of Cauchy's theory of limits, but this would certainly be a mistake, since Newton's theory of limits is referred to a geometrical rather than a numerical model. The objects to which Newton applies his 'synthetic method of fluxions' or 'method of first and ultimate ratios' are geometrical quantities generated by continuous flow. Cauchy referred limits to a numerical model, had an algebra of inequalities and developed convergence tests. First and ultimate ratios are referred to a geometric model, they are determined via geometric reasonings and their convergence is guaranteed by our intuition of continuous motion.

Relation with Seventeenth-Century Calculus

There is an ambiguity in relating Newton's geometric limits to 'calculus'. What do we mean with calculus? After having rejected Cauchy's calculus as a possible 'equivalent' to first and ultimate ratios, we have also to reject Leibniz's calculus. Leibniz conceived his calculus in a way which differs from Newton. Most notably, while he was conscious of the possibility of translating the calculus in limit terms, he did not pursue this programme but rather encouraged the use of infinitesimals. Newton's first and ultimate ratios should be rather related to Newton's method of fluxions. In fact they were conceived by Newton as a synthetic version of his analytical method of fluxions. They are the result of attempts carried on in the 1670s aimed at expressing the results of the *De analysi* and the *De methodis* in geometric terms. Terms which were considered by Newton better ontologically founded and reminiscent of the Greek geometric tradition.

Translations

The issues raised in this paper teach us that translation between mathematical languages can shift the meaning of mathematical terms considerably. There is permanence of meaning as well as modification. We can see the structure of the $\epsilon - \delta$ method in Lemma 1. The permanence of this structure, from Newton to Weierstrass, is a fact in the history of calculus. As Lakatos has observed, mathematical formal theories are always understood by mathematicians as translations of some pre-formal theory.[13[2:36]] Lemma 1 can be seen as a pre-formal version of Weierstrass's definition of limit. However, in this translation we lose aspects which were crucial for Newton: reference to geometric objects 'which have an existence in rerum natura' and to the methods of the ancients.

A translation which would be more historically founded is from first and ultimate ratios to the analytical method of fluxions. This can be done, again going against Newton's choices and values. In trying to carry on this translation one has to pay attention in avoiding to project into the analytical method of fluxions an abstraction and a degree of symbolisation which were achieved only in eighteenth-century calculus, say in Euler's calculus. Geometry played a leading role in the *De analysi* and the *De methodis.*

But again one has to be cautious. An analysis of the structure of the demonstrations in the *Principia* reveals that Newton's geometrical limits are, in many significant cases at least, to a certain extent independent from the analytical method of fluxions. This independence should be neither overstressed nor overlooked. Defining Newton's use of limits in dynamics is thus a complex task, not only because of the plurality of Newton's geometrical techniques, but also because of their contiguity with the calculus. In some cases, the geometrical demonstrations of the *Principia* lend themselves to an almost immediate translation into the analytical method of fluxions, in other cases this translation is complicated, unnatural, or even problematic. Needless to say, notwithstanding Newton's rhetorical declaration of continuity between his methods and the methods of the Ancients, his geometrical dynamics is a wholly seventeenth-century affair.

REFERENCES

1. Niccolò Guicciardini,the Principia: The Debate on Newton's Mathematical Methods for Natural Philosophy from 1687 to 1736, Cambridge University Press, Cambridge (1999).
2. Michel Blay, *La Naissance de la Mécanique Analytique: la Science du Mouvement au Tournant des XVIIe et XVIIIe Siècles*, Presses Universitaires de France, Paris (1992).
3. Isaac Newton, *The Principia: Mathematical Principles of Natural Philosophy*, A New Translation by I. Bernard Cohen and Anne Whitman, Assisted by Julia Budenz, preceded by a Guide to Newton's Principia by I. Bernard Cohen, University of California Press, Berkeley, Los Angeles, London (1999).
4. S. K. Stein, Exactly how did Newton deal with his planets?, *The Mathematical Intelligencer* 18:2 (1996).
5. Subrahamyan Chandrasekhar, *Newton's Principia for the Common Reader*, Clarendon Press, Oxford (1995).
6. Vladimir I. Arnol'd, *Huygens and Barrow, Newton and Hooke*, Birkhäuser, Basel, Boston, Berlin (1990).
7. Isaac Newton, *The Mathematical Papers of Isaac Newton*, 8 vols., D. T. Whiteside *et al.* (eds.), Cambridge University Press, Cambridge (1967–1981).
8. Helena M.Pycior, *Symbols, Impossible Numbers, and Geometric Entanglements: British Algebra through the Commentaries on Newton's Universal Arithmetick*, Cambridge University Press, Cambridge (1997).
9. Antoni Malet, *From Indivisibles to Infinitesimals: Studies on Seventeenth-Century Mathematizations of Infinitely Small Quantities*, Universitat Autònoma de Barcelona Servei de Publicacions, Bellaterra (1996).
10. Gottfried W. Leibniz, *De Quadratura Arithmetica Circuli Ellipseos et Hyperbolae Cujus Corollarium est Trigonometria sine Tabulis*, E. Knobloch (ed.), Vandenhoeck & Ruprecht, Göttingen (1993).
11. François De Gandt, *Force and Geometry in Newton's Principia*, Princeton University Press, Princeton (1995).
12. Henry Brougham & Edward J.Routh, *Analytical View of Sir Isaac Newton's Principia*, Longman *et al.*, London (1855).
13. Imre Lakatos, *Philosophical Papers*, 2 vols., Cambridge University Press, Cambridge (1978).

THE TWO FACES OF MATHEMATICAL MODELING: OBJECTIVISM VS SUBJECTIVISM, SIMPLICITY VS COMPLEXITY

Giorgio Israel

Dipartimento di Matematica
Università di Roma "La Sapienza"
Piazzale Aldo Moro, 2
00185 Roma, Italia
israel@mat.uniroma1.it

The use of mathematical models to describe real phenomena is a recent scientific practice, and had never been used prior to the present century. This may be illustrated in many ways, although perhaps the simplest is to compare two points of view separated by nearly three centuries.

Forty years ago John von Neumann referred to mathematical modelling as the very essence of doing science and clearly defined its characteristics:

« ... the sciences do not try to explain, they hardly even try to interpret, they mainly make models. By a model is meant a mathematical construct which, with the addition of certain verbal interpretations, describes observed phenomena. The justification of such a mathematical construct is solely and precisely that it is expected to work — that is, correctly to describe phenomena from a reasonably wide area. Furthermore, it must satisfy certain esthetic criteria — that is, in relation to how much it describes, it must be rather simple. I think it is worth while insisting on these vague terms [...] One cannot tell exactly how "simple" simple is. [...] Simplicity is largely a matter of historical background, of previous conditioning, of antecedents, of customary procedures, and it is very much a function of what is explained by it.»[1,2]

Science —von Neumann thus emphasized — does not set out *to explain*, to discover the inner essence of phenomena. It does not seek *the truth*, it does not aim to be a *mirror* for phenomena. It merely supplies the mathematical images — the models — that are evaluated exclusively according to criteria of effectiveness, that is, according to how well they "work", and thus allow certain effects to be predicted or at least to get some idea, albeit partial, of the facts. It does not matter how *true* they are, as long as they are *useful*.

Let us now remind that Isaac Newton, in a famous passage from "query" twenty-eight of his *Opticks*, asserts that the main task of natural philosophy is to argue on the basis of the phenomena without framing hypotheses, and to infer causes starting from the effects until we reach the First Cause, which is certainly not mechanical. Natural philosophy — Newton adds — has the task not only of *explaining the mechanism of the world*, but is fundamental to solve issues such as: what is there in places that are almost completely empty of matter, and whence it derives that the sun and the planets gravitate towards each other, without there being any dense matter between them? Whence comes it that Nature does nothing in vain; and whence derives all that order and all that beauty that we see in the world? What is the purpose of the comets, and whence comes it that the planets all move in one and the same way in concentric orbits, while the comets move in all kinds of ways in very eccentric orbits; and what prevents the fixed stars from falling upon each other? How is that the bodies of animals are designed so artfully, and what is the purpose of their numerous parts? Is it possible that the eye was constructed without a knowledge of optics, the ear of acoustics? How is it that the movements of the body derive from the will, and whence come the instinct of animals? ... Once all these things have been established clearly, it will be clear that an incorporeal, living, intelligent, ubiquitous Being exists, who in the infinite space of the sensorium, sees the things themselves intimately and perceives them completely, and understands them fully. Although every progress made in this philosophy does not bring us immediately to knowledge of the First Cause, yet it brings us close to it, and for this reason must be held in great consideration.

Not much discussion is required to be convinced that these words conjure up an image of science — "natural philosophy", as expressed in the language of the time — which is diametrically opposed to that illustrated by von Neumann. For Newton, the guiding criterion is not usefulness but *truth*, the discovery of the causes, the *explanation* of the causes, indeed for the purpose of attaining the First Cause. We know that Newton insisted on the fact that scientific analysis seeks causes starting from effects without "framing hypotheses", whether physical or metaphysical, that is, without making use of *ad hoc* conceptual constructions lacking any necessary relationship with the facts. We shall mention the masterly analysis made by Alexandre Koyré of Newton's famous aphorism "Hypotheses non fingo"[3] merely to say that it suggests that the aphorism should be paraphrased as follows: "I don't make models"... I don't use images or arbitrary conceptual constructions, but seek the intimate truth of the facts.

For about three centuries (starting from Kepler and Galileo) this was the credo of science, in particular, of its fundamental core – physical mathematics. It hinged on the idea of Galileo's that the book of nature was written by God in mathematical language. And even though nature may have a mathematical structure, mathematics is not a mere descriptive tool – a panoply of models – but contains within itself the intimate essence of the phenomena.

We have said — and noted in von Neumann's words — that this view was abandoned in our century. But how reluctantly! Suffice it to consider that even Einstein claimed that «we have the right to believe that nature is the realisation of what may be imagined as the simplest from the mathematical point of view», or that the astronomer Jeans asserted that the Creator must have been a mathematician. And to discover a great set of unified laws that can *explain* the mechanism of the world still seems to be the secret quest of each theoretical physicist.

Be that as it may, in spite of this more or less secret nostalgia, contemporary science appears to be increasingly addicted to making models, *mathematical models*.[4]

A mathematical model is a conceptual scheme intended to represent a set of phenomena using the language of mathematics. As it does not mirror a phenomenon exactly, a model does not claim to be the only possible representation: there is no one-to-one equivalence between models and phenomena. The same phenomenon may be represented by several models, some of which may be deemed more effective than others even though they are not necessarily in competition, as each may offer different but compatible perspectives. Conversely, one and the same model may be used to represent different phenomena, among which it establishes a kind of structural "homology". The latter aspect defines a typical method of mathematical modelling, namely the method of *mathematical analogy*. This consists in identifying features shared by phenomena that may even be distant from each other, and thus discover unexpected links between them. If one of these phenomena lends itself to an effective mathematical description, the latter may be considered as a *mathematical model* of all the other similar (or "homologous") phenomena. For example, in modern non linear dynamics it has been possible to develop a unified treatment of the phenomena of hydrodynamic turbulence, chemical kinetics, the behaviour of certain electronic circuits used in radio engineering and many other processes: the most satisfactory mathematical model is derived from the description of certain oscillatory processes that find their most conspicuous and best known expression in such electronic circuits.

A mathematical model may essentially be considered as a fragment of mathematics that can be used to represent a phenomenon, without this representation being unique and with the same fragment of mathematics being suitable for describing several quite independent phenomena. A single phenomenon may be observed through many different "lenses" (or models) and the same lens may account for different phenomena.

We have deliberately used the ambiguous term *representation*, because a model may have either a *descriptive* or a *prescriptive* function. For example, a model of meteorological phenomena is descriptive (insofar as it sets out to predict the weather as accurately as possible). On the other hand, a motor traffic model or a model for allocating scarce resources among consumers are prescriptive or control models. In this case, in fact, the interest does not lie in the description but in the establishment of procedures for achieving a given end: in our case, to ensure a rapid and fluid traffic flow and a distribution of resources that guarantees optimality or efficiency. However, these two different points of view can coexist. For instance, a flying aircraft involves both descriptive aspects and aspects of control related to human action to modify the trajectory. It is also clear that the increasing effect of human activities on meteorological phenomena is leading to the purely descriptive approach being abandoned in order to take into account the effects of this external influence and to study them in order to counter their negative aspects. The growing number of prescriptive or regulatory models is further evidence of the original nature of modelling as a scientific research practice as opposed to the almost exclusively descriptive-explanatory approach followed by classical science.

There is another aspect of mathematical modelling that represents a great innovation after several centuries of using mathematics to study various phenomena: its application to "non physical" contexts, such as biological, social, economic and psychological contexts. Prior to this century, there was a strong resistance to the use of mathematics in these "non physical" contexts, which subsequently weakened rapidly. Starting in the 'twenties, the application of mathematics to biology, initially spasmodic, rapidly developed into three mainstreams of research: population dynamics; population genetics (which made a significant contribution to the synthesis between Mendelism and Darwinism, that is, to the actual bearing structure of modern biology); the mathematical theory of epidemics; the development of many mathematical models of the physiology and pathology of human organs. The two decades between 1920 and 1940 have been defined as the "Golden Age" of mathematical biology. A similar explosion was observed in the socio-economic field, in which the more or less unsuccessful attempts to found a mathematical economics made by Léon Walras and Vilfredo Pareto in the nineteenth century were resumed by Abraham Wald and John von Neumann and then re-elaborated in the context of the theory of games, which was to produce results of fundamental importance.

Of course, one may wonder about the causes of the collapse of the two barriers standing in the way of the development of the modelling approach: the realist and objectivist view of phenomena, and the lack of faith in the mathematization of non physical phenomena.

There is no doubt that one of the causes lies in the processes that modified the image of physics at the beginning of the century. Indeed, research into the atomic and sub-atomic world led physics to abandon realism and to wonder less and less about the nature of the objects with which it was concerned, contenting itself to provide an effective and consistent formal representation of them: it is precisely in this context that the term "model" begins to spread, above all with reference to the various "models of the atom" aimed at representing its properties and behaviour without claiming to reflect its exact structure. Even the introduction of the concept of electron actually represents the creation of a model, as no one had clear ideas about what this concept represented in real terms.

Another extremely important factor in the development of the modelling approach is the rise of technology (that is, technical expertise based on science) in which it is quite natural for a closer and closer relationship to grow up between mathematics and the relevant applications. And these applications are not easy to restrict to the traditional confines of physics even when their subject matter is purely physical. Examples of this are problems linked to turbulent physical phenomena, as well as a series of topics suggested by new technological developments – for instance, the propagation of radio waves, the behaviour of non conventional electrical circuits and devices involving servomechanisms, engineering problems related to the flight of aircraft. The mathematical models constructed to handle these topics – which often follow on from non conventional topics in classical mathematical physics, such as the study of non linear differential equations — suggest new applications and hitherto undreamt of connections: from the problem of economic cycles to that of the numerous cyclic phenomena that occur in biology, from the study of economic equilibrium to that of the equilibria of animal populations competing inside the same ecosystem, to the way an epidemic tends to develop.

Such a wide variety of heterogeneous topics does not facilitate the task of incorporating the new developments of modelling into the rigid structure of classical science. They are being studied by groups of scholars working at the boundaries of the official scientific community who include (or are supported by) several highly prestigious figures. One example of a scholar as rich in fertile ideas as he was isolated from the official scientific community is Alfred J. Lotka, who wrote *Elements of mathematical biology* in the 'twenties. This book was misunderstood at the time but subsequently exerted a strong influence on one of the greatest applied mathematicians of this century, Norbert Wiener. Another top-ranking scientist involved on the new front of applied mathematics and modelling was Theodore von Kármán (who later migrated to the United States where he did important work at the California Institute of Technology). The distinguished Italian mathematical physicist Tullio Levi-Civita may be considered one of the main supporters and promoters of the new trend.

The spread of mathematical modelling during World War II was stimulated by problems of an applied nature. They are too numerous for a full list to be given. We shall take a few of the more important topics at random: problems of weather forecasting, problems of optimizing the allocation of resources (ranging from military logistics to problems related to the organization of the Marshall Plan, to problems of economic planning and programming and industrial management), problems of aerodynamics and ballistics (optimization of anti-aircraft fire, models of the atom and thermonuclear bombs, missile technology), industrial technology problems (from oil extraction to production organization methods). The applications of mathematics are so numerous that the spread among the mathematicians of the 'fifties of an approach tending towards abstraction and a pure and disinterested view of research — promoted by the group of French mathematicians gathered under the pseudonym of Nicolas Bourbaki — merely represents an anomalous, albeit influential, parenthesis. However, modelling was not limited solely to the technological and industrial aspects mentioned above. It revived the topics that had been central to biological and socio-economic modelling in the 'twenties: after a pause lasting several decades, population dynamics and epidemiology returned to the centre of interest and, in the 'seventies, underwent astonishing growth; and, starting in the 'fifties, mathematical economics again became focused on the study of the theory of economic equilibrium, accompanied by a growing interest in the theory of games, the foundations of which had been laid in the 'forties in Von Neumann and Morgenstern's famous book.[5] In more recent years, the mathematical study of weather problems led to the discovery of the phenomenon of "deterministic chaos", which some consider as a conceptual revolution of far-reaching importance. The study of "chaotic" phenomena, such as that of "complex" phenomena, can be traced back directly to the interest in turbulent processes and led to a radical change in the way phenomena were viewed mathematically: the long-standing conviction of a world structured into simple, orderly forms, which were believed to be reflected in the clear and harmonious structures of mathematics, was replaced by the idea of an irreducible complexity and an extraordinarily complicated web of relationships that could only be represented by means of the "new" mathematics of chaos or of fractal structures or else described using the more sophisticated instruments of probability theory.

It goes without saying that one crucially important factor in the development of mathematical modelling is the computer and the increasingly widespread use made of it for research purposes. Beginning in the 'eighties, the construction of increasingly powerful, fast and flexible machines led to the explosion of numerical calculus. Furthermore, the introduction of computer graphics and the consequent replacement of numerical representation by a geometric representation of solutions for the systems studied allowed a direct qualitative assessment to be made of the behaviour of the solutions themselves. The result was a radical change in research practice. Numerical analysis was no longer a support in the study of solutions that ordinary analytical tools could not render explicit, but became a tool for the simulation of the behaviour of the mathematical systems studied and consequently of the real systems that the latter claimed to represent. Computers could thus be used to *simulate* the behaviour of a real system in the hope that, at least in some cases, this approach could be replaced by direct empirical verification.

One further aspect characterizing modelling practice is the gradual abandonment of a fundamental principle of classical science: the idea of the *simplicity of nature*. This idea is rooted in the distant past but it was in the work of Galileo Galilei that it becomes associated with the notion that the world has a mathematical structure, and therefore a simple mathematical structure. The fascination wrought by this principle on scientist's minds is proved by its persistence: despite the much water that passed under the bridge of physics between Galileo and Einstein, one pillar has remained firmly intact and that is the conviction that (to repeat the already cited phrase of Einstein) «nature is the realization of what may be imagined as the simplest from the mathematical point of view».

The principle of the simplicity of nature was given an explicit mathematical translation at the turn of the eighteenth century, above all in the work of Joseph Louis Lagrange. The basic idea consisted in admitting that, in a differential equation representing a real process, the contribution of the linear part of the functions involved was that which expressed the characterizing aspects of the dynamics of the process itself. Therefore these functions can be replaced by their linear component without introducing any significant difference (the first term in the development of a Taylor series). The part that is neglected (the non linear part) certainly represents certain aspects of the process, although these consist of "perturbations" of the fundamental process: their omission leads to a slightly modified description that does not however alter the fundamental characteristics of the "true" nature of the process. *Linearization* means using a simpler description in which only the marginal aspects are lost. In return we gain in simplicity, as the process becomes amenable to easy mathematical treatment. In other words, linearization is the mathematical process used to penetrate the heart of a physical process and to reveal the simple inner structures by which it is governed.

Of course this approach is meaningful if, and only if, the hypothesis is true that omitting the non linear terms of the equation merely amounts to neglecting inessential or in any case secondary perturbations. This hypothesis has never been demonstrated and has even proved to be false, at least in principle. Nevertheless in mathematical physics it has been considered self-evident for at least a century, thus providing excellent proof of the fact that science is governed by metaphysical hypotheses in its conceptual choices and that it was accepted not because it had proved to correspond to the facts but because it expressed faith in the idea that nature is what is simplest to imagine from the mathematical point of view.

At the beginning of this century it was realized that the linearization method could be a source of serious errors, and had no general foundation. In many important cases in was precisely the non linear terms of the equation that contained the component needed to describe the physical process, that is, which expressed its most characteristic features. The development of mathematical modelling was linked to the rise of a paradigm of *non linearity* as opposed to the conventional paradigm of linearity, and this contrast was accompanied by the growing idea that natural processes were generally not simple but *complex*: simplicity (and thus linearity) became the exception and *complexity* (that is, non linearity) the rule.

The role of modelling in the rise of the paradigm of non linearity and complexity can be appreciated by taking into account the fact that both classical and more modern physics were dominated by the dogma of linearity. Indeed, quantum mechanics is completely structured around the concept of the harmonic oscillator: it is the apotheosis of the linear paradigm in view of the systematic use made in it of the theory of linear operators in a Hilbert space. It is thus easy to understand why it was precisely in physics that the greatest resistance to adopting a different point of view was found. The attention focused on non linearity (and the corresponding attention focused on the complex nature of a very large number of phenomena) thus emerges in non conventional areas of physics and technology (theory of oscillations in applied mechanics, turbulence problems, radio engineering, self-regulating processes, bifurcation) and in the applications of mathematics to the biological and socio-economic sciences. We have already mentioned the areas in which these interests developed. It should be noted in this connection that Russian science played a special and leading role in the development of the paradigm of non linearity, right from the beginning of the century, under the weight of its powerful tradition in the fields of engineering, applied science and mechanics.

In order to illustrate these issues more clearly let us refer briefly to the development of the concept of oscillation, which begins to take shape in the first few decades of the century. There is no point in insisting on the centrality of the notion of oscillation in science: a large number of oscillatory and periodic phenomena are found in nature and in artificial processes, ranging from mechanical vibrations to economic cycles, from acoustics to population dynamics, from the contractions of the heart muscle to the reproduction cycle of plants. When it is considered that a fundamental mathematical tool, Fourier's transformation, allows each evolution to be broken down into the sum of the periodic contributions, it is easy to grasp the extraordinarily general role played by oscillatory phenomena in the scientific (and mathematical) description of phenomena. The simplest and most effective mathematical structure for describing an oscillatory phenomenon is given by the linear harmonic oscillator: this explains the almost universal role it has taken on in mathematical physics. Nevertheless, many periodic oscillatory processes cannot be linked to this model. For instance, it is not sufficient to describe an ancient mechanism such as the clock insofar as it does not take into account the latter's capacity to "automatically" strike a balance between the input and dissipation of energy: the clock — like numerous other devices or natural processes— "self-regulates" its oscillations by means of a feedback mechanism. A better name for this kind of oscillation would be "auto-oscillation" and its most appropriate description is given by a non linear mathematical model that represents a variant of a famous equation introduced by the Dutch radiophysicist Balthazar L. Van de Pol in 1926.[6,7] Van der Pol introduced his equation in order to represent the behaviour of electric circuits in which triodes or neon tubes are inserted and applied it successfully to the description of phenomena such as the heartbeat, giving rise to the procedure we have called "mathematical analogy". Van der Pol's equation is essentially non linear and opens up the way to the study of a large number of self-oscillating processes governed by feedback processes.

The foregoing illustrates the profound change undergone by the relationship between mathematics and technology in this century. The consideration of the simplest system in which this kind of non linear relationship is present is indicative of the appearance of behaviours that cannot be explained in terms of the sum of the behaviour of the individual parts. Consequently, in a system comprising only two components, if a feedback relationship exists between the first component and the second, it is impossible to explain the overall behaviour of the system as the result of the behaviour of the two single components (in mathematical terms, the eigenvalues of the solution of the system cannot be obtained by simple combination of the eigenvalues of the two components). This already attests to a definite link between complexity, non linearity and several non conventional technological processes (that is, that are extraneous to previous technology), such as feedback processes, which are linked to the introduction of self-regulating devices on machines.

What has been said above leads to a more general line of thought. The classical conception of machines is based on a reductionist principle that may be expressed in the aphorism: "*the whole is the sum of the parts*". A machine, the components of which are linked by a linear causal connection, in which each element acts on the following one according to a scheme in which no branching or feedback is allowed, may be reduced to such a principle. On the other hand, the latter represents the paradigm of classical physics according to which each system is simply the aggregate of the elementary parts and the description of the evolution of the whole is obtained as the resultant of the description of the evolution of the elementary parts, and leads to the idea of simplicity and linearity. This mechanistic scheme based on the causal linkage of isolated or isolatable components proved inadequate in the case of the new technological conceptions. The first area to clash with this paradigm was that of feedback and servomechanisms. Norbert Wiener, in the late 'forties, starting precisely from the developments in computer technology, information theory and automatic machines, put forward the concepts of feedback, cybernetics and information as being of central importance also in much wider contexts than the strictly technological field. No less important a role in the gradual relinquishment of the rigid conception of mechanical reductionism was played by Claude Shannon's information theory and von Neumann's theory of games.[8]

Nevertheless, the inadequacy of mechanistic reductionism became apparent at a much deeper level. Modern technology raises much more complex issues than the description of an isolated device or of a single machine: one is always dealing with *systems*, in which different levels come into play and which often cannot even be reduced to purely physical factors. It is not only a question of the problems raised by technologies such as aerospace technology, in which mechanical, electronic, chemical and even biological problems are involved. Here we have to deal with problems such as the regulation of traffic or queuing problems, not to mention the scientific treatment of economic, production, financial and social questions. In this case a large number of different levels are involved which demand interaction among the skills of specialists who were previously considered to belong to different spheres: biologists, psychologists, economists, programmers, etc. Furthermore, the mathematical modelling of such processes does not require a descriptive approach and indeed must even explicitly preclude it in order to assure the conditions in which the process can achieve an optimal performance. The vast domain of control models represents a field in which mathematical-scientific description is called upon to play a hitherto unknown role.

In other words, twentieth century technology, already in its early embryonic forms at the beginning of the century and in a very obvious way at the present time, is bringing about the abandonment of the traditional and exclusive relationship between mathematics and physics, and of the constituent core of such a relationship, *mechanistic reductionism*. The problem remains of whether mathematics — in many respects profoundly shaped by its relationship with physics — has undergone such development as to be able to establish a new constitutive relationship with the non physical sciences, free from all residual linear causal mechanism, and thus truly suitable for reflecting the specificity of biological, social or economic problems. Indeed, as the systems theoretician Ludwig von Bertalanffy[8] observed, many modelling or systemic approaches appear more as *extensions* and *substitutions* of the mechanistic outlook and thus, although they offer an undoubtedly effective reinforcement of the mathematical analysis of physical phenomena, the same cannot be said for environments in which an original conceptual view is required.

These observations seem relevant in the case of an attempt to evaluate the present status of mathematical modelling. Of course, this kind of evaluation is very complex and must be carried out with great caution. However, it is not as difficult or intricate as the objective of tracing out an exhaustive historical picture of eighty years of mathematical modelling, which is complicated by the *intrinsically* scattered and fragmentary nature of this activity. On the other hand, this attempt to take stock leads us to an almost self-evident general observation, namely that the application of mathematics to questions that may be viewed as an extension of physical problems in which a mechanistic type approach persists is obviously effective, and acknowledged as such by the scientific community. No one denies the extraordinary effectiveness of mathematics in the theory of turbulence. The application to the biological and socio-economic sciences is quite a different matter, not to mention psychology, psychoanalysis and the behavioural sciences.

Outside the field of physics, the distinction seems to be between descriptive models and prescriptive ones: the former are generally the most disappointing. It is difficult to deny the effectiveness of models for the allocation of resources or the so called "activity analysis", of productive process models in industry or the dynamics of financial capital. On the other hand, the models of general economic equilibrium seem to be particularly mediocre. Nevertheless, the former produce an artificial image of reality for the purpose of setting the conditions in which certain desired results may be achieved. The latter instead have a descriptive purpose which must take account of the complexity of the economic process, and in fact it is observed that this complexity is reduced in an almost caricatural manner in the simplified mechanical scheme of the model.

A better example is that of the mathematical models of ecosystems. It is a known fact that the laboratory behaviour of several — i.e. two, three — animal species of prey and predators, corresponds reasonably well to the dynamics described by Volterra's classic equations. The same may be said for the case of a hypersimplified ecosystem such as that of farmland managed according to industrial principles. In both cases they are *artificial* situations, reduced to a simple mechanical scheme. But as soon as the problem becomes the *description* of a complex real ecosystem (the ecosystem of a rain forest) disconcerting discrepancies appear between the results of all the known models and empirical observation.

In this connection, particularly significant research was carried out by Robert M. May in the 'seventies for the purpose of examining the relationship between stability and complexity in ecosystems.[9] The problem addressed by May was the following: can it be said that, in Volterra's mathematical models and the generalizations thereof, an *increase in complexity* of the system (that is, an increase in the number of species and the interactions among them) leads to an *increase in stability* of the ecosystem (in the sense that its structure withstands perturbations from the exterior)? Ecologists are inclined to answer affirmatively, on the basis of known and investigated real phenomena. May's research produced opposite results in the case of mathematical models: *the more complex the model, the more unstable it is*. This extremely serious inconsistency between models and empirical observations is a problem that still remains to be solved.

A somewhat similar situation is found in the mathematical modelling of AIDS. In the past, considerable success was achieved in the mathematical modelling of "simple" epidemics, such as the plague, cholera or certain venereal diseases with a simple diffusion mechanism that can be reduced to a physical scheme similar to that of the "kinetic theory of gases". In the case of AIDS, however, things are different. Here the extreme complexity of the disease, the highly complex and only vaguely delimited patterns of its mechanisms of transmission, not to mention ideas about its period of incubation (or latency), which are uncertain, to say the least, have resulted in an enormous burgeoning of mathematical modelling which is however of dubious utility and validity. At a recent congress in the USA, mathematical model makers, biologists and physicians compared notes on this issue and the debate degenerated into a dialogue of the deaf: the latter expressed scepticism and their unwillingness to make all the empirical data available in the absence of any proof of the effective utility of the models; the former declared that they could do very little in the absence of a clearer quantitative scenario and a more extensive knowledge of the biological mechanisms of the disease.[10]

At the beginning of the great resumption of biomathematical studies that began in 1976, the mathematicians G. Oster and J. Guckenheimer pointed out:

«Biological systems tend to be considerably more complex than those studied in physics and chemistry. In analyzing models, one is frequently presented with two alternatives: either resorting to brute force computer simulation or to reducing the model further via such drastic approximations as to render it biologically uninteresting. Neither alternative is attractive. Indeed, the former alternative is hardly viable for most situations in ecology since sufficient data is rarely available to quantitatively validate a model. This contrasts starkly with the physical sciences where small differences can often discriminate between competing theories. The situation is such that many ecologists seriously question whether mathematics can play any useful role in biology. Some claim that there has not yet been a single fundamental advance in biology attributable to mathematical theory. Where complex systems are concerned, they assert that the appropriate language is English, not mathematical. A typical attitude among biologists is that models are useful only insofar as they explain the unknown or suggest new experiments. Such models are hard to come by.»[11]

These pessimistic conclusions did not hold back the explosion of biomathematics in the two decades 1974-1994, during which the publishing house Springer produced the series *Lecture Notes in Biomathematics*. In n° 100, published in 1994, which ended the publication of the series, the editor Simon A. Levin drew some «bittersweet» final considerations: there had been such a boom in this branch of science that, according to Levin, the series of books was no longer of any utility. No one was interested any longer in general overviews. Everyone had carved out his own specialist niche inside which he was busy digging away, completely oblivious to what was going on even in his immediate vicinity. Consequently, just when the number of researchers was increasing exponentially, the series had to close down for economic reasons – people had stopped buying the books!... In a word, the questions asked by Oster and Guckenheimer had found no answer at the end of the second "Golden Age" of biomathematics: they had simply been set aside as uninteresting by a research world unmindful of problems of general orientation.

A similar conclusion could be drawn in the case of mathematical economics. However, I shall merely say that, as in the case of May's studies in biomathematics, the mathematical results obtained in the theory of economic equilibrium in the 'seventies and 'eighties had a devastating impact on the theory itself. While the theory of the existence of economic equilibrium was an unqualified success, the research on stability and the uniqueness of equilibrium led only to negative or odd results.[12] Nevertheless, the making of models continued, gathering speed like a machine running out of control, without taking into account the critical or negative nature of some of the results, finding fuel and justification only within itself. It was merely declared that the topics of stability and uniqueness were in any case not all that important, after decades of efforts dedicated to them – like the sour grapes for the fox in La Fontaine's well-known fable.

From what has been said it may be concluded that the general outlook for mathematical modelling at the turn of the millennium is marked by strong contrasts — or, as Levin says, has a «bittersweet» flavour.

Mathematics appears increasingly as an all-pervasive tool, and is often presented as compulsory for any analysis of phenomena that claims to be scientific. Science appears to be increasingly pervaded by a "panmathematical" idea that apparently resurrects the ancient Galilean idea of nature written in mathematical language, although, as we have seen, without any pretence at realism and replacing the dogma of simplicity with the idea of complexity. However, while the mathematical representations of complexity seem to correspond quite well to the complexity of physical phenomena, the levels of complexity encountered in biological phenomena and, to an even greater extent, in economic, social and psychological phenomena, represent an extraordinarily tough and still open challenge for formal scientific analysis.

In this situation, mathematical modelling, insofar as it is a multiform activity scattered over a multitude of sectors and lacking any unifying criteria of a conceptual and methodological nature, throws out a fascinating challenge but also entails serious risks. The fascinating challenge obviously consists of the attempt to extend the domain of the scientific method to sectors from which it was apparently precluded. The risks are associated precisely with this extension. To the extent to which it is not supported by solid criteria of objective control and no longer claims to supply an objective image of the phenomena, modelling opens the way to arbitrariness, and comes closer to the very practice of those verbal and informal sciences of which the uncertain, questionable, unverifiable nature has so often been deplored. Very often the model is like a "*story*", a narrative, and while it may be endowed with the fascination of a story it also shares its subjectivism and arbitrariness.

One of the main difficulties encountered in the scientific analysis of the subjective world is well illustrated by a remark made by the distinguished Russian biologist Georgii F. Gause with reference to the principle of encounters on which Lotka and Volterra had founded population dynamics:

«Although Lotka and Volterra introduced this principle in the equations of their mathematical theory, they were fully aware that it could not be justified in practice. Lotka thus wrote in 1925 that many, and perhaps all, organisms possess some power of selection and are to some extent independent of pure chance. This introduces an extremely peculiar complication into the mechanism governing living beings, a complication which does not exist in the statistical dynamics of molecular physics. Not only is the living organism capable of accomplishing at the macroscopic level achievements similar to those that in the world of molecules are allowed to figments in the imagination, such as Maxwell's demon, but this power "to beat chance", as it were, is possessed to varying degrees by the different living organisms, and the mechanisms underlying the systems containing living matter must necessarily take into account, in addition to this capacity, also the degree to which this capacity is possessed, as it plays an important part in the determination of the place occupied by the various biological species in the scale of evolution.»[13, 14, 15]

What Gause is telling us here, very effectively, is that the behaviour of the biological subject in no way resembles the blind behaviour of a material particle — although his remark is even more pertinent to the social subject that, as Gause would say, occupies a higher level in the scale of evolution. It is aimed at warding off adverse circumstances, or in any case those imposed by an environment governed by blind principles and at *imposing its goals*. But how can this creative finalism be represented formally? At the risk of appearing tedious, it must be repeated that this is the consequence of re-proposing an unsolved problem: the creative choices of the biological or social subject cannot be described as causal or random processes without actually depriving them of their distinguishing characteristics. Practically everybody agrees on the fact that to describe creativity as a deterministic process is contradictory because it means relating it back to laws imposed in an absolute fashion and from outside the subject. But as probabilism is more fashionable, there is more resistance to admitting that the identification of creativity and randomness is no less absurd. As Gause said, the creative subject tends to «beat chance» and certainly not to submit to it: chance is indeed the antagonist in all planned activities. It is a «wretched loophole» to determinism — as René Thom observed —to try to «save the possibility of radical innovation, of creative invention» by using the «Brownian gait of a drunken sailor [...] forced unceasingly towards the burlesque alternative of Buridan's ass».[16]

The principle underlying the theory of games of Borel and von Neumann must be acknowledged as having one merit: it contained the awareness that the real problem is actually that of representing the behaviour of a free subject who, in an external context that is not completely controllable and not completely known, tries to "beat chance" and, within certain limits, to impose his goals. But then, when it came to formalizing this free and goal-oriented behaviour, it was necessary to content oneself with the panoply of tools available. There was no choice: the utility functions describing the "player's" behaviour are either ordinary functions of analysis or else random functions. This brings us back to the dilemma: to describe freedom and creativity by means of determinism or of chance. Both alternatives lead to grotesque situations: it is assumed that the actors all possess infinite foresight, or else that they behave like n drunken sailors. On the other hand, creative finalism does not have mathematical finalism as its counterpart, that is, the finalism of Maupertuis' principle of least action and of the variational principles of mechanics, which — as Ernst Mach and John von Neumann have clearly pointed out — offer an approach that although technically different is conceptually equivalent to the deterministic approach.

These considerations obviously lead us naturally to those made sixty years ago by Edmund Husserl concerning the irremediable contradiction that arises when it is attempted to tackle the problem of subjectivity via an objectivist approach. However, it is beyond our present scope to discuss such objections of principle. Let us consider the possibility of a more and more extensive mathematization, although let us evaluate its results critically and without any sham indulgence. From this point of view, it is perhaps impossible today to continue to be indulgent, as Husserl was when he acknowledged the scientific rigour of what he called the «concrete sciences of the mind». After eighty years of mathematical modelling we are now in a position to draw some preliminary conclusions. And it must be admitted that only too often do the mathematical models made of non physical contexts seem to come under the fierce but effective definition of Ludwig Von Bertalanffy who likened them to «Shakespearean tales told by an idiot». And Von Bertalanffy was probably right when he blamed the modest results obtained concerning the use of schemes slavishly borrowed from mathematical physics, or else from a totally different environment unsuitable for providing representative models in the field of the sciences of subjectivity.

We are of course still far from a mathematics suitable for investigating non physical phenomena. This is not surprising. John von Neumann used to warn the hasty that it had taken several centuries to achieve an effective mathematization of physical phenomena, and so it would take further centuries to achieve something satisfactory in the non physical field. Such a warning – expressed to boot by one of the staunchest supporters of the absolute mathematization of all categories of phenomena — should make those who are prepared to sacrifice scientific rigour and the quality of results on the altar of often undue haste pause for reflection.

REFERENCES

1. John von Neumann, Method in the Physical Sciences, in *The Unity of Knowledge*, L. Leary ed., Doubleday, New York (1955).
2. Giorgio Israel, Ana Millán Gasca, *Il mondo come gioco matematico. John von Neumann scienziato del Novecento*, La Nuova Italia Scientifica, Roma (1995).
3. Alexandre Koyré, *Études newtoniennes*, Gallimard, Paris (1968).
4. Giorgio Israel, *La mathématisation du réel, Essai sur la modélisation mathématique*, Éditions du Seuil, Paris (1996).
5. John von Neumann, Oskar Morgenstern, *Theory of Games and Economic Behaviour*, Princeton University Press, Princeton N.J., (1944, second edition 1947).
6. Balthazar L. Van der Pol, On Relaxation-Oscillation, *The London Edinburgh and Dublin Philosophical Magazine and Journal of Science*, 2: 978-992 (1926).
7. Balthazar L. Van der Pol, Jan Van der Mark, The Heartbeat considered as a Relaxation Oscillation, and an Electrical Model of the Heart, *The London Edinburgh and Dublin Philosophical Magazine and Journal of Science*, 6: 763-775 (1928).
8. Ludwig von Bertalanffy, *General System Theory*, Braziller, New York (1968).
9. Robert M. May, *Stability and Complexity in Model Ecosystems*, Princeton University Press, Princeton N.J. (1973).
10. Giorgio Israel, Analogie, metafore e verifica empirica nella biologia matematica contemporanea, in *Pristem/Storia, Note di matematica, Storia e cultura*, 1: 53-72, Springer Verlag-Italia, Milano (1998).
11. George Oster, John Guckenheimer, "Bifurcation Phenomena in Population Models", in J. E. Marsden, M. McCracken, *The Hopf Bifurcation and its Applications*, Springer, New York (1976).
12. Giorgio Israel, Bruna Ingrao, *The Invisible Hand. Economic Equilibrium in the History of Science*, The MIT Press, Cambridge, Mass. (1990).
13. Georgii F. Gause, *Vérifications expérimentales de la lutte pour la vie*, Hermann, Paris (1935).
14. Georgii F. Gause, *The struggle for existence*, Williams & Wilkins, Baltimore, 1934.
15. Ana Millán Gasca, Mathematical theories versus biological facts: A debate on mathematical population dynamics in the 1930, *Historical Studies in the Physical and Biological Sciences*, 2: 347-403 (1996).
16. René Thom, Préface, in P. S. Laplace, *Essai philosophique des probabilités*, Paris, Bourgois, 1984

THE SEARCH FOR THE MATHEMATIZATION OF THE SOCIAL DISCIPLINES

Sergio Monteiro

Department of Physics
Moorpark College
Moorpark, CA 93021

monteiro@physics.ucla.edu

ABSTRACT

The roots for the attempts to use mathematical language on the social disciplines can be traced as far back as 1687 with the publication of I. Newton's *Philosophiae Naturalis Principia Mathematica.* Newton's second law opened the possibility that not only all astronomy but also chemistry, even thoughts, and by extension all the complex unfolding of life, are deterministic. Accordingly, many tried to find the "equation of life", or how to compute the crystal ball. This *programme* arrived nowhere. A less ambitious *programme*, one that sprouted from John von Neumann's and Oskar Morgenstern *Theory of Games and Economic Behaviour*[1] succeeded in helping to analyze possible outcomes from a know current situation and given specific forces on society. We discuss the historical evolution of the *programme* of the search for the equation of life, show why it is unattainable, and analyze the limited possibilities as well as the strengths of game theory to help men to analyze human and other complex phenomena. Starting with the *Principia*, we end with an analysis of our time.

HISTORICAL PROLEGOMENA

The publication of Isaac Newton's Principia in 1687, had a profound intellectual impact. It is indeed hard for us, in the 20th Century, to correctly understand a world in which virtually all normal everyday natural events could only be understood as a result of either magic or of a hidden, mysterious, ubiquitous and omnipotent controlling being. As an indication of the plight of the 1600's men it suffices to observe how many educated people even in our own days, and people who should know it better, still hold believes on invisible beings that are totally against the available evidence. If we were to position ourselves in such a situation as a late 1600s man, the sudden revelation of a common cause for seemingly

The Application of Mathematics to the Sciences of Nature, edited by Cerrai *et al.*
Kluwer Academic/Plenum Publishers, New York, 2002.

dissimilar events as the falling of an apple on the ground and the orbiting of the moon around the earth would probably overwhelm us -- like it did them. There was suddenly one equation that controlled the motions of the stars, the planets, the moons and the falling apples. It is from the point of view of such a 1600 man that we should try to perceive the impact of the Principia.

THE EVOLUTION OF AN IDEA

In short time after the publication of the *Principia*, either its methodology or else only the form used by Newton was adapted to other areas. Coulomb[2] succeeded in framing electrostatics within the same mathematical framework created by Newton for mechanics, while John Craige[3] used the *form* of the *Principia* to calculate the next coming of the christian god. These two are examples among many. Coulomb succeeded in his task to perfection; it was his luck that the electrostatic force happens to obey a law of the same form as gravitation: inverse square with the distance between the elements involved, linear superposition, and the product of the relevant parameter of the elements, mass or electric charge. John Craige, sticking to a $1/r^2$, argued that the knowledge of god's coming to earth evanesces with the inverse square of the distance from Jerusalem, the specific point where John Craig believed he came down, and with the inverse square of the elapsed time from his tour of this particular planet. In line with his Newtonian leanings, Craige made use of much geometry to support his arguments. He is now only known as a bizarre curiosity. It was just quite a lot harder for John Craige to be right on his chosen topic than for Coulomb on his.

In between these extreme cases there are the introduction of the mathematical methods (languages) in slightly different forms to physics and to other branches of science. Examples in the field of chemistry is the work of the Swedish chemist Svent Arrhenius[4], in the field of biology the work of Gregor Mendel[5] and in the field of population genetics and the eugenic movement the work of Karl Pearson[6]. In each of these cases a different type of mathematics was used to describe a particular aspect of that field of knowledge.

These three examples were chosen not as the most representative of innovative mathematics to shine light on some particular aspect of knowledge. These were chosen rather to support the contention that some fields required only slight variations on existent mathematical formalisms, as the work of Arrhenius and Mendel, while others required more innovation on the mathematical formalisms, as the work of Pearson. Anticipating our conclusion, it turned out that the proper mathematics needed to analyze social and human affairs is rather different from the traditional mathematics used by physics, chemistry, geology, and other sciences. Partially because of this it took so long to bring mathematics into the study of human affairs.

In the *Principia* Newton disguised his real mathematical method (calculus) under a geometrical coat. In a way geometry was Newton's language, but only superficially so, because it appears that he used calculus to figure it all out, then just hid his method of approach under geometric arguments. This geometrical coat was already abandoned by all by the time of Newton's death, or shortly thereafter. Later on, particularly because of the work of Lagrange[7], the abstract mathematical language has been adopted as the language of choice, particularly in physics, to convey the description of nature.

Starting from the later part of the 19th Century more mathematical languages have being developed to deal with particular new problems posed not only by the traditional sciences, but by human analysis as well (social sciences, political sciences, etc.). The pace has been accelerating with the general increase of social investment in knowledge as the church power to keep in check humankind search for knowledge dwindled. Vector calculus (imbedded in the quaternion work of Hamilton[8]) which much improved the treatment of electromagnetism, then tensor calculus (Tullio Levi-Civita[9]), crucial for the development of the theory of relativity and an improvement for the treatment of electromagnetism as well, exterior differential calculus (Hermann Grassmann[10]), group theory, lie algebra, and others.

On the human and animal side of inquiry, investigations on animal behavior by Konrad Lorenz[11,12], Niko Tinbergen[13], Jane Goodal[14,15], Shirley Strum[16], and investigation on human behavior by B. F. Skinner[17,18], have been successfully carried without mathematical languages. One should keep in mind that because mathematics helps to carry a certain type of analysis of human affairs, it does not follow that *all* that is human is better explainable by mathematics. Indeed, as it will be shown later, though mathematics helps one to follow certain arguments, it is a human that must interpret the final result displayed by the mathematics, and then act on it. It is one of those mathematical tools for analysis of non-deterministic phenomena that we want to discuss in this article: theory of games.

GAME THEORY: FROM ITS ORIGINS TO ITS CURRENT APPLICATIONS

Starting in the 1920s John von Neumann began publishing a series of papers in which he laid down the foundations of what became later known as theory of games. Particularly important was his proof of the minimax theorem in 1928[19]. These papers draw little attention until von Neumann published in 1944, in co-authorship with Oskar Morgenstern, their influential book *Theory of Games and Economic Behavior*[1]. In this book they presented in an organized way a mathematical framework useful to analyze and understand the forces at play in an economic system and other interesting social interactions among free players, all of which are able to contribute for the outcome of a situation, while none of them could alone decide the outcome. Von Neumann himself was aware of the general applications of game theory to analyze social interactions, as per his own words in the book's preface[1].

Contrary to the naive interpretation of the title, which have caused it to be censored by the Brazilian military from at least one research institution, the book is not about games at all, except that the authors chose to use parlor games as examples of a new mathematics paradigm. The theory of games is a mathematical theory of decision-making with capabilities of accounting for both individual decisions by the participants in a competitive environment as well as for chance. The theory of games is applicable to problems where each participant may chose one of an array of options available to him/her, possibly changing his/her choice along the way, either out of whim or because of his/her observation of the other participants moves or because of chance events that happened to have occurred. These choices need not be the smartest in some way, nor they need to be consistent or indicative of mature behavior, in the common meaning of these words. Chance may also enter as a factor. Neither chance nor a single player can decide the outcome of the situation, called a game for simplicity. The theory of games is then a method or a language to organize the options available, the possible outcomes of each possibility, and a method to decide the outcomes of each possible evolution of events. It may be used by any player to maximize his payoff (his gain, in the theory's parlance).

As such, game theory can be considered as a language, a mathematical language like algebra or group theory. We will end this paper with a comparative analysis of algebra and game theory to analyze situations to which each applies. We do so with the intention of displaying the advantage of the use of mathematical languages to analyze particular problems. As it will be seen, different problems require different mathematical structures to analyze them.

Both game theory and algebra work better than men naked brain because the human mind has evolved to deal with a different set of situations than it is now being used for, and consequently human mind is prone to errors of judgement and all sorts of logical blunders (at least mine I know to be so). Our minds are limited and prone to errors. Mathematical languages, on the other hand, follow a rigid set of rules. These rules cause that each of the available mathematical languages is restricted in its use, but within their region of validity they are free of errors – assuming that the human user does stick to the rules of the language. Saying it in other words, mathematical languages, exactly because they are restricted by a rigid set of rules, are free of the errors of logic so common to human analysis. Another point that should be kept in mind is that the human mind did not evolve to follow complex, arbitrarily long arguments, as is needed to analyze many situations in economy, military strategy, political evolution of a society, and similar situations with a plurality of factors that influence the outcome, few of which, if any, may be precisely quantified or even identified. With time the superiority of game theory to help to analyze national economies and military strategies was recognized. Eventually game theory was also applied to other fields of human affairs, notably in political analysis.

It is important to understand from the beginning that game theory is not the crystal ball equation. It is a method that allows the user to collect the relevant facts together, then to follow the consequences of particular decisions of each player. It is also only superficially that this appears to be the same that common sense does, as common sense is unable to pursue an arbitrarily complex or arbitrarily long argument. This is indeed an important point to keep in mind: game theory may be considered as an extension of human common sense, but an extension that can be stretched to an arbitrarily long chain of arguments and that is free of the logical fallacies that the so-called common sense is so prone to display.

Game theory is a system for the analysis of possible scenarios and broader possibilities of evolution of current situations. Game theory does not attempt to make a deterministic prediction equivalent to the date and time of a lunar eclipse next year, let alone thousand of years into the future, an easy task for Newtonian mechanics -- though celestial mechanics position predictions fails if attempts are made to extend them to millions of years, this due to its now recognized sensitivity to the initial conditions. In this deterministic search the original *programme* failed, as the search for the mathematization of life is yet to find an equation for life. We also have indications that this problem is sensitive to the initial conditions, so it is likely that the equation for life, the crystal ball equation, if ever found will be useless for human predictions. The search has been successful, though, in finding what is possible to find: an algorithm that does allow one to follow an arbitrarily complex and long chain of arguments, leading to the contemplation of possible final outcomes. From there men still have to take a human (or sometimes an inhuman) decision. It is then the task of every generation to prepare the next one to be a better decision maker than the current one. Professors, politicians and sound technicians are all together in this endeavor; it is the task of all of us.

TWO EXAMPLES OF MATHEMATICAL LANGUAGES: ALGEBRA AND GAME THEORY

Edna Kramer, in her beautiful book *The Nature and Growth of Modern Mathematics*[20], quotes a numerical epigram about Diophantus, quite appropriately one of the greates algebraist of antiquity: "God granted him youth for a sixth part of his life, and adding a twelfth part to this, he clothed his cheeks with down; he lit him the light of wedlock after a seventh part, and five years after his marriage he granted him a son. Alas! lateborn wretched child; after attaining the measure of half his father's life, cruel Fate overtook him, thus leaving to Diophantus during the last four years of his life only such consolation as the science of numbers can offer." The epigram is supposed to tell Diophantus' life span. It is rather difficult to extract this information from the English language version of the epigram, yet it is a trivial task to get it from the algebraic version of the same information. So, let us rewrite the English language form of the epigram in the algebra form of it:

$$\frac{1}{6}x+\frac{1}{12}x+\frac{1}{7}x+5+\frac{1}{2}x+4=x$$

from which it trivially follows that x = 84; Diophantus lived 84 years. Algebra is the language of choice to analyze this information. Standard English is not, differential equation is not, neither is group theory. Nor is game theory either.

The next problem comes from an article by George Tsebelis[21] and references therein. The problem is modified from an American setting to fit an European one, otherwise being equivalent. Consider the problem of an average citizen that needs to get somewhere in a hurry, perhaps for an important meeting. He leaves his monthly subway pass at home, because he left in such a hurry, and by the time he enters the subway station he realizes that his train is coming. He has the option of continuing, boarding without a valid ticket and perhaps getting on time at his meeting (if there is no inspection), or else to stop to buy a daily pass, missing his train and certainly arriving late to his meeting. If he chooses the first option (to board without a valid ticket), he risks that the inspector happens to check that wagon he boarded, he will be fined, get exposed to public shame and arrive even later at his meeting than if he had stopped to buy a daily ticket, because of the delay incurred in paying the fine to the inspector, being identified, etc. But he may also not be inspected and get along just fine. If he chooses the latter option (to buy a daily ticket), he risks no fines, no shame, but he is assured of arriving late to his meeting. There are four possibilities for him, with payoffs a_1, b_1, c_1, and d_1: he hops into the train without a valid ticket and is inspected (a_1), he hops into the train without a valid ticked and is not inspected (b_1), he buys a ticket and is inspected (c_1), or he buys a ticket and is not inspected (d_1). A high negative value will be assigned to a_1 because it is the most undesirable situation with nothing good to it, while a high positive value will be assigned to the most desirable possibility b_1. which contains only good returns. Option c_1 is a mixed bag, as it involves arriving late at the meeting but he would have arrived late anyway because of the inspection, though our friend would be wondering if there would have occurred inspection in the preceding train, in which case the ticket purchase was a useless exercise. Finally option d_1 involves the loss of arriving late at the meeting but is not as bad as the loss incurred in a_1, because in the later case besides arriving late he also got arrested for a minor offense. Note that the actual numerical values assigned to these possibilities depend on the life constraints of the particular individual, besides being modulated by his / her possible near-term psychological fluctuations. A student, for example, does not look at being inspected without a valid ticked in the same way as his / her professor, who in turn is not as negatively impacted as a politician is. Though all three would assign a negative value to a_1 , the student would assign a number of smaller absolute value, the politician would assign a very large absolute value to it, while the professor would be in between. On the other hand, a large fine for riding without a valid ticket would normally be considered more negative for the poor student than for the rich politician, the professor again falling in the middle. The amount of fine may, in some cases, affect the payoffs for a subset of the public and consequently the amount of fine influences some of the values discussed, with indexes "1". Game theory analysis begin with the what is called the payoff matrix:

$$M_1=\begin{bmatrix} a_1 & b_1 \\ c_1 & d_1 \end{bmatrix}$$

If our friend were the only player in this game, and if he knew the probability of inspection he could make an intelligent decision maximizing his payoff (his benefit, defined in some way according to the values he ascribes to the four parameters). In particular one should remember that the amount of the fine may influence some of these numbers. But he is not the only player, as there is also the subway inspection agency, which will make or not make inspection according to the probability of finding offenders. Because if there were no offenders, certainly all inspectors would be assigned to other jobs, while as more people were found to be free riding, more and more inspectors would be added to the trains. The inspecting agency has then four options (here labeled with subindexes 2): inspecting while there are offenders – good for the agency (payoff a_2), fail to inspect while there are offenders – bad for the agency (payoff b_2), inspect in the absence of offenders – bad public relations and waste of resources (payoff c_2), and fail to inspect in the absence of offenders – good (payoff d_2). If one assumes that the collected fines go to the government coffers and is not used to pay the dinners of the inspectors or their supervisors, which is a reasonable assumption, none of the inspecting agency payoffs should depend on the amount of fine, since it does not profit from the fines. Therefore the entries with subindexes "2" are independent of the amount of fine. But note that these values may depend on variable external factors. The adverse public relations of inspecting in the absence of offenders would be highly undesirable in a functioning democracy with a free press, while close to neutral in a dictatorship.

In this case the payoff for the passenger is a product of his own payoff by the inspecting agency payoff. Common sense tell us that an increase in the fine (penalty for been found in the train without a valid ticket) ought to decrease the number of free-riders. This is a conclusion that most everybody would agree with. Yet this obvious conclusion is false, as can be easily uncovered by game theoretical methods. Indeed, the payoff for this game is the matrix

$$M_{12} = \begin{bmatrix} a_1a_2 & b_1b_2 \\ c_1c_2 & d_1d_2 \end{bmatrix}$$

From which game theory methods shows that the probability of free riding is[21]

$$p = \frac{d_2 - c_2}{a_2 - b_2 + d_2 - c_2}$$

which, depending only on the payoffs of the inspecting agency (subindexes 2), is not a function of the amount of the fine. We arrived at the most unexpected conclusion that the amount of fine does not affect the number of free riders! The fallacy, which is beyond the ability of most anybody's mind to evaluate, is that an increase in the fine amount would indeed cause an *initial* decrease in the free riders, but this would be followed by an adjustment by the inspecting agency, that would reassign inspectors to other duties, given that the new smaller number of free riders would no longer justify that many inspectors creating bad public relations for the agency. This would in turn be followed by a readjustment of the incentive for free riders, causing the number of free riders to go up again, in adjustment to the smaller number of inspectors. In the end, there would be the same number of free riders, these would be inspected less times and would pay more each time that they were caught, paying on the average as much as with a smaller fine paid more times with more inspections.

Naturally that legislation could also be passed, forcing the inspecting agency to keep the old number of inspectors, in which case the number of free riders would indeed go down. In this case the payoff matrix for the inspecting agency would be different, as it had to reflect external imposed conditions too. But this does not take away from the point that I want to make here, which is that as much our brains are unsuited to analyze Diophantus epigram in English language (at least my brain is), they (or at least mine) are unsuited to analyze complex interrelated situations, the outcome of each depending on several decisions taken by a multitude of players, each one having his/her own agenda to pursue, acting independently and without knowledge of how his/her opponent will act, because they are also free to chose any of an array of possibilities.

CONCLUSION

Is then game theory a crystal ball that can be used to predict the future? Did the original search for the equation of life succeeded? The answer is no to both, something that surprises no one, and is also a welcome conclusion to most everybody, if not to all humans. All that game theory can do is to guide the thoughts of a human being through the analysis of complex situations, presenting him/her with the possibilities of each one of the available choices at any given situation. Game theory enhances human analytical capabilities, as much as algebra does, but ultimately a human (or sometimes an inhuman) decision has to be made still based on the prevailing human values of the particular society in which the decision makers are inserted.

Has game theory changed our world, or at least contributed for some change? In the United States, the Department of Attack uses game theory as part of its strategic planning (Melvin Dresher[22]). Rand Corporation, one of the so-called think tank groups[22], has developed much expertise in game theory which is used for the several attack contractors and for similar groups as well. The evidence it that the answer to this question is yes. It appears also that it has been mostly right wing groups that have taken advantage of the method, except for the academic environment, where many political sciences departments in the United States include game theory in the course work and routinely use game theory to support their analysis.

It is still possible for all of us, from all walks of life, to make choices that will lead to a better or to a worse world when we leave it. Game theory can help to analyze the possibilities, but it is still humans that are making decisions that are not predictable by any mathematical formalism. This is good, because the conclusion is that it is now possible to see deeper before making decisions that depend on complex factors, factors too numerous to be kept organized by ordinary reasoning. We can then work smarter to pass on a better world to the next generation. But we still have to fight, whether we are politicians, professors or audio technicians, for this better world to come, exactly like we had to do before John von Neumann gave us game theory.

ACKNOWLEDGMENTS

I am indebted for fruitful discussions with Professor Claudio Pellegrini and Professor George Tsebelis, both from UCLA, and to Professor Ximenes A. da Silva, from Centro Brasileiro de Pesquisas Fisicas, who introduced me to game theory. I am also indebted to Professor Claudio Pellegrini for the congenial environment provided by him during a sabbatical leave I spent at UCLA.

REFERENCES

1. John von Neumann and Oskar Morgenstern, *Theory of Games and Economic Behavior* Princeton University Press (1944).
2. Duane Roller and Duane H.D. Roller, *The Development of the Concept of Electric Charge – Electricity from the Greeks to Coulomb,* Harvard University Press (1954).
3. John Craige, *Theologiae Christianae Principia Mathematica* ("The Mathematical Principles of Christian Theology"), London (1699).
4. Svante Arrhenius, *Theories of Solutions,* Yale University Press (1912).
5. Gregor Mendel, *Experiments in Plant Hybridisation* Translation made by the Royal Horticultural Society of London, Harvard University Press (1950), original published in the *Verh. naturf. Ver. in Brunn*, Abhandlungen, iv (1865), which appeared in 1866.
6. Theodore Porter, *The Rise of Statistical Thinking – 1820 – 1900,* Princeton University Press (1986).
7. J.-L. Lagrange, *Mecanique Analitique*, Gauthier-Villars, 4eme edition, Paris (1888). I am thankful to Prof. C. Pellegrini for having called my attention to this.
8. William Rowan Hamilton, *Elements of Quaternions* third edition, Chelsea Publishing, New York (1969).
9. Tullio Levi-Civita, *Lezioni di Calcolo Differenziale Assoluto*, Rome (1925).
10. Hermann Grasssmann, *Die Ausdehnungslehre von 1844*, Otto Wigand, Leipzig (1878).
11. Konrad Lorenz, *On Agression,* (translated by Marjorie Kerr Wilson) Harcourt, New York (1963).
12. Konrad Lorenz, *Here I Am – Where Are You?,* (Translated by Robert Martin) Harcourt (1991).
13. Niko Tinbergen, *The Animal and Its World,* Harvard University Press (1972).
14. Jane Goodal, *Through a Window - My Thirty Years With the Chimpanzees of Gambe*, Houghton Mifflin, Boston (1990).
15. Jane Goodal, *In the Shadow of Men,* Houghton Mifflin, Boston (1971).
16. Shirley Strum, *A Journey Into the World of Baboons,* Random House, New York (1987).
17. B. F. Skinner, *About Behaviorism,* Knopf, New York (1974).
18. B. F. Skinner, *The Behavior of Organisms – An Experimental Analysis,* Appleton, New York (1938).
19. John von Neumann, Zur theorie der Gesellschaftsspiele, *Math. Annalen* **100**, 295-320 (1928).
20. Edna Kramer, *The Nature and Growth of Modern Mathematics,* Hawthorn Books, New York (1970).
21. George Tsebelis, The abuse of probability in political analysis: the Robinson Crusoe fallacy, *Am. Polit. Sc. Rev.*, **83**, # 1 pg 77-91 (March 1989).
22. Melvin Dresher, *Games of Strategy,* Prentice-Hall (1961).

MATHEMATISATION OF THE SCIENCE OF MOTION AND THE BIRTH OF ANALYTICAL MECHANICS: A HISTORIOGRAPHICAL NOTE

Marco Panza

University of Nantes
Nantes, France
Marco.Panza@irem-hst.univ-nantes.fr

Usually, one speaks of mathematization of a natural or social science to mean that mathematics comes to be a tool of such a science: the language of mathematics is used to formulate its results, and/or some mathematical techniques is employed to obtain these results.

It seems to me that the notion of mathematization is misleading when used in this sense, however. Instead, we should distinguish what I would like to call "mathematization" from something quite different I would call "application of mathematics".

We should speak of application of mathematics to a natural or social science when a natural phenomenon that such a science is studying is described or explained using mathematics, or when a problem this science aims to solve is answered by means of mathematical techniques. By contrast, we should speak of mathematization only when the object of this science becomes a mathematical object, i.e. mathematics provides a model or a scheme of a natural or social phenomenon and this model or scheme becomes the real object of studying. I shall illustrate this distinction by means of the example of the science of motion from Aristotle to the middle of 18th century. The same example will also be used to illustrate another topic, namely when analytical mechanics was born.

Concerning my first topic, the standard view maintains that the mathematization of the science of motion begins within 17th century, specially thank to Galileo. Moreover, it maintains that the mathematization of the science of motion is a distinctive character of the "scientific revolution". But, I do not think that something as a "scientific revolution" occurred in 17th century (or in some other date in the history of science). I shall not discuss this point here however. I will confine myself to argue, according to my previous distinction, that when application of mathematics to the science of motion is a usual practice within Aristotelian physics, no mathematization of this science occurs before Newton. From such point of view, Galileo's theories of motion share with Scholastic theories an essential aspect: they do not deal with a mathematical object, but they merely use a mathematical theory (that is actually a suitable version of classical theory of proportion) to deal with some natural phenomenon.

Concerning my second topic, there is no standard view. That analytical mechanics was born in 18th century is a well-grounded historiographical judgement. When it is used in

The Application of Mathematics to the Sciences of Nature, edited by Cerrai *et al.*
Kluwer Academic/Plenum Publishers, New York, 2002.

historiographical contexts, the term "analytical mechanics" is quite imprecise, however. This made the previous judgement open to different interpretations, each of them ascribing to this term a different meaning. I shall consider two examples of such different conceptions that seem to me to be opposite end extreme.

TWO OPPOSITE CONCEPTIONS ABOUT THE BIRTH OF ANALYTICAL MECHANICS

In his many and fundamental essays on the history of mechanics, Clifford Truesdell seems to use the term "analytical mechanics" in a very strict sense. According to him this term has to be used, in historiografical contexts as well in scientific ones, to designate a part of mechanics and specifically of rational mechanics. The following quotation is typical of this attitude:

> Mechanics appears [in Lagrange's *Méchanique analitique* [1]] as a physically closed subject, a branch of the theory of differential equations.
>
> [...] As its title implies, the *Méchanique Analitique* is not a treatise on rational mechanics, but rather a monograph on one method of deriving differential equations of motions, mainly in the special branch now called, after it, analytical mechanics. [2]

A similar judgement is also advanced by Truesdell in the preface of his *First Course in Rational Continuum Mechanics*:

> [...] that book [Lagrange's *Méchanique analitique*] covers only certain aspects o the rational mechanics created by Lagrange's great predecessors [...].
>
> In 1788 [the date of publication of Lagrange's *Méchanique analitique*] the mechanics of deformable bodies, which is inherently not only subtler, more beautiful, and grander but also far closer to nature than is the rather arid special case called "analytical mechanics", had been explored only in terms of isolated examples, brilliant but untypical. [3]

Thus, according to Truesdell's conception, the term "analytical mechanics" has to be used in historiographical context, in a sense that is very close to the sense it takes in the present-day treatises of classical mechanics, where it refers to that part of mechanics that treats of the equations of motion for dynamic systems, and particularly of the equations connected to the Lagrange's ones.

Of course, in Truesdell's interpretation the adjective "analytical" qualifies the term "mechanics" by limiting its scope. This limitation entails a number of different aspects. The first aspect is quite obvious, being connected with the fact that, by speaking of analytical mechanics, we are implicitly speaking of classical (rather than relativistic or quantum) mechanics. By contrast, the second aspect is only apparently obvious. Though it is connected with the plain fact that, by speaking of analytical mechanics, we are implicitly speaking of mathematical mechanics, it is not so obvious because of the way in which Truesdell means this fact. He seems to think first that mechanics' being mathematical is the same that its being rational. Then he assumes that, since it is mathematical, rational mechanics is, as such, a part of mathematics [4]. Finally he maintains that rational mechanics "is that part of mathematics that provides and develops logical models for the enforced changes of positions and shape which we see everyday things suffer" [5]. In the first of these attitudes, Truesdell follows Newton who in the preface to the *Principia* (that is for Truesdell the first treatise of rational mechanics [6]) maintained that mechanics is either rational or practical and, when it is rational, it "accurate procedit" and thus "ad geometriam referatur" [7]. The third aspect is only apparently obvious too. For it is concerned with the fact that mathematics which enters analytical mechanics is a particular one: it is just analytical mathematics. The difficulty is here concerned with the different sense of this last term: "analytic mathematics". I confine myself here to observe that for Truesdell this means that mathematics is here merely a tool to provide a formalistic exposition of mechanics, where this was not the case, for example, for Jean Bernoulli or Euler which used mathematics to make clear the fundamental concept of rational mechanics [8]. Finally, the fourth aspect is in no sense obvious at all. It comes from Truesdell's attitude to transpose on historical contexts the present-day practice to consider analytical mechanics as concerned only with a certain sort of mechanical problems and a certain type of formalism. In the already quoted preface of the *Principia*, Newton says that "*Mechanica rationalis* erit Scentia Motuum qui ex viribus quibscunque resultant, et virium quæ ad motus quoscunque requiruntur, accurate proposita ac demonstrata" [9]. Truesdell seems to assume that rational mechanics as such understands this definition in a large sense, while analytical mechanics understand it in such a limiting sense to constraint itself to the consideration of the motion (and equilibrium) of discrete systems of bodies (also considered under certain restrictive conditions).

Starting from a similar conception, Truesdell has to assume that analytical mechanics is essentially the result of Lagrange's formalistic conception of mathematics, even though other works, as Euler's *Mechanica* [10] for example [11], are part of its early history.

The second example refers to a more widespread attitude among historians of mechanics and mathematics. This attitude has been very well exposed by Michel Blay in his book concerned with the *Naissance de la mécanique analytique* [12]. Blay assumes that the term "analytical mechanics" in the title of his book refers to an analytical (rather that a geometrical) way to do mechanics. I quote the *Avant-propos* of Blay's book:

> [...] ce livre [...] est construit autour de deux questions précises:
> — Comment la conceptualisation differentielle de la science du mouvement a-t-elle été obtenue?
> — Comment cette concetualisation différentielle a-t-elle pu favoriser la réorganisation et le développement de la science de mouvement?
> Partant de ces deux questions, nous voudrions déterminer aussi exactement que possible par quelles démarches et au moyen de quels concepts une science analytique du mouvement a réussi à se constituer. [13]

Thus, for Blay, as well as for Truesdell, the adjective "analytical" qualifies the term "mechanics" by limiting its scope, but the only aspect of such a limitation, on which Blay seems to agree completely with Truesdell, is the first of the previous ones: by speaking of analytical mechanics, we speak of classical mechanics. Of course, there is also a sense in which Blay agrees with Truesdell on the second and third aspect, but it seems to me that this is just the sense that makes these aspects obvious: analytical mechanics is a mathematical theory and it is concerned with a particular sort of mathematics. Starting from his premises, Blay obviously maintains that analytical mechanics origins when Newtonian mechanics is translated from its original geometrical style in an analytical (that is a differential) language. According to Blay, this is initially done by Pierre Varignon who devoted to such an aim a number of papers between 1698 and 1711 [14,15]. To discuss these papers is the principal matter of Blay's book.

In his short article on classical mechanics in Grattan-Guinness' *Companion Encyclopaedia*, Craig Fraser accepts a perspective very similar to Blay's [16]. Here is what he writes under the title "Analytical Mechanics":

> The eighteen-century progress of theoretical mechanics occurred almost entirely on the Continent, using the mathematics of the Leibnitian calculus. The *Principia* became historically influential once its theory had been absorbed by researches working in the scientific academies of Paris, Berlin and St Petesburg. The French priest Pierre Varignon wrote a series of memoirs beginning in 1700 on orbital dynamics that employed Leibnitian analytical methods. He gave an elegant demonstration of *Principia*, Book I, Proposition 11 that replaced Newton's involved geometric reasoning by a simple algorithmic procedure. Varignon contemporaries John I Bernoulli and Jakob Hermann also contributed to his subject, providing a uniform mathematical treatment of the so-called 'inverse problem' in which it was required to determine the orbit from a knowledge of the force law. [17]

Fraser insists, however, on the essential role, in the development of the analytical mechanics, of Euler's differential formulation of the second Law of Newton, occurred only in 1750 [18,19].

I do not want to deny the legitimacy of the two previous interpretations of the term "analytical mechanics" in historical contexts. On the one hand, their difference can be finally reduced to a merely terminological one and it is also possible that the distinctions introduced by Truesdell are accepted by someone that decides to use the term "analytical mechanics" in historical contexts as Blay and Fraser do. On the other hand, it is clear that neither Truesdell, nor anyone else can deny the existence of a relevant difference between the geometrical style of Newton's *Principia* and the analytical style of the most part of treatises and memoirs devoted to mechanical subjects in the 18th century. Since Blay and Fraser use the term "analytical mechanics" just to refer to such an important difference, that I am obviously far to deny, it is clear that I can not disagree on any important point with them. Besides, even though I can follow Truesdell neither in his idiosyncratic insistence in diminishing the role and importance of Lagrange's contributions to mechanics and in reducing them to a mere formalistic compilation, nor in his implicit elimination of Aristotelian tradition from the history of mechanics, I think his distinctions are correct and I agree with him in stressing that 18th century mechanics was not only (and perhaps not mainly) a theory of motion and equilibrium for discrete systems of bodies.

Thus, the only incompatibility between the thesis I will defend here and the previous ones is terminological. But, since the term which is concerned—that is just the adjective "analytical"—is a crucial term in a very large number of scientific texts in the 18th century, this terminological incompatibility has a more fundamental meaning and is strictly connected with my first distinction between the application of mathematics to a certain science and the mathematization of this science. I think that when the adjective "analytical" is used in 18th century to refer to mechanics, it takes a meaning which is both narrower than that Blay and Fraser ascribe to it and larger than that Truesdell ascribes to it. My only aim is to suggest a possible reconstruction of this meaning.

As I have already anticipated, I shall consider here only one example, that is that of the science of motion, and even more particularly, the theories of motion of one or more (indeformable and distinct) bodies. Of course, mechanics is more than that. Nevertheless, I maintain that this example is historically relevant as such.

Before presenting my point (and justifying my historical judgements) let me sketch what I mean as the main lines of the development of the theory of motion from Aristotle to Newton. This shall allow me to illustrate partially my first distinction.

CLASSICAL FRAMEWORKS IN THE SCIENCE OF MOTION FROM ARISTOTLE TO NEWTON

Even though Aristotle provides, in the third book of *Physics*, a logical definition of it as "the fulfilment of what is potentially, as such" [20], he [21] seems to consider motion (κίνησις) as a primitive phenomenon. The definition need not introduce or fix the notion of motion, but provides only a logical analysis of it, in terms of the fundamental notions of the science of being, or metaphysics. We could even claim that with such a definition Aristotle legitimates the notion of potentiality, "en lui faisant représenter conceptuellement une réalité physique évidente" [22]. Aristotelian science of motion is thus the science of such a reality and not a science of an abstract object defined in terms of the primitive notions of fulfilment and potentiality. Moreover, Aristotle appeals to motion, as a real phenomenon, in order to characterise both nature and the science of nature, since, according to him, nature is "a principle or cause of being moved" [23], and the science of nature is essentially the science of motion. The reality of motion is a composite reality, however, since there are different sorts of motion, according to the different modalities of fulfilment or potentiality. Such a difference is also a primitive reality and it is as such perfectly evident to us. Namely, we are able to distinguish displacement (φορά)—that is the fulfilment of the potentiality of taking certain different positions or occupying certain places—from other kinds of motion.

By contrast, according to Aristotle, time is not a primitive or external reality, but only a phenomenological form of our perception of motion, or an abstract representation of it. Logically, it is the "number of motion in respect of 'before' and 'after'" [24], where the term "number" refers to what is numbered, rather than to what we use to number. However, the time, understood as the number of a certain motion, can be used to measure other motions. Namely, the number of the first (eternal, necessary and continuous) motion—that is the circular displacement of the first heaven—can be used to measure particular motions.

These are nothing but local and accidental consequences (or even local expressions, or accidental particularisations) of the first motion due to particular causes. As such, they are not only characterised by their respective measures, according to the time of the first motion, but also by the particularity both of the potentiality of which they are the fulfilment and of their subject. These positions change from a point to another point by following a continuous path, while time* (continuously) evolves from an instant to another instant†. Thus, by abstracting from the particular bodies which are moving and from the causes of their motions, different particular displacements can be compared—as quantities are compared (that is according to an order relation, including the possibility of equality between distinct objects)—by comparing their paths and the respective times. As this is a double comparison and paths are not always comparable as such, Aristotle can not get in such a way a total order of displacements and he is only able to compare certain particular displacements to certain other ones.

In fact, though Aristotle speaks about speed, when two particular displacements are compared in such a way, we have not to imagine that he compares the displacements by comparing their respective speeds. Even though any particular displacement has for Aristotle a certain speed, he never says exactly what speed is, and how it is measured in terms of the path of this displacement and of the time taken to cover it. Aristotle could not have been able to define speed as the ratio of space and time. This does not only depend on its general conception of nature and of its first principles, it is also connected to the absence in Greek mathematics of something similar to our conception of ratio between two quantities, particularly if two non-homogeneous quantities—as a path and a time—are considered. As long as it can not be a ratio (or any sort of entity defined in terms of space and time), speed is, according to Aristotle, a specific and primitive property of particular displacements, that is a quality of real phenomena. Of course, this property changes when the covered path and the corresponding time change and vice versa, but it is not determined as such when this path and this time are determined: it is not reducible to them.

In short, for Aristotle motion is a natural primitive phenomenon that can be characterised in general in term of the fundamental metaphysical categories of fulfilment and potentiality and described in its particular manifestation by specifying the particular body which is moving, the causes of its motion, the path this body covers, the time it takes to do it, and the speed with which it does it. From such a point of view no mathematization of the science of motion is of course possible. No obstacle forbids to apply mathematics to the study of motion, however. Though Aristotle did not write, as long we know, any mathematical treatise on motion, he referred, in a number of occasions, to some

---

* In what follows, I shall use the term "time" to refer both to the time of the first motion, that is the time as such, and to a fragment of the time of the first motion, that is a temporal interval. The context will make clear what of these meanings is intended.

† Even though such a description of a particular displacement in term of a moving body, a path and a certain time seems to be accepted by Aristotle as perfectly legitimate, it is far to be unproblematic in Aristotelian framework. As a matter of fact, the exigency to make it clearer is one of the main motivations for scholastic discussions about motion. I dealt with an example of these discussions in [21]: Panza (1994), where a large bibliography of secondary literature is also provided

mathematical notion in order to speak about motion. In particular, he did it when he considered the relations different displacements have with each other.

If two displacements take the same time (or equal times) and the paths are comparable—i.e. we can speak about distances, rather than paths in general—the order of distances is also the order of speeds. If the paths of two displacements—or the distances covered by them—are equal, the order of times is opposite to the order of speeds. If neither the paths covered by two displacements, nor the corresponding times are equal, these displacements can not be ordered in the Aristotelian framework. Nevertheless, if their paths are comparable, such a framework is at least compatible with the possibility of stating certain relations between two hypothetical distinct motions.

In the chapter VII.5 of *Physics*, Aristotle states quite clearly four rules or theorems about motion and the possible relations of two hypothetical distinct motions‡. He provides, of course, quite a discursive formulation of these rules, but their content can be formulated in a more formal language. In order to do it, I suppose that:

— $s(x, y, z)$ is the space covered by the mobile x upon which the force y is applied during the time z;

— $t(x, y, z)$ is the time taken by the mobile x upon which the force y is applied to cover the space z.

Then, the rules of Aristotle consist of the previous implications (where n is any positive integer; I put "*n*" where, to be faithful to Aristotle's text, I should put "2", since it is clear that Aristotle is here making an exempla of any integer multiple or sub-multiple):

R.1 if $s(nM,F,T) = S$ then $s(M,F,T) = nS$;

R.2 if $t(nM,F,S) = nT$ then $t(M,F,S) = T$;

R.3 if $t(M,F,nS) = nT$ then $t(M,F,S) = T$;

R.4 if $t(nM,nF,S) = T$ then $t(M,F,S) = T$

Aristotle claims that it is so because "in this way, the proportion is preserved" [25]. This is not really a prove, however, and, really, it would be difficult to understand how he could really prove his rule, without starting from any definition or postulate fixing the relations between mobiles, distances covered, forces applied upon a mobile, and time during which this is applied. Probably, he thought that these rules were evident enough to provide a description of the relations that would take place between two real distinct motions if they satisfied the conditions indicated in these rules. Thus, he does not define in general the notions he is employing in stating his rules, just because he thinks that these notions simply refer, in general, to empirical entities we can distinguish in reality when considering a particular motion. Nevertheless, it is easy to imagine something like some general laws that can be formulated in the framework of the classical theory of proportion, starting from which the rules of Aristotle, or other similar rules, can be proved. In Aristotle's rules, speed is not mentioned. It could occur in these general laws, however, and be used in the proof of Aristotle's rules as an intermediate term. The following are two groups of laws fixing respectively the relations between space, time and speed and between force, resistance and speed. I use the term "resistance" to refer, in the Aristotelian spirit, to something that opposes the motion: force brings about motion and resistance opposes the latter. If S_i, T_i, V_i, F_i, and R_i $(i = 1,2)$ are respectively the space (or distance) covered by two distinct motions, the times these motions take, their speeds, the forces that bring about them, and the resistances that oppose them, while K, L, H are three magnitudes taken in order to satisfy the antecedents of the implications, then these laws are:

L.I.1 if $S_1 : S_2 = K : L$ and $T_2 : T_1 = L : H$, then $V_1 : V_2 = K : H$

‡ Cf. [20] Aristotle, *Physics*, 249b, 30 – 250a, 7.

L.I.2 if $V_1 : V_2 = K : H$ and $S_1 : S_2 = K : L$, then $T_2 : T_1 = L : H$

L.I.3 if $V_1 : V_2 = K : H$ and $T_2 : T_1 = L : H$, then $S_1 : S_2 = K : L$

L.II.1 if $F_1 : F_2 = K : L$ and $R_2 : R_1 = L : H$, then $V_1 : V_2 = K : H$

L.II.2 if $V_1 : V_2 = K : H$ and $F_1 : F_2 = K : L$, then $R_2 : R_1 = L : H$

L.II.3 if $V_1 : V_2 = K : H$ and $R_2 : R_1 = L : H$, then $F_1 : F_2 = K : L$

If we the mobile occurring in the Aristotle's laws is intended as a resistance, the proof of his rules starting from these laws is easy. I will just consider R.1. If one put $R_1 = nM$, $R_2 = M$, and $F_1 = F_2 = F$, then the two premises of L.II.1 become respectively $F : F = 1 : 1$ and $M : nM = 1 : n$; thus, according to this law we have: $V_1 : V_2 = 1 : n$. If one pose then $T_1 = T_2 = T$, the two premises of L.I.3 become respectively $V_1 : V_2 = 1 : n$ and $T : T = n : n$, thus, according to this law we have: $S_1 : S_2 = 1 : n$. From this it follows that if $S_1 = S$, then $S_2 = nS$, that is exactly what R.1 states. The proofs of the other rules are similar.

I do not want to argue either that Aristotle actually asserted the laws L.I.1-3 and L.II.1-3, or that he referred to them implicitly. I simply argue that these rules formulate in a mathematical language, which was perfectly at disposal of Aristotle and the scholastics, some relations between some aspects or properties of actual motions that seem to be implicitly accepted by the latter. The laws of the first group compare two distinct motions, whatever they are, by stating some relations between the mutual ratios of the distance they cover, the time they take and the speed the have. In these way they fix the relations between the distance covered by any motion, the time it takes and its speed. By using a typical language of theory of proportions, the first law of this group states that the speeds of two motions are between them in the composed ratio[§] of the distances they cover and the times they take to do it. The other two laws state the converse of the same. The laws of the second group compare two distinct motions, whatever they are, by stating some relations between the mutual ratios of the forces that bring about them, the resistances that oppose to them, and their speeds. In these way they fix the relations between the force that brings about any motion, the resistance that opposes it and its speed. By using the language of theory of proportions again, the first law of this group states that the speeds of two motions are between them in the composed ratio of the forces that bring about them and the ratio of the resistances that oppose them [26]. In this case as well, the two other laws state the converse of the same.

These are two different ways to compare motions, the first one according to their effects, the second one according to their causes. Thus, the previous laws and deduction from them or the rules of Aristotle prove that the Aristotelian conception of motion, both of its effects and if its causes, is compatible with an application of mathematics to the science of motion, even if it is far away from any sort of mathematization.

These two distinct ways of comparing and characterising displacements, either according to their effects or according to their causes, are both part of the Aristotelian science of displacements. Though the second one is typical of the official Aristotelism, at least at 17th century—so much that we could define Aristotelian theory of displacements, as a "physical theory of displacements and their causes" [27]—the first one provides the basis for the main result (from a modern point of view) of this science, that is the distinction between uniform, uniformly difform and difformly difform displacements, introduced in the 14th century by Oresme and the Mertonians [28]. Such a distinction does not need any reference to the notions of speed and acceleration as variable ratios. Speed is nothing but a quality, whereas acceleration, or better difformity, is nothing but the phenomenon of variation of the

§ Note that the term "ratio" does nor refer here to a specified object, as it does today; it is only a technical term occurring in the description of a certain system of proportions.

intensity of speed. The previous distinction is thus a particular application of a more general distinction, concerning the modalities of variation of the intensity of a quality. A certain body moving according to a certain displacement is thus a particular subject with a certain quality, that is just speed. The intensity of such a quality can change in different ways. In particular, it can change with time and hence be taken as an instantaneous quality. This change can be either uniform or difform; in the first case the displacement is also uniform, while in the second case it is difform. This is a very natural distinction, but it can work only if we know, by an independent source, what the quality of speed is or, at least, how it can be measured. The basic idea both of the Mertonians and Oresme is that of measuring speeds by using the rectilinear spaces covered in a certain time by the moving bodies which are affected by these speeds, taken as constant with respect to this time. Thus [29], the displacement of a certain body is uniform, if the spaces covered by this body in equal times are always equal, otherwise it is difform. Moreover, if it is difform, it is uniformly difform, if "in quancunque equali parte temporis, equalem acquirit latitudinem velocitatis."

The later definition is, as such, quite obscure, since spaces are not mentioned, However, if we represent time by using a certain axis and different speeds at the extremes of equal times by means of equidistant perpendicular segments to such an axis, such a definition is perfectly sufficient for concluding that when the displacement is uniformly difform the extremities of these segments fall all on a straight line oblique with respect to the axis, while they fall on a straight line parallel to the axis, if the displacement is uniform. This is the basis of Heytesbury's and Oresme's [28, 30] diagrammatic representation of uniform and uniformly difform displacements and of the consequent proof of the "theorem of Merton college" which appeals to the comparison of these diagrams. This theorem states that the distances covered by a uniformly difform displacement at any given time is the same as that which is covered by a uniform rectilinear displacement, with a constant speed equal to the speed of the uniformly difform motion in the middle of this time. Though it explicitly deals with speed as an instantaneous (or punctual) quality, the proof of this theorem is not founded on any explicit definition, either of speed in general, or of instantaneous (or punctual) speed. It simply works because of a diagrammatic formalism associated to the metaphysical idea of speed (or quality, in general). This formalism is far for providing a mathematical object modelling the real phenomenon of motion. It is only a convenient way to speak about it and represent certain of its properties. Once again it makes possible a convenient application of mathematics to the science of motion, but is far from being the result of mathematicisation.

Different interpretations are possible of both Heytesbury's and oresme's proof [31]. All of them agree on an essential point, however: these proofs have nothing to do with the classical paradigm of the theory of proportions. Under certain possible interpretations, they even break the most fundamental clause of this theory—that is, homogeneity—by considering a certain surface both as a representation of a linear distance and as the result of an infinite sum of segments representing instantaneous speeds. By contrast, for Galileo (who nevertheless proves twice the theorem of Merton College, by two distinct arguments, both analogous, even if not equivalent, to those of Heytesbury and Oresme [32, 33]), as long as the science of motion is a mathematical science, it is founded on the theory of proportion. At most, this theory should be reformed, for the new aims of the science of motion [34], but it can not be substituted with alternative formalisms.

According to Galileo**, there is no unitary science of Aristotelian motion. Even though he still speaks of "local motion" (that is, displacement) to specify the object of his research, for him the science of motion is nothing but the mathematical science of displacements, since motions are nothing but displacements. Moreover, this science does not deal with the causes

** The *locus classicus* for Galileo's science of motion is obviously [33]: Galilei (1638), 3th and 4th *Giornata*.

of motions—it is not a form of dynamics (or a theory of gravity and lightness)—but it deals, like in Oresme's and the Mertonians' approach, with motions as long as they are characterised in terms of spaces that are covered and times that are spent for that—it is properly a kinematics (a theory of space and time) [26, 35]. Nevertheless, even though this science has to be, according to Galileo, a mathematical science, it does not introduce its objects as mathematical ones. Rather it is founded on a mathematical characterisation of physical phenomena. Once again, Galileian science of motion results from an application of mathematics to the study of motion and not from a process of mathematization. Galileo does not study, for example, uniformly accelerated motion, rather he studies the "natural motion" of a falling body. He recognises—by experimental and rational arguments—that this motion is such that speed uniformly changes in it, characterises it as a motion which satisfies a certain mathematical property in the context of the theory of proportions, and then looks for a deduction of its laws starting from such a mathematical characterisation. Both the first and the second stage of this research were far from being easy. As a young man, Galileo believed that the natural motion of a falling body was essentially uniform, afterwards he was quite sure that instantaneous speed was proportional to the covered space, rather than to the elapsed time [36].

Still, even when Galileo recognised that the natural motion of a falling body is "uniformly accelerated", he did not characterise it as a motion whose acceleration is constant, since acceleration was not a quantity for him, but just a variation of speed (or a cause of such a variation). Moreover, he did not reduce speed to space and time, by means of an explicit mathematical definition. As long as it is a quantity, speed is as such, independent from space and time. As long as it is possible to provide correlative definitions of these three quantities, these definitions are just metaphysical characterisations and do not provide any sort of mathematical reduction. The problem of fixing the mathematical relations between these quantities is thus independent from the problem of definying or characterising them. This is a very natural situation in the context of the theory of proportion, where the condition of homogeneity prevents two non homogeneous quantities from operating with each other [37, 38].

While Galileo bases his "new science" on the theory of proportion, Newton [7] has completely absorbed the new Cartesian geometry. What was essentially new in Cartesian geometry [39, 40, 41] was the definition of internal product and ratio between two continuous quantities: according to such a definition two continuous quantities could be multiplied and divided one for the other, the result being a third continuous quantity. This quantity is not understood by Descartes as produced by multiplication or division but it is only selected as the result of these operations in a domain of existing quantities, according to the previous choice of a unity. If the multiplied or divided quantities are homogenous and the chosen unity is also homogenous to them, the resulting quantity is necessarily homogenous to its factors. This is the case implicitly considered by Descartes in his definitions. Since these definitions use classical proportions, and a proportion does not require the homogeneity of the four quantities occurring in it, such a complete homogeneity is not necessary, however.

Let be u the unity and a, b and c three continuos quantities such that $a : c = u : b$. If a and b are given and they are homogenous with each other, u must also be homogenous with a and b and the Cartesian product c of a and b is necessarily homogenous with its factors. However, if a and b are given and they are not homogenous with each other, such a proportion requires only the homogeneity of u with b and of c with a. Because of the commutativity of multiplication, this means that the Cartesian product of two non homogeneous quantities a and b can be homogenous either with a or with b.

Let us imagine now that c and b are given and they are not homogeneous. The Cartesian ratio a of c and b is necessarily homogenous with c. But a can be also intended as the ratio of two other (homogenous) quantities C and B that are both not homogenous with it, and thus whit c, and this ratio can be, at its turn, intended as equal to a new quantity A non homogeneous both with c and b. The Cartesian ratio of the two non homogeneous quantities c and b can then be equal to a quantity A non homogeneous with both c and b. Thus—even without introducing measures of these quantities by means of segments or any other elements of a universal class of quantities [42]—it is possible to imagine, in the Cartesian framework, that speed is the ratio of space and time and that it is not homogeneous with both the former and the latter. The same is true of acceleration, that is now the ratio of speed and time, even though it is not, as such, homogeneous with both speed and time.

This is a very nice situation from the point of view of definitions, but it is quite impossible to imagine how these definitions could be translated as such in the principles of a mathematical science of motion. The difficulty is obvious: Cartesian definitions of algebraic operations on continuous quantities are not, as such, constructive; they provide the conditions under which a certain quantity is the product or the ratio of two other given quantities, but they do not provide, as such, this product or ratio, even though the unities are fixed. Thus, they do not provide as such any way to compare, respectively, two different speeds or two different accelerations with each other.

It is to solve such a difficulty (and not to provide suitable definitions) that Newton has to assume the reducibility of any mechanical quantity to its "measure" in terms of the elements of the universal class of segments. Segments are, in fact, the only continuous quantities for which Descartes provided not only a definition of internal multiplication and ratio but also showed how the results of these operations (as well as of power and of any sort of integral root) could de constructed. This means that if a certain motion is given, by giving the measures, in terms of segments, both of the space it covers and of the time taken to do it, then the measures of its speed and acceleration can be easily constructed under the form of suitable segments. It is just this construction that provides both a sense for algebraic (or even numerical) calculation of these quantities, and makes their algebraic expressions able to denote a proper object [43].

But a new difficulty arises. If the space covered by any motion is represented by a segment, Cartesian geometry cannot be used for representing trajectory of non rectilinear motions. Newton's central idea to solve such a difficulty is directly connected to the Newtonian conceptions of *inertia* and force, since it appeals to the composition of motion.

The motion of any sort of body is represented by the curve described by a point reacting in a specific way to certain forces in the three-dimensional space where the motion takes place. If no force acts upon a moving body, it moves of a rectilinear uniform motion; other sorts of motions are due to the action of forces that accelerate the body and change its direction. The rectilinear direction of original uniform motion combines itself with the rectilinear directions associated with any force acting on it. If a force acts punctually upon the body, it changes its direction and speed, but as the force acts only in one point, after this point the body moves of a uniform rectilinear motion, as long as no other force occurs. Curvilinear trajectories are thus only possible if one or more forces act continuously upon the body and change continuously its direction. A single non rectilinear motion is then understood as a continuous composition of continuous changing virtual and rectilinear motions. Thus, at any point of the curvilinear trajectory of this motion, distinct virtual motions are associated, expressing the punctual speed of this motion and the forces acting in this point upon the moving body. Even though speed and forces are variable, these punctual virtual motions are not only rectilinear, but also uniform, since they express a punctual value of speed and forces. Hence, they are represented by suitable segments: the former by an oriented segment taken on the tangent to the curve (since the direction of this tangent is just the punctual direction of the curve), with its origins in it; the latter by any oriented segment with the same origin. Of course, two constant speeds or forces φ and ψ (or the punctual values φ and ψ of two speeds or forces) have to be respectively represented by two segments A and B verifying the proportion $\varphi : \psi = A : B$. In general, Newton interprets these segments as a representation of the rectilinear spaces that would be covered in a certain time by a body moving according to the rectilinear uniform motion expressing these speeds or forces (or their punctual values). Under this interpretation, they are measures of speeds and forces, since they represent spaces that measure motions, expressing speeds and forces. As long as speeds and forces are measured by these segments, they compose each other according to the law of the parallelogram applied to them. Finally, speeds and forces are related by the condition of proportionality between force and acceleration (where the constant of proportionality is intended as the mass of the moving body and it is consequently the same for any force acting upon the same body).

Newton mechanics develops, according to these premises, as a geometry of trajectories—referred to two or three Cartesian coordinates and described by points with a constant mass—to each position of which suitable vectors representing punctual speeds and forces are associated. Time is in general understood as an external common parameter, fixing the measure of uniformity for the variation of any considered quantity. The aim is to determine trajectories, if forces and initial speeds and positions are given, or forces, if trajectories and initial speed and positions are given [44..52]. If only one body is considered, its mass can be taken as unitary. But it is also possible that the motions of two or more bodies are connected with each other, either because the moving bodies are linked, or because the forces acting upon each of these bodies depend on the masses and positions of the other bodies. If it is the case, we are faced with a an n-bodies problem.

Since motions are considered as the effects of forces (or *inertia*), Newton mechanics is, as Aristotelian science of motion, a dynamic, rather than a kinematics. Nevertheless, as long as speeds and forces are no more understood by Newton as constant or variable qualities of motion, the latter is nothing else but a mathematical object. Thus, as long as it is studied by means of the mathematical theory introduced by Newton, motion is not given as a real external phenomenon. Rather, it is given as a mathematical object, once a certain system of connected geometrical entities is given. Hence, the Newtonian theory of motion is not only a mathematical science, i.e. the result of the application of mathematics to the study of motion, but it is also a science of mathematical objects, i.e. the result of a process of mathematisation. Mathematics is not just a tool which is able to provide suitable characterizations of natural phenomena, as it was for Galileo; rather it is now a science inside which suitable models for natural phenomena are constructed and studied.

As long as motion is, for Newton, a mathematical object, it is essentially a geometrical one. It is a relation between two magnitudes, represented by a certain curve, which has to be studied by a new sort of geometry. One of the aims of analytical mechanics in 18th century is that of transforming the Newtonian science of motion in an analytical science, i.e. to pass from a geometrisation of the science of motion to a new theory of motion where the latter is just an analytical object. I shall try to describe briefly this transition in the next paragraph.

ANALYTICAL MECHANICS

As long as speed is defined in general as a ratio between a certain space and a certain time, it could seem very natural to express punctual speed as the differential ratio ds/dt and consequently punctual acceleration by $dv/dt = d^2s / dt^2$. However, in order to understand these ratios as proper objects, that is quantities which could be exhibited, post-Cartesian mathematicians had to interpret them as the ratios of two segments, measuring respectively differentials or fluxions of space and time. Of course, they could assign a numerical value to them, but as long as continuity were not considered as a property of the field of real numbers (like in the modern conception based on Cantor-Dedekind's construction), this value can only be understood as a numerical expression of a continuous quantity, like a segment[††]. Thus, an analytical expression of mechanical quantities in terms of differentials

[††] Before Cantor defined a real number as the limit of a Cauchy's succession of rational numbers, no justification was possible for the practice of using numbers in order to express continuous quantities. In

could not be understood in the end of 17th century as independent from a geometrical interpretation of them. Moreover, since the essential law of composition of motions (that provided also compositions of forces) was intrinsically geometrical [53 .. 57], as well the Newtonian representation of trajectories of motions as curves referred to two or three Cartesian coordinates, no analytical formulation of the new science of motion in terms of differential calculus could be independent from a geometrical basic treatment. Namely, such an analytical formulation could not be anything else than an analytical description of a geometrical model. Even though such a description would have simplified many Newtonian proofs, by using algebraic formalism where these proofs advanced by considering geometrical relations, it is a matter of fact that Newton never appeals to it. A number of reasons have been advanced to justify such an apparently strange attitude. However, I take it that one of them is that such a treatment would be probably appeared to Newton as an unnecessary translation of an intrinsically geometrical science (where simplicity could have been gained at the expense of a loss of rigor, globality and intelligibility).

As I have already said, the mathematical object of the Newtonian science of motion (that is just motion as a mathematical object) was intrinsically a geometrical object. With the term “intrinsically” I mean that this geometrical nature did not depend merely on an expository choice of Newton, but it depended on Newton’s conceptualisation of the objects of his science.

Nevertheless, once Newton's *Principia* had been studied and understood, an analytical formulation of its principal definitions, results and proofs in terms of differential calculus, should have not been quite easy to obtain for a good mathematician of the end of 17th century. As I have already said, the first programmatic effort in this direction was made by Varignon. According to Blay’s reconstruction, this effort could be considered as the third stage of a progressive process along which—in the years following the publication of the *Nova Methodus* [58]—Leibnitian calculus appropriates the theory of motion.

The first stage of this process [59] is marked by a number of differential solutions for problems concerned with the determination of a certain curve responding to certain cinematic or dynamic properties, like the isochrone, the brachistochrone or the curve of equal pressure. What is typical of these solutions (proposed for the most part by Leibniz and Jacob and Johann Bernoulli) is that they proceed all from a preliminary reduction of the problem to a purely geometric condition, and apply then the differential calculus to such a condition. Thus, as Blay remarks, they “ne present finallement aucune conceptualisation différentielle spécifique de la science du mouvement” [60]. Moreover, the preliminary reduction does not depend on any geometrical representation (by means of the Newtonian vectors or anything else) of speed or forces, being founded on known relations between these latter entities and certain entities entering the usual geometrical representation of the searched trajectory.

The second stage [61] is completely due to Leibniz and consists in the establishment of a differential expression for the relations occurring in any motion between the instantaneous speed and both the elements of space and time and the elements of this speed itself. Leibniz reaches, for example, the determination of a “general principle”, where he generalises the law asserting that the speeds of two uniform motions are between them in the composed ratio of the ratio of distances and the inverse ratio of times, by stating a relation of proportionality between ds and vdt [62]. As Blay remarks, this not yet a differential characterisation of the instantaneous speed. Rather, instantaneous speed is geometrically represented by certain (infinitesimal) segments, that are thought as rectilinear spaces covered

---

such a situation, to ascribe to a segment (or any other continuous quantity) a numerical value was only a practical simplification without any justification.

in a given time, in such a way that it is impossible to calculate the times when spaces and speeds are given [63].

Finally, the third stage [64] consists in Varignon's differential characterisation of the principal notions of Newton's theory and of their general relations. This is done by means of a geometrical representation not only of certain linear spaces or trajectories, but also of times speeds and accelerative forces. This representation enables Varignon to eliminate any constraint of homogeneity, since it allows him to compare spaces, times, speeds and accelerative forces by means of a comparison of segments and leads him to the crucial identities $v = \frac{ds}{dt}$, $dv = \frac{d^2s}{dt}$ and $\gamma = \frac{dv}{dt}$ (γ being the accelerative force). These identities make the solution of a number of cinematic and dynamic problems independent of a geometric analysis: if the data are expressed by means of suitable equations, the solutions can also be expressed by other equations derived by using the algorithm of the calculus. As Blay says:

> [Varignon] elabore la solution de[...][ces] problèmes, non plus comme ses prédécesseurs, à partir des éléments géométriques représentés dans telle ou telle figure, mais directement à partir des relations entre les symboles. Dans cette perspective [...], la figure, essentielle pour le développement et l'organisation de la pensée géométrico-infinitésimale, devient progressivement avec Varignon simple diagramme. C'est-à-dire que cette figue perd sa valeur traditionnelle d'intellection pour prendre une valeur toute secondaire d'illustration [65].

Varignon's approach was largely improved by the treatment of mechanical problems in the Leibnitian school, in particular by Jean Bernoulli [66], and is the basis both of Euler's *Mechanica* and of the footnotes of Le Seur and Jacquier to Geneva's edition of the *Principia* [67], where a number of Newton's proofs and arguments are reformulated in analytical terms, according to the algorithm of the calculus.

As I have already said, there is certainly a sense according to which these important results constitute the first stage of analytical mechanics, they mark the birth of it, as Blay says. However, there is also a sense according to which the birth of analytical mechanics is not there, it does not only depend only on the translation of Newton's definitions, principles and proofs in the analytical language of differential (or fluxional) calculus.

A first remark is obvious: all the differentials entering Varignon's formulation of mechanical relations are infinitesimal differences between segments or portions of curves; moreover, they measure (or express) mechanical entities, just because they are infinitesimal differences. As long we deal with a cinematic or a simple dynamic problem, this is nothing but a question of interpretation of algorithmic procedures, but when a more difficult problem is considered the geometrical nature of Varignon's approach becomes manifest from an operative point of view too.

Let us consider the example of the expression of the punctual central force that is responsible for a curvilinear motion [68, 69]. The following is a simplified version of Varignon's argument drawn from Blay's reconstruction [70, 71]. Let QL be the trajectory described by a moving body because of a central force with centre in C. Let AH = x and HL = y be the coordinates (respectively linear and circular, with centre in C) of the running point L. Varignon considers a point l infinitesimally close to L and assumes that the accelerative force γ acting on L during the time dt is proportional to the infinitesimal segment LR. Then, he decomposes this forces into two orthogonal components taken respectively on the direction of the tangent LM and the normal LN to QL in L. Since γ and its orthogonal components can be taken as constant during dt, we can assume the identity $\frac{\gamma_t}{\gamma} = \frac{\mathrm{LP}}{\mathrm{LR}}$, where γ_t is the tangential component of γ. Now, because L and l are two points infinitely close to each other, the infinitesimal curvilinear triangle LRl, can be considered as a rectilinear one, which is similar to the infinitesimal triangle LRP.

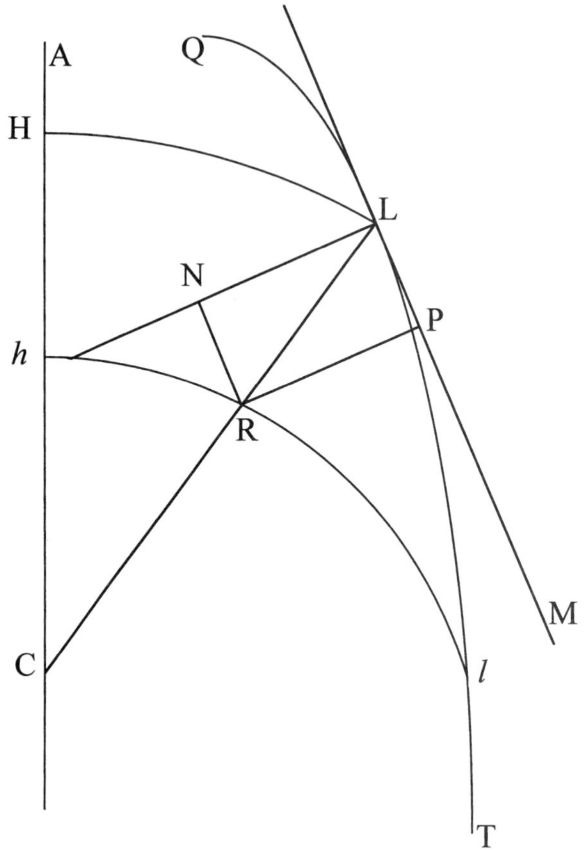

Figure 1. Example of the expression of the punctual central force

Hence, we will have also $\frac{\text{LP}}{\text{LR}} = \frac{\text{LR}}{\text{L}l}$ and thus: $\frac{\gamma_t}{\gamma} = \frac{dx}{ds}$ or $\gamma_t = \gamma\frac{dx}{ds}$. Consider now the (virtual) rectilinear motion along LM, due to the force γ_t. Since this motion is rectilinear, it satisfies the identities $d^2s = dvdt$ and $dv = \gamma_t dt$, which entail: $d^2s = \gamma\frac{dx}{ds}dt^2$ or $\gamma = \frac{dsd^2s}{dxdt^2} = v\frac{dv}{dx}$.

Even though it uses the differential algorithm, this argument consists of a Newtonian analysis of the central force acting upon the body. This is a very simple exemplification of my point: Varignon's differential expression of mechanical quantities and relations does not enable him (when the problem is a dynamic one and non rectilinear motions are concerned) to avoid any appeal to the Newtonian method of analysis of forces, which is founded on the consideration of intrinsically geometrical models. Now, according to the meaning of this term I am considering here, what is essential in order to speaking about "analytical mechanics" is neither the expression of speed and acceleration by means of differential or fluxional ratios, nor the expression of the link between the different variables occurring in a geometrical model of motion by means of suitable equations where analytical forms occur. Rather, it is just the substitution of Newtonian geometrical models with equations of motion as a basic description of a certain mechanical situation. When simple situations are considered, the difference between these two approaches is not always evident and often vanishes. In a sense, Newton himself uses his principles to determine equations of motions (even though these equations are always understood as a description of a geometrical model), and the basic idea of determining standard forms for equations expressing the data of a mechanical problem of a certain sort and containing its solution is already explicit in Varignon's memoirs—particularly when resisting motion is considered [72, 73]—Bernoulli's solution of the problem of shock of bodies [74] and, overall, Euler's *Mechanica*. Nevertheless, in all these texts, equations of motions are the result of an analysis of forces—a real Pappusian analysis, applied to geometrical models of mechanical problems—and their general form results from codified methods for such an analysis, which are applied to different sorts of problems.

In the original sense of such a term—that Viète and Descartes had inherited by Pappus—this is a very analytical procedure, since it starts with the assumption of what is searched as it were given [75, 76]. But it is not so, if the term is taken in the new sense it will take in 18th century. In 18th century, mathematical analysis is neither the first stage of the Pappusian "method of analysis and synthesis", nor a particular procedure for solving mathematical problems—as it is in classical language. Moreover, it is not a branch of mathematics—as it is today . It is a mathematical theory—the new theory of functions—but it is not a particular theory among many other, since its aim is a reformulation of mathematics as a whole: a function is not a particular mathematical object, among a number of other mathematical objects, rather it is an abstract representation of a quantity, and mathematics is just the science of quantities [77]. Considered as functions, quantities need not to be distinguished between numbers (or discrete quantities) and magnitudes (or continuous quantities), and magnitudes need not to be qualified, according to their specific nature. As long as it is represented by a certain function, a quantity need not to be exhibited as such, and it is not necessary to give specific criteria of identity for it. General criteria for functions work in any circumstance and if we are studying a particular sort of quantities, we must only provide a suitable interpretation of a consistent formalism. Thus, 18th century mathematical analysis is a unitary formulation of mathematics. In fact; it is the first unitary exposition of the science of quantities. More generally, in 18th century science and philosophy, the ideal of analysis is that of a logic organisation of knowledge, according to an internal and unitary order coming from general principles or elements [78], and transmitting to particular applications of them. The place of mechanics, in such an organisation, is not simply that of a collection of standard methods for solving problems concerning with motion, or, as Truesdell says, with "the enforced changes of positions and shape which [...] things suffer". Rather, it is the place of the science of these changes, understood as general and universal phenomena expressed by mathematical objects and governed by general laws of nature.

In such a framework, the program of analytic mechanics could not have been merely a program of exposition of Newtonian mechanics by means of a certain language. It was a program of edification of a more general science, founded on general principles expressing the natural laws of motions (rather than on Pappusian analysis of geometrical models), where Newton's results are particular cases of general results. Of course, the general theory of functions could not provide, as such, the principles of this science. However, these principles were expressed by general equations where suitable functions occur, in such a way that mathematical deduction could operate starting from them according to the methods and results of the general science of functions (that is analysis) in order to provide the particular equations characterising a particular mechanical situation.

Let us consider the example of the problem of motion of one or more bodies again. The essential aim of what I would like to call "analytical mechanics", as long as it was concerned with this problem, was to determine a little number of general principles—like those of the least action, the conservation of the *vis viva* or the virtual velocities—according to which it were possible to express the dynamic characters of a system of n bodies, upon which any sort of forces act, by means of suitable equations [79, 80]. As long as these equations were general, they referred to any sort of mechanical system, independently from the number of bodies and forces acting on them. *Stricto sensu*, they were forms or schemes of equations. But, as long as forms or schemes of equations are equations, they worked, at the same time, as the starting point of deductive proofs of general results relative to any sort of mechanical system, and as the guidelines for writing the equations of motion of particular systems. From such a point of view, motion is no longer merely a mathematical object, rather, it is an analytic object, i.e. a functional relation expressed by an analytical form.

This approach was not only different from the Newtonian one for epistemological reasons. It made also possible address problems that no analysis of force could solve, as those concerning a great number of bodies. Moreover, from such a point of view, the different systems of bodies became elements of an infinite domain of objects of a common and universal form that could be easily expressed in general. As Kant said, it made possible to treat motion as a "universal in *concreto*" [81, 82].

Understood in this way, i.e. as a system of deductive consequences of general principles expressed by suitable equations (or forms of equations), 18th century analytical mechanics seems to be completely independent of Newtonian geometrical models and analysis of forces.

However, this is not the case, if the question of justification of the general principles is considered. In fact, even though these principles were often formulated in metaphysical terms and supported by metaphysical arguments, they could only be translated in a certain equation when they were applied to something such as a simple, and easily generalisable, mechanical model, which is, necessarily, in 18th century, a Newtonian geometrical model. This is even clearer if mathematical arguments are considered. These arguments were either *a priori* or *a posteriori* [83]. The former tried to deduce the equations expressing the general principles directly from something as a Newtonian analysis of forces applied to a general system; the latter merely consisted in the verification of the equivalence between the consequences of the general principles when they were applied to simple systems and the consequences of Newtonian analysis of force applied to the same systems. In the same way, and for obvious reasons, 18th century mechanics could not be independent of Newtonian geometrical models, if the aim was to give a proper mechanical content both to the analytical description of a mechanical situation and to the different sorts of mechanical equations.

Thus, an unsolved tension, between the philosophical need of elimination of Newtonian geometric analysis of forces and the mathematical need of appealing to them is intrinsic in the birth and evolution of analytical mechanics in 18th century: in my sense, no history of this birth and evolution is possible if it is not also a history of such a tension, which is also a tension between two distinct way of accomplishing a mathematization of the science of motion.

ACKNOWLEDGMENTS

The present paper has been completed during a visiting scholarship at the Faculty of Sciences (Departamiento de Matematica) of the University of Mexico-City (UNAM). I thank it for its logistic and financial support. I thank also Pierre Souffrin for useful discussions concerning a part of my paper.

REFERENCES

1. J. L. Lagrange, *Méchanique analitique*, La veuve Desaint, Paris (1788); IInd ed., *Mecanique analytique*, Courcier, Paris, 2 vols. 1811-1815

2. C. Truesdell, *Essays in the History of Mechanics*, Springer-Verlag, Berlin, Heidelberg, New York (1968). p. 134. ("A Program toward Rediscovering the Rational Mechanics of the Age of Reason", first publication in 1960, *Archive for History of Exact Sciences*, **1**, 1960-1962, 3-36).
3. C. Truesdell, *A First Course in Rational Continuum Mechanics*, vol. 1, p 9, Academic Press, New York, San Francisco, London (1977).
4. Ref 2, p. 335 ("Recent Advances in Rational Mechanics", first published, in a different version in *Science*, **127**, 1958, pp. 729-739).
5. Ref 3, vol. I, p. 4. As Truesdell explicitly remarks the term "things" refers here to natural things (as animals, plants, seas, minerals, planets, etc.) and man's artifices.
6. Ref 2, p. 335 ("Recent Advances [...]", *op. cit.*).
7. I. Newton, *Philosophiæ naturalis principia. mathematiaca [...]*, Jussu Soc. Regiæ ac Typis J. Streatner, Londini (1687); *Præfatio ad lectorem*, first page [not numerated], 2nd. ed., without publisher, Catabrigiæ 1713; 3th ed., apud G. and J. Innys, Londini, (1726).
8. Ref 2, p. 133-134 ("A Program [...]", *op. cit.*).
9. Ref 7, *Præfatio ad lectorem*, first page [not numerated].
10. L. Euler, *Mechanica, sive motus scientia analytice exposita*, ex typ. Acad. sci. imp., Petropoli, 2 vols. (1736).
11. Ref. 2 p. 106 ("A Program [...]", *op. cit.*).
12. M. Blay, *La naissance de la mechanique analytique. La science du mouvement au tournat des XVIIème et XVIIIème siècle*, PUF, Paris (1992).
13. Ref 12, p. 8.
14. Ref 12, pp. 153-221 and 277-321, where a complete bibliography of Varignon's papers on this subjects is given.
15. D. Bertoloni-Meli, *Equivalence and Priority: Newton versus Leibniz*, pp. 201-205Clarendon Press, Oxford (1993).
16. C. Fraser, "Classical mechanics", in I. Grattan-Guinness (ed. by), *Companion Encyclopedia of the History and Philosophy of the Mathematical Sciences*, particularly pp. 973-975 Routledge, London and New York (1994), 2 vols. with continued pagination, pp. 971-986
17. Ref 16., p. 974.
18. L. Euler, "Découverte d'un nouveau principe de mécanique", *Mém. Acad. Sci. Berlin*, **6** (1750, publ. 1752), pp. 185-217. Note that in this paper, Euler presents the Nawton's law (in his new formulation) as a "new principle of mechanics" that he pretends to have discovered. Lots of papers have been written on the subject of Newton's second Law where distinct interpretation of it and of its role and meaning in the *Principia* are discussed. Among the more recent ones, cf. ref 19: Verlet where a large bibliography is given.
19. L. Verlet, "*F=ma* and the Newtonian Revolution: an Exit from Religion through Religion", *History of Science*, **34** (1996), pp. 303-346.
20. Aristotle, *Physics* Φυσικῆς ἀκροάσεως, *in Aritoteles graece*, ex recensione I. Bekkeri, editit Academia Regia Borusica, apud G. Reinerium, Berolini, 1831, 2 vols., vol. I, pp. 184-267
21. M. Panza, "Dalla metafisica del moto alla scienza matematica della natura. Considerazioni critiche a proposito di alcuni problemi cinematici trecenteschi", *in* L. Bianchi (ed. by), *Filosofia e teologia nel trecento. Studi in ricordo di Eugenio Randi*, édité par la Fédération Internationale des Instituts d'Etudes Médiévales, Louvain la Neuve (1994), pp. 413-478.
22. H. Carteron, La notion de force dans le systhéme d'Aristotle, p. 211Vrin (Paris), (1923).
23. Ref 20, 192b 21
24. Ref. 20, 219b 2.
25. Ref 20, 250a 4.
26. E. Giusti, "Galilei e le leggi del moto", *in* G. Galilei, *Discorsi e dimostrazioni matematiche intorno a due nuove scienze...*, p. XIII ed. by E. Giusti, Einaudi, Torino (1990), pp. IX-LX.
27. Ref. 26, p. XXIII.
28. M. Clagett, *The Science of Mechanics in the Middle Ages*, Univ. of Wisconsin Press, Madison, Wisconsin (1959).
29. Ref 28 pp. 238 e 241-42 [doc. 4.4. ("William Heytesbury, *Rules for Solving Sophisms*"), 14-17 e 89-97].
30. M. Clagett, *Nicolas Oresme and the Medieval Geometry of Qualities and Motion. A Treatise on the Uniformity and Difformity of Intensities Known as Tractatus de configurationibus qualitatum et motum* (Edited with an Introduction, English Transaltion and Commentary by M. Clagett), The Univ. of Wisconsin Press, Madison, Milwaukee, London (1968).
31. Ref 21 for discussion of Heytesbury's proof.
32. G. Galilei, *Dialogo [...] sopra i due Massimi Sitemi del Mondo [...]*, G. B. Landini, Fiorenza (1632) [*in* G. Galilei, *Le Opere*, Ed. Nat., a cura di A. Favaro, Barbera, Firenze 20 voll., 2th *Giornata* (Ed. Nat., pp. 254-256) 1890-1909, vol. VII]

33. G. Galilei, *Discorsi e dimstrazioni matematiche intorno a due nuove scienze...*, Appresso gli Esevirii, Leida, 3th *Giornata* (Ed. Nat., pp. 208-209). (1638) [*in* G. Galilei, *Le Opere*, *cit*., vol. VIII].
34. E. Giusti, *Euclides reformatus. La teoria delle proporzioni nella scuola galileiana*, Bollati Boringhieri, Torino (1993).
35. E. Giusti, "Aspetti matematici della cinematica galileiana", *Bollettino di storia delle scienze matematiche*, **1**, p. XXIII (1981), 2, pp. 3-42.
36. Ref 26 pp. XIII-XLVII.
37. Ref 26, pp. XXIX-XXX
38. Ref 35, p 19.
39. R. Descartes, *La Géométrie*, *in* R. Descartes, *Discours de la Méthode [...]*.pp. 297-298 *Plus la Dioptrique. Les Météores. Et la Géométrie qui sont des essaies de cette Méthode*, I. Maire, Leyde (1637), pp. 295-413
40. N. H. Bos and K. Reich, "Der Doppelte Auftakt zur frühneuzeitlichen Algebra: Viète und Descartes", *in* E. Scholz (ed. by), *Geschichte der Algebra*, Wissenschaftverlag, Mannheim, Wien, Zürich (1990), pp. 183-234
41. V. Jullien, *Descartes. La* Géométrie *de 1637*), pp. 75-76, PUF, Paris (1996).
42. Ref 34 p. 169.
43. M. Panza, "Quelques distinctions à l'usage de l'historiographie des mathématiques", *in* J.-M. Salanskis, F. Rastier, R. Sceps (éd.), *Herméneutique: textes, sciences*, P.U.F., Paris (1997), pp. 357-383.
44. J. W. Herivel, *The Background of Newton's Principia. A Study od Newton's Dynamical Researches in the Years 1664-1684*, Clarendon Press, Oxford (1965).
45. D. T. Whiteside, "Newtonian Dynamics", *History of Sciences*, **5** (1966).
46. D. T. Whiteside, "The Mathematical Principles underling Newton's *Principia Mathematica*", *Journal for the History of Astronomy*, **1** (1970), pp. 116-138.
47. I. B. Cohen, *Introduction to Newton's* Principia, Harvard Univ. Press, Cambridge, Mass., (1971).
48. F. De Gandt, "Le style mathématique des *Principia* de Newton", *Revue d'Histoire des Sciences*, **39** (1987).
49. F. De Gandt, *Force and Geometry in Newtons's* Principia, Princeton Univ. Press, Princeton, (1995).
50. *Les* Principia *de Newton. Question et commentaires*, special issue of the *Revue d'histoire des sciences* (RVS), **40**, 3/4 (1987).
51. G. Barthélémy, *Newton Mécanicien du cosmos*, Vrin, Paris (1992).
52. M. Panza, "Eliminare il tempo: Newton, Lagrange e il problema inverso del moto resistente", *in* M. Galuzzi (ed. by), *Giornate di storia della matematica* (proceedings of the meeting held in Cetraro, September, 8th-12th 1988), Editel, Commenda di Rende, Cosenza (1991), pp. 487-537 (see a discussion of an example of Newton treatment of motion of an isolated body, by comparing it to Lagrange's solution to the same problem
53. J. Dhombres, "Un style axiomatique dans l'écriture de la physique mathématique au 18ème siècle. Daniel Bernoulli et la composition des forces", *Sciences et Techniques en Perspective*, **11** (1986-1987), pp. 1-38.
54. J. Dhombres and P. Radelet De Grave, "Contingence et necessité en mécanique. Etude de deux textes inedits de Jean d'Alembert", *Physis*, **28** (1991), pp.35-114.
55. P. Radelet De Grave, "Daniel Bernoulli et le parallélogramme des forces", *Sciences et Techniques en Perspective*, **11** (1987), pp. 1-20.
56. P. Radelet De Grave, "Les forces et leur loi de composition chez Newton", *Revue Philosophique de Louvain*, 86 (1988), p. 505-522.
57. P. Radelet De Grave, "La composition des mouvements, des vitesses et des forces", *in* H.-J. Hesse und F. Nagel (hrsg. von), *Der Ausbau des Calculus durch Leibniz und die Brüder Bernoulli [...]*, *Studia Leibnitiana*, Sonderheft 17 (1989), pp. 25-47.
58. G. W. Leibniz, "Nova methodus pro maximis et minimis [...]", *Acta Eruditorum* (1684), pp. 467-473.
59. Ref 12 part. I, ch. 3, pp. 63-109, where the necessary bibliographical references are given.
60. Ref 12, p. 109.
61. Ref 12 part. II, ch. 1, pp. 113-152, where the necessary bibliographical references are given.
62. Ref 12 p. 129.
63. Ref 12, pp. 150 and 152.
64. Ref 12, part II, ch. 2, pp. 153-221, where the necessary bibliographical references are given.
65. Ref 12, p. 161.
66. Ref 15, pp. 208-215, where a bibliography is given for the famous controversy on the inverse problem of central forces.
67. I. Newton, *Philosophiæ naturalis principia mathematiaca*, perpetuis Commentariis illustata, communi studio PP. T. Le Seur & F. Jacquier, Barrillot & Filli, 3 toms in 4 vols., 1739-1742 (G).

68. P. Varignon, "Du mouvement en général partout sorte de courbes, et des forces centrales, tant centrifuge que centripètes, nécessaires aux corp qui les décrivent", *Hist. Acad. Roy. Sci. [de Paris], Mém. Math. Phy.*, (1700, publ. 1703; pres. 31/3/1700), pp. 83-101.
69. Ref 12, pp. 193-213.
70. Ref 68, p. 85.
71. Ref 12, pp 197-198.
72. P. Varignon, "Des mouvements faits das des milieux qui leur résistent en raison quelconque", *Hist. Acad. Roy. Sci. [de Paris], Mém. Math. Phy.* (1707, publ. 1708; pres. 13-17/8/1707), pp. 382-476.
73. Ref 12, pp. 280-299.
74. Jean Bernoulli, *Discours sur les loix de la communication du mouvement [...]*, C. Jombert, Paris (1727).
75. M. Otte and M. Panza (ed, by) *Analysis and Synthesis in Mathemetics. History and Philsophy*, Boston Studies in the Philosophy of Science, 196, Kluwer, Dordrech, Boston, London, (1997).
76. Ref. 75, M. Panza, "Classical sources for the concepts of analysis and synthesis", pp. 365-414.
77. M. Panza, La forma della quantita. Analisi algebrica e analisi superiore: il problema dell'unità della matemetica nel secolo dell'Illuminismo, Cahiers d'Histoire et de Philosophie des Sciences, **38** and **39**, Ch. II.2. (1992).
78. J. B. l. R. d' Alembert, "Eléments des sciences" (1st part), *in Encyclopédie, ou dictionnaire raisonné des sciences, des arts et de métiers*, Briasson, David l'aîné, le Breton, Durand, Paris, 1751-1780, 35 vols., vol. V, 1755, pp. 491a-497a.
79. M. Panza, "De la nature épargante aux forces généreuses : le principe de moindre action entre mathématiques et métaphysique. Maupertuis et Euler, 1740, 1751", *Revue d'histoire des Sciences*, **48** (1996), pp. 435-520.
80. M. Panza, "Die Entstehung der analytischen Mechanik im 18. Jahrhundert", *in* H. N. Jahnke (Hrsg.), *Geschichte der Analysis*, Spektrum Akad. Verl., Heidelberg, Berlin (1999), pp. 171-190.
81. I. Kant, *Kritik der reinen Vernunft*, Hartknoch, Riga, 1787; *ibid.*, vol. III , 1911, pp. 1-552 (B).
82. I. Kant, *Kritik der reinen Vernunft*, Hartknoch, Riga, 1781; the pages 1-405 are in *Kant's gesammelte Schriften* (herausg. von de Königlich Preukischen Akademie der Wissenschaften; after: Deutsche Academie der Wissenschaften zu Berlin; after: Academie der Wissenschaften der D.D.R.), G. Reimer and (from 1922) W. De. Gruyter, Berlin (1900-...), vol. IV, 1911, pp. 1-252 (A).
83. M. Panza, "De la nature épargante aux forces généreuses : le principe de moindre action entre mathématiques et métaphysique. Maupertuis et Euler, 1740, 1751", *Revue d'histoire des Sciences*, **48** (1996), pp. 435-520.

MODELS, ANALOGIES, AND STATISTICAL REASON, 1760-1900

Theodore M. Porter

Department of History Box 951473
University of California, Los Angeles
Los Angeles, California 90095-1473
Tporter@history.ucla.edu

INTRODUCTION

In our own time, statistics refers first of all to mathematical tools of inference and estimation, and it is natural that these concerns play a central role in the histories. Stephen Stigler [1], in his distinguished study, used the subtitle "the measurement of uncertainty" for his history of statistics before 1900. His account places the theory of errors, worked out in the first decade of the nineteenth century, at the center of the early development of statistical methods and conceptions. In relation to the background of eighteenth-century probability, least squares involved a synthesis of several eighteenth-century traditions: rules and customs for combining observations in astronomy, the extension of basic probability to estimate uncertainties for large numbers of random events, and the inverse methods of probability developed by Thomas Bayes and Pierre Simon Laplace. The method of least squares was developed mathematically by Laplace and Carl Friedrich Gauss around 1810, and thereafter became exemplary for efforts to analyze data from various fields using probability mathematics. Yet, as Stigler observes, the spread of least squares in the nineteenth century was quite narrow. Other sciences of the observatory, such as meteorology, began soon to use it, and in fact the geodetic campaign to measure lines of longitude and thereby to fix the meter provided the context for Adrian Marie Legendre's initial formulation of least squares in 1805 [2]. In social science, despite the enthusiasm of such persons as Adolphe Quetelet, probabilistic measures of variation and uncertainty were never really successful until after 1900. The experimental field of psychophysics, championed by Gustav Theodor Fecnher, figures in his account as a partial exception, but the real breakthrough had to await the biometric school of Francis Galton, Karl Pearson, and G. Udny Yule.

An alternative understanding of the history of probability and statistics can be found in the two edited volumes produced by a research group at Bielefeld, Germany, bearing the title *The Probabilistic Revolution* [3]. Their *problématique* was defined not so much by changing methods of data analysis as by new ways of understanding the world. Although their account begins in earnest about 1800, the contributors to these volumes write rather little about least squares. On the whole, they treat it as part of the philosophical background against which modern probabilistic conceptions were defined. Least squares

The Application of Mathematics to the Sciences of Nature, edited by Cerrai *et al.*
Kluwer Academic/Plenum Publishers, New York, 2002.

was one of the more important, or at least most successful, achievements of classical probability, but it was part of the old world in which probability was first of all about errors and uncertainty, not about variation and a world of chance. Perhaps the most important outcome of the probabilistic revolution, for them, was what Ian Hacking calls "the erosion of determinism" Hacking [4]. Within this conception, the English biometricians remain central, but perhaps more for their incorporation of variability into the structure of the world than for their methods of managing error and uncertainty. An equally decisive role is assigned to some developments that are virtually omitted from Stigler's account, the rise in physics of statistical mechanics and quantum theory. Here, the deterministic view of the world gradually broke down, and was replaced by one in which chance had a fundamental role.

The history of probability and statistics can of course be approached in many ways. Here I will contrast two broad perspectives: one that examines statistics as a tool of reasoning or of inference, and another that develops statistical analogies, models, and theories of the natural and social worlds. Without in any way diminishing the significance of the history of estimation and inference, I will here discuss the significance of models and analogies for the history of statistical reason.

MATHEMATICAL MODELING: DANIEL BERNOULLI ON SMALLPOX

The language of "mathematical modeling" seems to have developed only in the twentieth century, and one should be wary of anachronisms, but models have in some sense been used in science since ancient times. Alistair Crombie [5] discusses modeling as one of the basic styles of scientific reasoning, and supplies abundant examples from the medieval and early modern periods. Its typical purpose was to provide a visualizable, macroscopic image or representation of what cannot be observed. Model-making reached a peak of influence under the "mechanical philosophy" of the seventeenth century, in which hooks and eyes explained chemical bonds, springs gave bounce to the air, and sharp-pointed particles accounted for the experience of sourness. Often these were primarily aids to the imagination, put forth as theories but often with the understanding that even if nature contained something like these curious objects, its mechanisms were not quite the same as the springs and levers of common experience. This was no reason to despair of them. Sometimes a physical picture permitted very precise reasoning, as in Descartes' mathematical analysis of the refraction of light in water droplets, his explanation of the rainbow.

In the late eighteenth century, it was not uncommon to use probability mathematics to understand and propose reforms for human institutions. In one of the best-known cases, it served as the basis for theories of testimonies, elections, and judicial procedures. Work of this kind was, as Lorraine Daston [6] points out, altogether typical of classical probability theory, which was commonly conceived up to the early nineteenth century as good sense mathematized. In some cases, probability went beyond the capacity of good sense to provide counsel on questions presented as purely quantitative. Mere judgment could not work out the complex consequences of enlarging or shrinking a jury and altering the majority required for conviction (Baker [7], Brian, [8]). Probability calculations could provide more precise estimates of quantities that untutored intelligence could never reduce to numbers.

A particularly remarkable instance of mathematical modeling is to be found in a 1760 memoir by Daniel Bernoulli [9], which explored the effects of smallpox on mortality, and the potential gains if it could be eliminated. In policy terms, he wished to demonstrate the rationality of inoculation, which since Voltaire had been a favorite Enlightenment campaign. Bernoulli used Edmond Halley's mortality table, slightly modified for the first year of life, as the standard of the death rate for every year of life. Since there were no comparable data for incidence by age of smallpox, nor for deaths from smallpox, he had recourse to simplifying assumptions, the unmistakably hypothetical elements in what I am calling his model. Each year, he supposed, one eighth of those who have not yet suffered smallpox will contract it, and one eighth of these, in turn, will die from it. He wrote and solved a differential equation to determine how many at each age remained susceptible in the "natural state" where smallpox remained present, and how many of these became sick, and how many died. This permitted him to calculate a table that distinguished annual deaths due to smallpox from those attributable to other causes and thence, finally, to calculate the life table that might reasonably be anticipated if smallpox could be eliminated.

Bernoulli insisted on the impartiality of his own position. He would neither defend not reject smallpox inoculation, but only explore its consequences. Still, he developed the implications of his argument in a way that made clear his view that inoculation can be very useful. No more than one in 200 will die as a result of surgical inoculation, and a child or young adult can prolong his expectation of life by up to three years in this way. For the state, the arguments are even more forceful. Most deaths from smallpox, according to his calculations, take place during childhood. Those who die before age 16 consume resources without contributing significantly to wealth. By raising the ratio of adults to children, inoculation adds to prosperity and power.

Bernoulli's memoir was an impressive piece of mathematical reasoning, but it did not initiate a tradition. As it happened, a refutation appeared even before the Parisian Académie Royale des Sciences finally printed it, in the *Mémoires de mathématique et de physique* for 1766 [10]. The critique was by Jean d'Alembert, a rival of Bernoulli's and a famous skeptic regarding probability theory. He objected on several levels, including the psychological. Bernoulli's reasoning, even if it were ironclad in terms of mathematics and evidence, might not yet appear decisive, especially to a mother, who may well refuse to buy a longer expectation of life at the cost of increasing the chance that she will lose her baby within the next month. The state can rely unambiguously on statistics and averages, but for an individual there is always the possibility that his particular life will be shortened (on the different perspectives of state and individual see Desrosières [11]) For our purposes, it is particularly important to note that d'Alembert rejected the modeling aspects of Bernoulli's paper. He wanted data, not assumptions (d'Alembert [10]):

In the first place, the assumption which the illustrious mathematician makes about the number of persons of each age who catch smallpox and about the number who die of it would seem to be absolutely unfounded. It is by no means certain—it is, indeed, to say the least, more than doubtful—that smallpox constantly attacks one eighth of those who have not had it, at whatever age; and it is still more doubtful that it constantly destroys, at whatever age, an eight of those whom it attacks. Moreover, it would be necessary to know whether inoculation always carries off, as is assumed, the same fraction of those inoculated, irrespective of the age at which they were inoculated.

Bernoulli had in fact introduced modeling assumptions as a substitute for data that could not be obtained. Where possible, he compared the predictions of the theory with the available data. D'Alembert rejected the whole approach, without any effort to explore how the conclusions might change if the simplifying assumptions were altered. And this was in the eighteenth century, the apogee of classical probability. The nineteenth century was much less tolerant of using probabilistic conjectures as a substitute for empirical information, and less sympathetic also to mathematical versions of human reasonableness.

STATISTICAL SOCIAL SCIENCE AND STATISTICAL THINKING

A single model underlay much nineteenth-century statistical reasoning. It was not, in the first instance, a mathematical model, but rather a social one. It respected the dedication to precise measurement that increasingly characterized the sciences after about 1750 (Wise [12]). It was, however, no mere consequence of scientific investigation, but equally of bureaucratic expansion and of the production and circulation of social information. It was official publications of French criminal statistics that first provoked the astonishment of Quetelet and some of his contemporaries. Crimes in every category—murder, suicide, theft—occurred in almost constant numbers from year to year. Quetelet wondered in his first published response to these numbers, in 1829, if the moral acts of man were driven by some strange fatality, leaving no room for individual freedom (Porter [13]).

The regularities of "moral statistics" led to a considerable intellectual ferment in ensuing decades. They were diffused most widely in the wake of Henry Thomas Buckle's immensely successful *History of Civilization in England*, published in 1857, whose influence was particularly impressive in Germany and Russia as well as Britain. Throughout this period, there was much uncertainty and disagreement about what exactly the regularities of statistics meant. Buckle himself associated them with scientific necessity, if not some strange fatality. Quetelet had taken in the end a less rigorous line, insisting that regularity at the level of the mass still left a modest space for free will within any particular individual. Given, however, that statistics provided little or no assistance in predicting the actions of individuals, it was far from clear at first just what they actually described. Quetelet resolved this problem in favor of the reality of the social. "Statistical laws" were characteristics not of any particular individual, but of the human species *en masse*, of that special body we call "society." Or, more precisely, it characterized not individual men, but *l'homme moyen*.

The idea of society, a progressive entity whose development could not in the end be controlled by political authorities, was an important one in mid-century Europe. It gave support to a discourse of liberalism, since it tended to narrow the potential role of the state. At the same time, the study of disease and crime in terms of mass regularities rather than of the failings of individuals was seen by advocates and opponents alike as weakening or diffusing moral responsibility. While an individual criminal might yet be justifiably condemned, "social physics" focused its attention instead on collective phenomena, on crime rates, which could be compared with other times and places to learn something about the social causes that lay behind them. This was the most important implication of what I am calling the statistical model. Large numbers of individuals, acting independently within a system, produce regularities that do not depend on coordination or direction of all these persons. It is possible to reason about collectives without any detailed knowledge of the individuals.

This liberal conception of society, and of the spontaneous production of social order, provided a model also for a form of mathematical reasoning whose subsequent career was enormously successful. It is thus reasonable that a word for administrative and social investigation, "statistics," should eventually have come to name a field of applied mathematics. It was, after all, in the social field that the quantitative study of mass phenomena was first practiced, and its fundamental conceptions worked out. The mathematical field of statistics came into being about 1900, drawing its formalism from mathematical probability, and especially from nineteenth-century developments in the method of least squares. Beneath it, however, was a conception of how collective order was produced, and a scheme for detecting higher level causes that remained opaque at the level of individuals.

SOCIAL PHYSICS

Quetelet had envisioned that statistics might provide the needed materials for that Newton of the social sciences who was to come, and who would develop a *mécanique sociale* or *physique sociale* on the model of Laplace's *mécanique celeste*. The mantle of social physics was claimed from time to time in the later nineteenth century. Ironically, though, it was most effectively realized in the physical sciences. Here is how Ludwig Boltzmann introduced his landmark memoir of 1872 on the convergence of the distribution of molecular velocities in a gas to the squared exponential distribution later named for him and for James Clerk Maxwell:

> "It is to be ascribed exclusively to the circumstance that even the most irregular processes yield the same average values when placed under the same conditions, that we perceive wholly determinate laws in the behavior of warm objects. For the molecules of a body are indeed so numerous, and their movements are so rapid, that nothing ever becomes perceptible to us except average values. The regularity of these mean values may be compared to the astonishing constancy of average numbers furnished by statistics, which are also derived from processes in which each individual occurrence is conditioned by a wholly incalculable collaboration of the most diverse external circumstances. The molecules are like so many individuals, having the most various states of motion, and the properties of gases only remain unaltered because the number of these molecules which on the average have a given state of motion is constant. The determination of averages is the task of the calculus of probability."

Or, in a popular exposition on the same subject in 1886, he wrote:

> "As is well-known, Buckle demonstrated statistically that if only a sufficient number of people is taken into account, then not only is the number of natural events like death, illness, etc., perfectly constant, but also the number of so-called voluntary actions—marriages at a given age, crimes, and suicides. It occurs no differently among molecules." (quotes from Porter, [13])"

Maxwell began using this analogy at almost the same time. In an address at the British Association for the Advancement of Science in 1873 [14], he explained that the molecules of a gas are far too numerous, and their interactions too complex, to be solved using equations of dynamics.

> "The modern atomists have therefore adopted a method which is, I believe, new in the department of mathematical physics, though it has long been in use in the section of Statistics. When the working members of Section F get hold of a report of the Census, ... they begin by distributing the whole population into groups, according to age, income-tax, education, religious belief, or criminal convictions. The number of individuals is far too great to allow of their tracing the history of each separately, so that, in order to reduce their labour within human limits, they concentrate their attention on a small number of artificial groups. The varying number of individuals in each group, and not the varying state of each individual, is the primary datum from which they work....
>
> "The equations of dynamics completely express the laws of the historical method as applied to matter, but the application of these equations implies a perfect knowledge of the data. But the smallest portion of matter which we can subject to experiment consists of millions of molecules, not one of which ever becomes individually sensible to us. We cannot, therefore, ascertain the actual motion of any one of these molecules; so that we are obliged to abandon the strict historical method, and to adopt the statistical method of dealing with large groups of molecules. (Maxwell [14])

The combination of complexity and disorder at the level of individuals with statistical order at the level of populations was analogous in the social and physical domains. The social model provided a way of visualizing a system of infinite colliding molecules as orderly, and capable of supporting the lawlike regularities that had already been formulated in thermodynamic terms. The analogy provided an attractive mode of exposition for introducing the subject to popular audiences. It provided also an analogy for exploring the limitations of statistical knowledge in the kinetic gas theory. Maxwell, who disliked deterministic claims about the human will, followed a line of critique developed by social statisticians when he pointed out that collective molecular regularities are not binding on individuals. Even the macroscopic relations of temperature and pressure do not have the same status as laws of dynamics, but reduce at bottom to statistical regularities.

SOCIAL MODELS OF BIOLOGICAL PROCESSES

In physics, as in social science, statistical thinking had somewhat paradoxical consequences. It arose from the discovery of a novel form of scientific order, yet led to the recognition of an unexpected uncertainty. In evolutionary biology, the complexity of the social model took a different form. The problem of homogeneity—a central one, as Stigler observes, for the entire century—was very much at issue in the biometric study of "natural inheritance." Biometry did not, however, solve it. Francis Galton's social model, for example, permitted distinct alternative understandings of the mechanisms of heredity, and alternative frameworks for statistics.

Charles Darwin, Galton's cousin, had offered his "provisional hypothesis of Pangenesis" in 1868 to give a theory of heredity that would support his theory of evolution. It involved the budding of tiny genetic elements, called "gemmules," from every unit—perhaps every cell—of the body. Sexual reproduction concentrates enough gemmules to permit the formation of an embryo. Galton elaborated and modified Darwin's hypothesis through a "simile" that he published in his *Hereditary Genius* [15]. His version, in contrast to Darwin's excluded use-inheritance. More crucially, it permitted selection of gemmules by affinities, and not only through random selection.

Galton's model of spontaneous organization derived from a model reflecting a liberal political economy. The formation of gemmules into an individual, he proposed, "may be compared to the typical appearance always found in different descriptions of assemblages.... The assemblages of which I speak are such as are uncontrolled by any central authority, but have assumed their typical appearance through the free action of the individuals who compose them, each man being bent on his own immediate interest, and finding his place under the sole influence of an elective affinity to his neighbors." (Galton, [15]). The invocation of elective affinities is crucial here. The formation of a town involves not a random sampling of the surrounding region, but a very special selection. In some cases, as with a fishing town and a tourist resort, the elements combine easily, since the coastline is picturesque and visitors happily purchase the day's catch. A noisy, smoking factory and its rough laborers, by contrast, would drive the tourists away. In the same way, the genetic elements from heterogeneous parents might not combine into a viable offspring. Rather, some particular type would prevail.

The statistical model of natural inheritance depended on an assortment of gemmules that was congruent, at least, with what would happen under a regime of random selection. Galton himself used the analogy of drawings from an urn to characterize these processes. When he began his experimental and statistical studies of heredity, he anticipated that for any given trait the offspring (after correcting for sex differences) would be distributed around the mean of the parents. He discovered that they tended rather to "revert" in fixed proportions toward the mean for the whole population. The term "reversion" had a specific biological meaning, and his use of it suggested that this tendency to return to the mean owed to the presence of gemmules from more remote ancestors that had not been expressed in the parents. An assortment of genetic elements from many generations would naturally represent a nearer approximation to the mean for the whole population than did those expressed in any particular individual.

In the mid-1880s, he changed his terminology from "reversion" to "regression." This latter term could also refer to a move backwards in time, but Galton indicated a different meaning. In his 1885 presidential address to the anthropology section of the British Association, he explained that: "The type is an ideal form towards which the children of those who deviate from it tend to regress.... The stability of a type would, I presume, be measured by the strength of its tendency to regress." (Galton, [16]). That is, he now considered regression not so much as a move to an ancestral form as to an ideal type, the point of stability. He now believed that the "effective heritage of the child" is "less than the accumulated deviates of his ancestors." (Galton, [17]). On these matters, he never fell in line with what Karl Pearson would put forth as the view of biometrics, that evolutionary change is always continuous. Galton and Pearson shared an unwavering dedication to statistics, but Galton's understanding of heredity and the formation of the individual required discontinuities in the evolutionary process. So long as evolution takes place within the confines of the type, deviations will always tend to return in subsequent generations to the point of stability, the type for the population. True evolutionary change requires sports—changes so radical that they form a new point of stability, a new hereditary type.

Whereas "reversion" and "regression" referred to the relations between generations, "correlation" meant for Galton the interrelations among the traits of a single individual. It was in his guise as criminologist, seeking a way to identify criminals, that the man who later defined a method for categorizing fingerprints was drawn to the problem of correlation. In what proportions does unusual height imply a particularly long arm or finger? Following the hypothesis of Pangenesis, this was a question of elective affinities. Somehow, gemmules associated with length of arm must be related to those that determine legs and torso. Curiously, though, the relationships conformed to the mathematics of reversion, so that unusual height implied a longer cubit for that individual in just the way that it implied greater than normal height for his or her progeny. Yet correlation of parts was not easily reconciled with the model of independent drawings of balls from an urn that underlay the mathematics of reversion and regression. Correlation was linked rather to the logic of elective affinities, of instabilities, and of distinct hereditary types. If it were to be modeled in terms of drawings from an urn, these would have to be biased and not random.

The disharmony of Galton's models was linked, perhaps, to his shortcomings as a mathematician. Yet his refusal to be confined by the strictest conditions of mathematical derivation contributed indispensably to his statistical achievement. As Stigler (1986, p. 274) remarks, he set out consciously to soften the restrictive assumptions that had gone into Laplace's derivation of the normal curve of error. Correlation and regression meant dividing up the variation into a portion that could be attributed to the correlated variables and a random portion. That partitioning did not follow logically from his model, but emerged out of its ambiguities, in the context of his detailed work with pea seeds, criminal records, and other biological measurements. He deployed models of random sampling on some occasions, and invoked the correlation of parts on others, with more attention to practical fruitfulness than to logical consistency.

CONCLUSION

Galton was highly critical of statisticians who could see variation only as error. The objects of his scorn were not mathematicians, but census officers and their ilk, who too rarely went beyond presenting mean values. Yet the mathematical tradition of error theory was limited in its subject matter and in its perspectives. Here I have pointed to a few salient moments in an alternative tradition, which can be characterized in terms of its models. On occasion, as with the kinetic gas theory, the statistical model involved very sophisticated mathematics, but the image of regularity and social stability underlying it all was mathematically very elementary. Nineteenth-century statistics involved, in a way, a retreat from the bold assumptions deployed by Bernoulli. The social statisticians preferred not go far beyond the empirical data. Yet statistics was not merely a set of tools for analyzing data; it provided the rudiments of a theory of society. It came to function also as a model in the sense that its basic conceptual structure could be transported into other fields, such as gas physics and biological evolution, where it was developed in novel ways. Theories of errors and measures of uncertainty, then, are only a part of this story. The migrations and elaboration of a characteristic statistical model form a key element in the history of statistical mathematics during the nineteenth century.

REFERENCES

1. Stephen M. Stigler, *The History of Statistics: The Measurement of Uncertainty before 1900*, Harvard University Press, Cambridge, Mass. (1986).
2. Stephen M Stigler,. *Statistics on the Table: The History of Statistical Concepts and Methods*, Harvard University Press, Cambridge, Mass. (1999).
3. Krüger, Lorenz, Gerd Gigerenzer and Mary Morgan, (eds.), *The Probabilistic Revolution, volume 2: Ideas in the Sciences*, MIT Press, Cambridge, Mass. (1987).
4. Ian Hacking, *The Taming of Chance*, Cambridge University Press, Cambridge, U.K. (1990).
5. Alistair Crombie, *Styles of Scientific Thinking in the European Tradition*, 3 vols., Duckworth, London. (1994).
6. Lorraine Daston, *Classical Probability in the Enlightenment*, Princeeton University Press, Princeton. (1988).
7. Keith Baker, *Condorcet: From Natural Philosophy to Social Mathematics*, University of Chicago Press, Chicago. (1975).
8. Eric Brian, *La mesure de l'Etat: Administrateurs et géomètres au XVIIIe siècle*, Albin Michel, Paris. (1994).
9. Daniel.Bernoulli, "An Attempt at a New Analysis of the Mortality Caused by Smallpox," (1766) in T. Bradley, (ed). *Smallpox Inoculation: An Eighteenth-Century Mathematical Controversy*, Adult Education Department of the University of Nottingham, Nottingham, (1971).
10. Jean le Rond d'Alembert,. "On the Application of the Calculus of Probabilities to Inoculation against Smallpox," (1760) in T. Bradley, (ed)., *Smallpox Inoculation: An Eighteenth-Century Mathematical Controversy*, Adult Education Department of the University of Nottingham, Nottingham, (1971).
11. Alain Desrosières, *La Politique des grands nombres: Histoire de la raison statistique*, Editions la Découverte, Paris. (1993).
12. M. Norton Wise, , (ed).. *The Values of Precision*, Princeton University Press, Princeton. (1995).
13. Theodore M Porter,. *The Rise of Statistical Thinking, 1820-1900*, Princeton University Press, Princeton. (1986).
14 .James Clerk Maxwell, "Molecules," (1873). in W. D. Niven, (ed)., *Scientific Papers of James Clerk Maxwell*, 2 vols., Cambridge University Press, Cambridge, U.K., (1890).
15. Francis Galton, *Hereditary Genius*, Macmillan, London. (1869).
16. Francis Galton, "Address to the Section of Anthropology of the British Association," *Nature*, 32, 507-510 (1885).
17. Francis Galton, "Family Likeness in Stature," *Proceedings of the Royal Society of London*, 40, 42-73. (1886).

IS MUSIC RELEVANT FOR THE HISTORY OF SCIENCE?

Tito M. Tonietti

Dipartimento di Matematica - Università di Pisa
Via Filippo Buonarroti, 2.
56127 Pisa
Italia
tonietti@dm.unipi.it

SUMMARY

I maintain the following thesis: music was the original mathematical model for natural sciences in the West. To support this, arguments and numerous examples could be presented, from the Pythagoreans to Kepler, ..., from Newton to d'Alembert. It is to be noted that the original idea of harmony included the whole of nature and did not leave out human beings, medicine, or life sciences.

THE OLDEST OF ALL QUANTITATIVE PHYSICAL LAWS

My aim here is simply to illustrate a thesis, which, in its strongest and clearest form, may be stated as follows:

Music was one of the original mathematical models for the sciences of nature in the West.

This thesis is suggested by one of the most ancient events of which man has memory, so ancient that it has entered the realms of myth, and become lost amid the desert wastes. A relationship exists between the length of a string that emits sounds when it is plucked and the pitch of those sounds, as perceived by the ear. This relationship was described by means of a precise mathematical formula, that of inverse proportionality. After Descartes, this was to be written in the form

$$\nu \propto \frac{1}{l}$$

where the pitch, after Mersenne, is measured as the frequency ν and the length is l.

This relationship has remained unchanged until today, for about 2500 years. Is it the only mathematical natural law which is still considered to be valid? Whereas

The Application of Mathematics to the Sciences of Nature, edited by Cerrai *et al.*
Kluwer Academic/Plenum Publishers, New York, 2002.

perhaps others may have been modified over the years several times? "... possibly the oldest of all quantitative physical laws" wrote C. Boyer[1]. If so, that "possibly" could be deleted.

The sounds emitted by instruments, that is to say, the musical notes, as heard by the ear, could thus be classified by the proportions 1 : 2, 2 : 3, 3 : 4. These corresponded to the *diapason* interval (the octave, *do - do*), the *diapente* (the fifth, *do - sol*) and the *diatessaron* (the fourth, *do - fa*). The proportion thus established that the greatest importance was to be attributed, not to the single isolated sound, but to the relationship between sounds. At this point the story became even more fascinating, and also relatively well documented, because during the whole of the subsequent evolution of the sciences, controversies continually arose over two main problems. How was the octave to be divided into notes? Which of the relative intervals were to be considered 'consonant', that is to say, 'pleasant', and thus to be included in pieces of music, and which were 'dissonant'? And why? The constant presence of contrasting answers to these questions allows us to consider the sciences as an integral part of our culture, which regularly thrives on the various connections between debates of this kind.

Anyway, in view of the surprising success of our original model of a mathematical law, it was linked with other regular features, identified here and there, and posited as the explanation for other phenomena. The most famous of these was undoubtedly the movement of the planets and the heavenly bodies, thus giving rise to the so-called music of the heavenly spheres. This original seal, this original aporia of its foundation, remained visible for a long time. All, or almost all, the figures that we usually take into consideration in the evolution of the mathematical sciences wrote about these problems. Sometimes they made original contributions. At other times, they simply repeated, with a few personal variations, what they had learnt from tradition. It might be called Pythagorean tradition, from the name of the legendary founder to whom the original discovery was attributed, or Platonic or neo-Platonic, or else it might be placed alongside the rival tradition of Aristoxenos. In any case, many felt they were obliged to render homage to it, in the form of commentaries, compendia, diverse quotations, or actual theories.

Without stopping to think at length, and as a result forgetting others, Euclid[2], Plato[3,4], Ptolomy[5] and Aristoxenos[6] come to mind. These were followed, skipping the medieval era, before and during the scientific revolution of the seventeenth century, by Maurolico[7,8], Benedetti, Stevin, Kepler, Galileo Galilei, Mersenne, Huygens, Descartes and Wallis. And the interest in the division of the octave into a certain number of notes, the interest in explaining the consonances, passed intact, or almost intact, through the epoch-making substitution of the Copernican system in the place of the Ptolemaic one. It could be interpreted variously as a theory of music, or as acoustics, as the music of mathematics or the mathematics of music. Under whatever guise, it recurred incessantly in Leibniz, Newton, a member (members?) of the Bernoulli family, D'Alembert and Euler. We shall present only a few examples here, referring the reader to the bibliography for the others.[9,...,22]

KEPLER AND LEIBNIZ

In 1619, Johannes Kepler published his *Harmonices Mundi Libri Quinque.*[23] Starting from the curious idea of continuing Claudius Ptolemy's *Harmonica*, he wrote a complex treatise on geometry, music, astrology and astronomy. His theory of the division of the octave and his justification of consonances is highly original, because it is based on an idea linked with the 'constructibility' of regular polygons with a certain number of sides. In his fifth book, he included his third law of the movement of planets, stating that a sesquialter proportion exists between their periods of rotation T and their mean

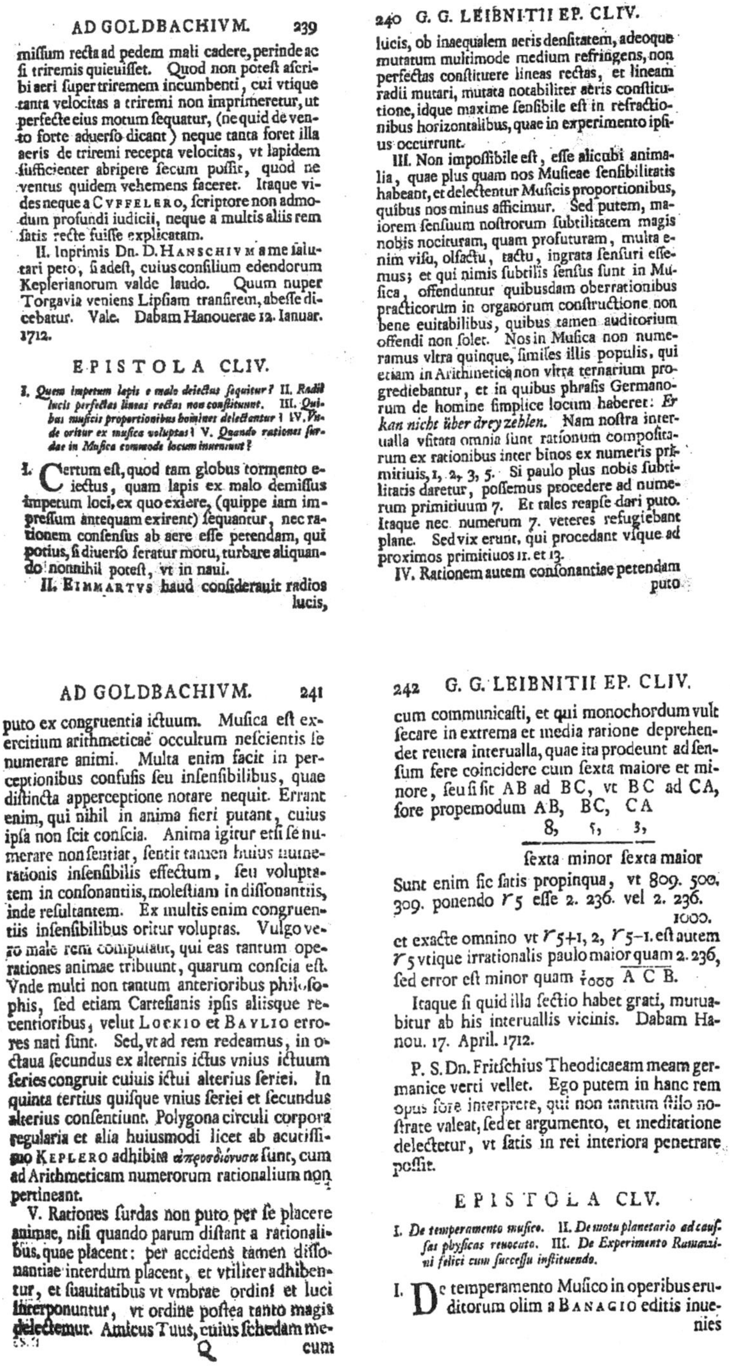

AD GOLDBACHIVM. 239

miſſum recta ad pedem mali cadere, perinde ac ſi triremis quieuiſſet. Quod non poteſt aſcribi aeri ſuper triremem incumbenti, cui vtique tanta velocitas a triremi non imprimeretur, ut perfecte eius motum ſequatur, (ne quid de vento forte aduerſo dicant) neque tanta foret illa aeris de triremi recepta velocitas, vt lapidem ſufficienter abripere ſecum poſſit, quod ne ventus quidem vehemens faceret. Itaque vides neque a CVFFELERO, ſcriptore non admodum profundi iudicii, neque a multis aliis rem ſatis recte fuiſſe explicatam.

II. Inprimis Dn. D. HANSCHIVM a me ſalutari peto, ſi adeſt, cuius conſilium edendorum Keplerianorum valde laudo. Quum nuper Torgavia veniens Lipſiam tranſirem, abeſſe dicebatur. Vale. Dabam Hanouerae 12. Ianuar. 1712.

EPISTOLA CLIV.

I. Quem impetum lapis e malo deiectus ſequitur? II. Radii lucis perfectas lineas rectas non conſtituunt. III. Quibus muſicis proportionibus homines delectantur? IV. Vnde oritur ex muſica voluptas? V. Quando rationes ſurdae in Muſica commode locum inueniunt?

I. Certum eſt, quod tam globus tormento eiectus, quam lapis ex malo demiſſus impetum loci, ex quo exiere, (quippe iam impreſſum antequam exirent) ſequantur, nec rationem conſenſus ab aere eſſe petendam, qui potius, ſi diuerſo feratur motu, turbare aliquando nonnihil poteſt, vt in naui.

II. EIMMARTVS haud conſiderauit radios lucis,

240 G. G. LEIBNITII EP. CLIV.

lucis, ob inaequalem aeris denſitatem, adeoque mutatum multimode medium refringens, non perfectas conſtituere lineas rectas, et lineam radii mutari, mutata notabiliter aeris conſtitutione, idque maxime ſenſibile eſt in refractionibus horizontalibus, quae in experimento ipſius occurrunt.

III. Non impoſſibile eſt, eſſe alicubi animalia, quae plus quam nos Muſicae ſenſibilitatis habeant, et delectentur Muſicis proportionibus, quibus nos minus afficimur. Sed putem, maiorem ſenſuum noſtrorum ſubtilitatem magis nobis nocituram, quam profuturam, multa enim viſu, olfactu, tactu, ingrata ſenſuri eſſemus; et qui nimis ſubtilis ſenſus ſunt in Muſica, offenduntur quibusdam oberrationibus practicorum in organorum conſtructione non bene euitabilibus, quibus tamen auditorium offendi non ſolet. Nos in Muſica non numeramus vltra quinque, ſimiles illis populis, qui etiam in Arithmetica non vltra ternarium progrediebantur, et in quibus phraſis Germanorum de homine ſimplice locum haberet: *Er kan nicht über drey zehlen.* Nam noſtra interualla vſitata omnia ſunt rationum compoſitarum ex rationibus inter binos ex numeris primitiuis, 1, 2, 3, 5. Si paulo plus nobis ſubtilitatis daretur, poſſemus procedere ad numerum primitiuum 7. Et tales reapſe dari puto. Itaque nec numerum 7. veteres refugiebant plane. Sed vix erunt, qui procedant vſque ad proximos primitiuos 11. et 13.

IV. Rationem autem conſonantiae petendam puto

AD GOLDBACHIVM. 241

puto ex congruentia ictuum. Muſica eſt exercitium arithmeticae occultum neſcientis ſe numerare animi. Multa enim facit in perceptionibus confuſis ſeu inſenſibilibus, quae diſtincta apperceptione notare nequit. Errant enim, qui nihil in anima fieri putant, cuius ipſa non ſcit conſcia. Anima igitur etſi ſe numerare non ſentiat, ſentit tamen huius numerationis inſenſibilis effectum, ſeu voluptatem in conſonantiis, moleſtiam in diſſonantiis, inde reſultantem. Ex multis enim congruentiis inſenſibilibus oritur voluptas. Vulgo vero male rem computant, qui eas tantum operationes animae tribuunt, quarum conſcia eſt. Vnde multi non tantum anterioribus philoſophis, ſed etiam Carteſianis ipſis aliisque recentioribus, velut LOCKIO et BAYLIO errores nati ſunt. Sed, vt ad rem redeamus, in octaua ſecundus ex alternis ictus vnius ictuum ſeries congruit cuiuis ictui alterius ſeriei. In quinta tertius quiſque vnius ſeriei et ſecundus alterius conſentiunt. Polygona circuli corpora regularia et alia huiusmodi licet ab acutiſſimo KEPLERO adhibita ἀπροσδιόνυσα ſunt, cum ad Arithmeticam numerorum rationalium non pertineant.

V. Rationes ſurdas non puto per ſe placere animae, niſi quando parum diſtant a rationalibus, quae placent: per accidens tamen diſſonantiae interdum placent, et vtiliter adhibentur, et ſuauitatibus vt vmbrae ordini et luci interponuntur, vt ordine poſtea tanto magis delectemur. Amicus Tuus, cuius ſchedam mecum

Q

242 G. G. LEIBNITII EP. CLIV.

cum communicaſti, et qui monochordum vult ſecare in extrema et media ratione deprehendet reuera interualla, quae ita prodeunt ad ſenſum fere coincidere cum ſexta maiore et minore, ſeu ſi ſit AB ad BC, vt BC ad CA, fore propemodum AB, BC, CA

	AB,	BC,	CA
	8,	5,	3,
	ſexta minor	ſexta maior	

Sunt enim ſic ſatis propinqua, vt 809. 500. 309. ponendo $\sqrt{5}$ eſſe 2. 236. vel 2. 236/1000. et exacte omnino vt $\sqrt{5}+1$, 2, $\sqrt{5}-1$. eſt autem $\sqrt{5}$ vtique irrationalis paulo maior quam 2. 236, ſed error eſt minor quam $\frac{1}{1000}$ $\overline{ACB}$.

Itaque ſi quid illa ſectio habet grati, mutuabitur ab his interuallis vicinis. Dabam Hanou. 17. April. 1712.

P. S. Dn. Fritſchius Theodicaeam meam germanice verti vellet. Ego putem in hanc rem opus fore interprete, qui non tantum ſtilo noſtrate valeat, ſed et argumento, et meditatione delectetur, vt ſatis in rei interiora penetrare poſſit.

EPISTOLA CLV.

I. De temperamento muſico. II. De motu planetario ad cauſſas phyſicas reuocato. III. De Experimento Ramazzini felici cum ſucceſſu inſtituendo.

I. De temperamento Muſico in operibus eruditorum olim a BANAGIO editis inuenies

Figure 1.

distances from the sun R:

$$T \propto R^{\frac{3}{2}}.$$

This famous result was thus included in a book about the theory of music.

On April 17, 1712, Leibniz wrote a letter to Christian Goldbach.[24] We here present a facsimile of the letter (Fig. 1), together with the translation of parts III, IV e V.

"III. It cannot be excluded that animals may somewhere exist, which have a greater musical sensitivity than us, and take delight at musical proportions which impress us to a lesser extent. I believe, however, that a greater sensitivity of our senses would be more harmful than profitable for us; we would have to hear many unpleasant things, together with our sight, our sense of smell and of touch; and those who have too fine a sensitivity to music are offended by certain deviations which cannot easily be avoided in the practical construction of organs, by which, however, the listeners are not usually offended. We do not count beyond five in music, like those peoples who have not even got beyond three in arithmetic, to whom the German saying about the simpleton is applicable: *er kan nicht über drey zehlen* [he can't count beyond three]. All our usual intervals are ratios based on two of the prime numbers, 1, 2, 3 and 5. If we were endowed with a little more subtlety, we might arrive at the prime number 7. And actually I believe the following ones are also given. Thus the ancients did not openly avoid the number 7. But hardly anybody proceeded as far as the following prime numbers, 11 and 13.

IV. On the contrary, I believe that the basis of consonance is to be sought in the congruence of beats. Music is a recondite arithmetical exercise carried out by the soul, which is unaware that it is counting. It performs many things with a confused, or insensitive perception, which it cannot discern with a clear understanding. Those who think that nothing can exist in the soul, of which the soul itself is not aware, are mistaken. Thus, although the soul does not realise that it is counting, it feels, nevertheless, the effects of this unconscious numbering, or a pleasure at the consonances and a distaste at the dissonances which are the result. The pleasure is aroused by several unconscious congruences. Those who attribute only conscious operations to the soul, in truth, usually make a bad judgement. This has given rise to many errors, not only with philosophers of the past, but also with the followers of Descartes themselves, and with others who are more recent, such as [John] Locke and [Pierre] Bayle. But, to return to our subject, in the octave, every other beat in one series of beats coincides with any beat in the other series. In the fifth, there is a concordance between every third beat in one series and the second of the other one. The polygons of the circle, regular bodies and others of this kind admitted by the highly intelligent Kepler are used $\alpha'\pi\rho o\sigma\delta\iota o'\nu\upsilon\sigma\alpha$ [inopportunely], seeing that the arithmetic of rational numbers does not apply to them.

V. I do not believe that irrational ratios are pleasing to the soul in themselves, except when they are very close to the rational ones which give pleasure. In some cases, however, dissonances are sometimes pleasing, and are used with profit, placed between attractive intervals, just as shadows are contrasted with order and light, so that even more pleasure may be taken in order. Your friend whose note you sent me, who chooses to divide the monochord into the extreme and medium ratios, would actually find intervals which proceed in such a direction as almost to coincide with the major and minor sixth, that is to say, if $AB : BC = BC : CA$, it will be approximately

$$AB - - - - - - - - BC - - - - - - - - CA$$

$$8 - - - - - - - - - 5 - - - - - - - - - 3$$

$$minor \quad sixth - -major \quad sixth.$$

These are as close, approximately, as 809 . 500 . 309, if we admit that $\sqrt{5}$ is 2.236, that is to say, 2.236 to the thousandth and exactly the same as $\sqrt{5}+1$, 2, $\sqrt{5}-1$, whereas $\sqrt{5}$ is wholly irrational and slightly higher than 2.236, but the error is less than $\frac{1}{1000}$. And thus, if that division held in it something pleasing, it would be taken from these close intervals."

EULER AND D'ALEMBERT

Euler's letter to Johann Bernoulli dated May 25 (June 5) 1731[25] is equally interesting.

"My long delay in writing to Your Excellency and in assuring you of my respectful homage, is due to the fact that for some time now, I have not worked at anything that I thought worthy of being sent to you. For almost all this period, I have dedicated my energies to the construction of a *Systematis Musici*, which I have now almost finished. I therefore take the liberty of sending Your Excellency what I believe I have thus achieved, because I imagine that you may be curious about it, as I recently heard from your esteemed son [Daniel] that you had received news of it from Herr [Jakob] Hermann. My aim, in this work, was to try to develop music as a part of mathematics, and to derive in an orderly fashion, from the right foundations everything that a combination and a mixture of notes may make pleasing. For this complete treatise, I have felt the need of a metaphysical foundation on which to base the reason why one may take delight in a piece of music, and why something is pleasant for one person, but unpleasant for another. I have demonstrated that with notes, the following situation arises: we appreciate many notes combined together if we recognise the ratio of high and low pitch between the notes, in other words *rationem intervallorum pulsuum* [the proportion of the beats of the intervals] given by the strings. And I have also devised the rules for this, to explain how the notes should be put together so that an intelligent ear may take delight in them. But there is yet another reason why we take delight in a piece of music, and this lies in the duration of the notes, and consequently we may even appreciate a piece of music only because we grasp the ratio between the durations of the notes. According to the first *principio*, it is the high or low pitch that we appreciate, but according to the other one, it is their different duration. In order to be perfect, music must combine both and thus we appreciate both *ratione acuminis et gravitatis* and *ratione duratione sonorum* [in view of the ratio between high and low, and that of the duration of the sounds]. The Choral is a kind of music which we appreciate only in the first manner, as it is based only on the harmony of voices, without considering the duration. But an *Exempel* of music which can give pleasure only in the second way is the sound of drums, where only the duration of notes is considered, and not their pitch. Initially I have described music in accordance with the first *principio*, and only subsequently in accordance with the second one, and finally I will deal with both of them together.

In the first part, which is undoubtedly the most *fürnehmste* [important?], the following elements are presented and examined. First, the *accorten* [chords], where I have demonstrated how two or more notes should be used in order to create a pleasant *Harmonie* when they are played together. Secondly, I have studied how two chords should be used, so that when they are played one after the other, they sound pleasant. Thirdly, I have considered a whole succession of chords, demonstrating what is needed for them to sound pleasant. The limits within which a succession of this kind may be found constitute the *modos* that I have subsequently described. In this same part, I have also shown what notes are needed on instruments for each *modum*, finding the *genus chromaticum* with 12 notes in an octave very attractive. In the following discussion, I have considered everything in accordance with the *Systemate chromatico*. After this, I have done something more particular than the preceding and I have not applied myself any more to *praxin*. And as I have at the same time prepared a *perfectam enumerationem modorum*, it may be seen that much more progress could be made in music, and that out of all the various *modis*, only two are at present in fashion. These are included in the music that is based on the *Genus chromaticum*, and if it were desirable to use other notes in addition to the 12 found in an octave, or in their place, then it would be possible to obtain infinitely many different types of music.

But all this I have found starting from the first *principio*. Let many different notes be taken, whose *numeri pulsuum* [numbers of beats], that occupy the same period of time, stand to each other as the whole numbers a, b, c, d etc., by which the same notes are usually expressed. Let A be the *minimus communis dividuus* [lowest common multiple]. I call this number the *exponentem* of the same notes, because it is on this basis that we recognise grace that is produced when the same notes are played either at the same time or in succession. I have thus devised a *gradus suavitatis*, the first of which includes the most perfect chord, that is to say, when all the notes are relatively equal. The following ones include the less perfect, according to their order. From the *exponente* it is possible to recognise the *gradus suavitatis* in the following manner: I break it down into its *factores simplicissimos* [most simple factors], I add these together, and then I subtract from their sum $n-1$ (n stands for the *numerum factorum* [number of factors]); the number that is thus obtained represents the *gradum*. For example, let the notes be 1, 2, 3, 4, 6, then the *exponens* of these notes will be 12. The *factores simplicissimi* of this are the following three: 2.2.3, the sum of which, less 2, gives 5. It is clear from this that the harmony of these notes will be pleasant to the 5th degree. In this way, it is possible to find the *exponentem* of a whole piece of music, if all the notes are expressed as whole numbers, and the *minimum commune dividuum* is taken. But this *exponens* is nothing other than the *modus musicus* of the piece itself. As all the notes are present on the clavichord, the *exponens* of a piece is $2^n.3^3.5^2$, where n depends on the number of octaves. Consequently, the following 12 notes must be present in an octave:

$$480, 512, 540, 576, 600, 620, 675, 720, 768, 800, 864, 900$$

the thirteenth note is one octave higher than the first, that is to say, 960.

This is what I wanted to let Your Excellency know about my *Musica Theoretica*. ....".

Johann Bernoulli's reply was dated August 11 (July 31).[25]

"It was a great pleasure for me to learn that you are working on the completion of a *Systematis Musici* (which must be almost ready); I have no doubt that it will be a fine work when it is published, and will offer a sufficient demonstration of the excellent *ingenium* of its *Autoris*; I can well imagine that nobody else will attempt a similar *opus*, in which everything is derived from a mathematical basis, seeing that there are few, if any, *Scriptores Musici* endowed with such a great and unlimited mathematical understanding as the *Herr Professor*. I therefore look forward to seeing your work. I certainly could not imagine what that basis may be which must be so metaphysical, as you say, and through which it may be possible to explain why a piece of music gives pleasure, and why something sounds pleasant to us, and unpleasant to another: of course, we have a general idea of harmony, that it is pleasant when it is well founded and the consonances are well organised; then, I note that also dissonances are useful in music, with the result that the beauty of the consonances which follow them is improved, in accordance with the common proverb, *opposita juxta se posita magis elucescunt* [opposites which are placed next to each other stand out more clearly], and this is also connected with the use of shadows in painting, which serves to let the light stand out. In *musica practica*, I believe it depends above all on the type and modification [modulation?] to which one is accustomed and this type depends largely on the nature and temperament of people; as a result, some think that this piece is sweet and pleasant, and others that other piece, and consequently, the type of Italian music does not agree with French music, and the latter does not agree with English music, etc. In a word, *de gustibus non est disputandum.* Thus, if one desires to base the pleasantness of a piece of music on its nature itself, it is necessary to define what one means by pleasantness and not say that this or the other piece is pleasant because I like it, another might not like the same piece; for example, Midas preferred the musical din of Pan to the sound of the Apollo's lyre. The *Herr Professor* say that it is possible to judge the pleasantness or unpleasantness of 4 notes played together, by examining the ratio between the high or low pitch of the notes, that is to say, the *rationem intervallorum pulsuum* [the ratio between the beats of the intervals] given by the strings. From this, you have derived the rule which establishes how the notes are to be combined, so that an intelligent ear can take delight in them. I think that this is appropriate for a musician who is more concerned about the accuracy of a piece of music than its effect, which satisfies the listener; a person of this kind will undoubtedly find enjoyment and delight, if you have written this down and examine it and find that it is well composed in accordance with the fundamental rules; but as a piece of music is usually played to ears that are devoid of understanding, and are not able to recognise the *rationem intervallorum pulsuum* of the strings, and are even less able to count, then I believe that the same ears will appreciate or refuse the same piece of music, depending on whether they are used to this or that kind of music. I greatly appreciate your project because at least in this way the *Theoria Musices* is improved, and it is demonstrated that a *Mathematicus* is able to deal with almost all the sciences, whereas, on the contrary, the other masters who are only *practici* are as capable of writing of their art as a blind man is of colours.

When this *Tractatus Musices* is finished, you will undoubtedly give serious attention to the *Mechanicam* that you have planned (of which I have been informed in a letter); I expect something special from this ...".

The *Tractatus Musices* was to be entitled *Tentamen novae theoriae musicae* and was finally published only in 1739,[26,27] after the *Mechanicam* of 1736.

In 1752, there was a brief exchange of letters between Euler and Jean Philippe Rameau, the famous musician and theoretician.[26,27] But it was rather Jean D'Alembert who was the main interlocutor of this French composer, who based (tonal) harmony on the harmonics of sound. In actual fact, the mathematician and encyclopaedist dedicated a whole book to the problem *Élémens de musique Théorique et pratique suivant les principes de M. Rameau,*[28] which was published in the same year, 1752.

The part that D'Alembert played in the famous *querelle des bouffons* in connection with Italian music is fairly well known.[29] I believe, however, that it will come as a surprise to many to learn what decisive argument he used to back up his thesis. In about 1752, he wrote: "I speak from experience. I have examined many recitative scenes; I have sung them in the Italian style, leaving out the long cadences and the held notes: the result was a song very similar to Italian recitative, and I would dare say that it would not be unpleasant, however little inclined one may be. I carried out this experiment in the presence of some other people, who reached the same conclusion as myself".[30]

Under the entry "Fondamental"[31] in the *Encyclopèdie*, he dealt with the basic sound and the first harmonics which form the perfect major chord. If the basic sound is a *do*, these will include the repetition of *do* in the higher octave, *mi* and *sol*. Consequently, he also criticised the solutions considered by Daniel Bernoulli and Euler for the equation of a vibrating string. In view of its importance in the evolution towards the notion of a function, as the solution of a differential equation, this controversy is one of the most widely studied by historians of mathematical sciences. However, in the best known current reconstructions, there is no longer any trace of music. And yet it is not necessary to read the 'frivolous' (frivolous?) entry in the *Encyclopèdie* to realise this.

The wave equation proposed by D'Alembert was to become the main explanatory model in mathematical physics. The dynamic phenomenon to be studied was translated into a differential equation (from now on also with partial derivatives); the solutions were calculated as functions of space and time, and from them the properties of the phenomenon could be read. This method of explaining and calculating was to be the main one used, at least up to Poincaré. Only in 1912, faced with new quantum phenomena, which appeared to impose the well-known discontinuities, did the latter comment disconsolately: "physical phenomena would cease to obey laws that can be expressed by means of differential equations, and this would undoubtedly be the biggest, and most wide-reaching revolution in natural philosophy since the time of Newton".[32]

D'Alembert opened the period dominated by differential equations in the modern sense in 1747, with his famous article, "Recherches sur la courbe que forme une corde tenduë mise en vibration".[33] This was followed in 1750 by an "Addition", at the end of which his polemical stance against the rival musical theory was quite clear.

"It is clear from the preceding formulae that, given an equal tension and thickness, the number of vibrations in the same time is inversely proportional to the length of the strings. As the higher or lower sound of the strings depends on their larger or smaller number of vibrations in the given time, it is undoubtedly for this reason that some very capable modern authors have considered it possible to represent the sounds by means of the logarithms of the ratios between the lengths of the strings. This idea is ingenious, and would appear to be based equally on figures of speech in acoustics and music, when we say that if four strings a, b, c, d are geometrically proportional, the interval formed by sounds a and b will be *equal* to the interval formed by c and d; hence it was considered possible to conclude that the logarithms of the relationships $\frac{a}{b}$ and $\frac{c}{d}$ represented the intervals between the sounds. But undoubtedly this conclusion was not claimed to be anyhting more than a purely arbitrary supposition; the words *interval* between sounds, *equality* and *difference of intervals* are only abbreviated figures of speech, which should not be given a wider meaning than they really have. Sounds are merely sensations, and consequently they do not in reality have any ratio with one another; sounds cannot be compared, any more than colours can; all that is needed is a little attention to hear this ...".

Thus the French mathematician and physicist believed that this controversy in the field of music was pertinent to mathematical studies on vibrating strings. (The *corpus callosum* of his brain had not been cut, and the right hemisphere, that of music, communicated perfectly with the left hemisphere, that of language.) It was not difficult for him to believe this, because the harmonic sounds, placed by him and by Rameau at the basis of music, appeared to undermine the millenary Pythagorean tradition based, instead, on numerical proportions. Euler, on the other hand, appeared to prefer to stay within this tradition, even if he wanted to extend its rigourousness and its application, making musicians count up to 7! The controversy about vibrating strings thus saw the participation of people who were fighting at the same time on behalf of the relative musical theories.

SINGULAR COINCIDENCES

We might be surprised at the singular coincidence between a scientific controversy and a musical one, which arose in the century of the Enlightenment. And yet this coincidence was not new, and was not the only one.

Let us come back to the original seal of our story, and let us carefully consider an important detail. Thanks to the proportions chosen for the octave, the fifth and the fourth, the Pythagorean musical school was rapidly able to calculate the tone $fa - sol$ as the difference between the fifth $do - sol$ and the fourth $do - fa$, and consequently as the ratio $\frac{3}{2} : \frac{4}{3} = \frac{9}{8}$. All the treatises on music at that time were taken up with the question whether it was possible to divide the tone into two equal parts (semitones). The Pythagorean tradition denied that it was possible, but the followers of Aristoxenos readily admitted that it was. Why? Dividing the Pythagorean tone into two parts would mean admitting the existence of the proportional mean between 9 and 8, that is to say, $9 : \alpha = \alpha : 8$, where $9 : \alpha$ and $\alpha : 8$ are the proportions of the required semitone. But how much was α to be? Clearly, $\alpha = \sqrt{9.8}$ and therefore $\alpha = 3.2.\sqrt{2}$! Thus the most famous controversy in ancient Greek mathematics, the one concerning irrational numbers, immediately acquired a musical tone. The discussion of the relative model of the numerical or geometrical continuum,[34] variously represented in ancient philosophy by means of the paradoxes of Zeno of Elea, could be directly transferred to music. Anybody who, like Aristoxenos, preferred to look for answers in the practical activity of the earthly world, and on musical instruments instead of in the heaven of Platonic ideas, could not doubt that it was possible to place his finger on the string at exactly the point which corresponded to the division into equal semitones.

As we know perfectly well, the story was to continue to evolve constantly, without ever succeeding in reaching a complete solution, at least not for everybody. It returned, for example, in the margins of a manuscript left by Francesco Maurolico, bearing dates between 1567 and 1570, in the form "9.r72.8" where r stood for root. And it was denied that consonant intervals suitable for music could be obtained by means of irrational proportions. Because "Only God is infinite".[7,8]

At the beginning of the twentieth century, Georg Cantor would have liked to grasp that continuum, but it always slipped through his fingers, with the result that he ended up by losing his mind. He, too, was forced to use the string of a violin to evoke it.[35]

I have found, very recently, only one book on the history of mathematics, published in 1997 by R. Cooke, in which the exercise is assigned of dividing the octave into two equal parts and discussing what the Pythagoreans thought of this idea.[36]

ACKNOWLEDGMENTS

I would like to thank Andreas Kleinert for kindly letting me have copies of the letters of Leibniz and Euler, Dr. Ronald Packham for his help in the English styling of the manuscript, and Isabella Capitani for her help in the Greek language.

REFERENCES

1. C. Boyer, *A History of Mathematics*, Wiley, New York (1989).
2. Euclides, *Rudimenta Musices*, Giovanni Pena (cur.), Andrea Wechelo, Paris, (1557). Italian tr. in Fabio Bellissima, "La sezione del monocordo di Euclide".
3. Plato, Italian tr. *La repubblica*, Franco Sartori, Mario Vegetti, Bruno Centrone (cur.), Laterza, Bari (1999).
4. Plato, *Timaeus*, English tr. *The Collected Dialogues of Plato*, E. Hamilton, H. Carins (eds.), Princeton U. Press, Princeton (1987).
5. Claudius Ptolemaeus, *Harmonicorum libri tres*, Johannes Wallis (cur.), Oxford (1682).
6. Aristoxenos, *Elementa Harmonica*, Rosetta Da Rios (cur.), Istituto poligrafico dello stato, Roma (1954).
7. Francesco Maurolico, *Musica*, Tito M. Tonietti (ed.), [Manuscrit latin 7462, 1567-1570], Paris, Bibliothèque Nationale de France. www.maurolico.it (2000).
8. Francesco Maurolico, Musicae Traditiones, in *Opuscula Mathematica*, Francesco De Franceschi, Venezia (1575).
9. H. F. Cohen, *Quantifying Music*, Reidel, Dordrecht (1984).
10. Paolo Gozza (ed.), *La musica nella rivoluzione scientifica del seicento*, Il mulino, Bologna (1989).
11. Victor Coelho (ed.),*Music and Science in the Age of Galileo*, Kluwer Academic Publishers, Dordrecht (1992).
12. Eberhard Knobloch, Rapports historiques entre musique, mathematique et cosmologie, in *Quadrivium, Musique et Sciences*, Éditions ipmc, Paris:123-167 (1992).
13. Patrice Bailhache, *Leibniz et la théorie de la musique*, Klincksieck, Paris (1992).
14. Patrice Bailhache, Cordes vibrantes et consonances chez Beeckman, Mersenne et Galilée, *Sciences et techniques en perspective*, n. 23:73-91 (1993).
15. Patrice Bailhache, Deux mathématiciens musiciens: Euler et D'Alembert, *Physis*, 32:1-35 (1995).
16. Eberhard Knobloch, Harmony and Cosmos: Mathematics Serving a Teleological Understanding of the World, *Physis*, 32:55-89 (1995).
17. Thomas Settle, Galileo's Experimental Research, preprint Max Planck Institute for the History of Science, Berlin (1996).
18. Patrice Bailhache, Sciences et musique: quelques grandes étapes en théorie musical, *Littérature, médicine, société*, n. 13 (1996).
19. Tito M. Tonietti, Albert Einstein and Arnold Schönberg Correspondence, *Naturwissenschaft Technik und Medizin*, 5:1-22 (1997).
20. Tito M. Tonietti, Verso la matematica nelle scienze: armonia e matematica nei modelli del cosmo tra Seicento e Settecento, in *La costruzione dell'immagine scientifica del mondo*, Marco Mamone Capria (ed.), La città del sole, Napoli:155-219 (1999).
21. Tito M. Tonietti, Does Newton's Musical Model of Gravitation Work?, *Centaurus*, 42:135-149 (2000).
22. Tito M. Tonietti, *Dalla musica alla matematica*, to be published.
23. Johannes Kepler, *Harmonices Mundi Libri V*, J. Plank, Linz (1619); repr. Forni, Bologna (1969); German tr. *Welt-Harmonik*, Max Caspar (ed.) München (1939), repr. R. Oldenbourg, München (1990); English tr. *The Harmony of the World*, E. J. Aiton, A. M. Duncan, J. V. Field (eds.), American Philosophical Society, Philadelphia (1997).
24. Gottfried Wilhelm Leibniz to Christian Goldbach (1712), in *Epistolae ad diversos*, Lipsiae:239-242 (1734-1742).
25. Leonhard Euler to Johann Bernoulli, Johann Bernoulli to Leonhard Euler (1731), in *Leonhardi Euleri Opera Omnia*, Birkhäuser, Basel Ser. IV A, vol. II:146-150 (1975).
26. Leonhard Euler, *Tentamen novae theoriae musicae*, St. Petersburg (1739), repr. *Opera Omnia* Ser. III, vol. I, Teubner, Leipzig:197-427 (1926).
27. Jean-Philippe Rameau, *Musique raisonnée*, Stock, Paris (1980).
28. Jean Le Rond d'Alembert, *Élémens de Musique Théorique et Pratique suivant les principes de M. Rameau*, Paris (1752); repr. Lyon (1779).
29. Enrico Fubini, *Gli enciclopedisti e la musica*, Einaudi, Torino (1971).

30. Jean Le Rond d'Alembert, Fragment sur la musique en général et sur la notre en particulier (1752), in *Oeuvres et Correspondances inédites*, Charles Henry (ed.), Didier Perrin, Paris:164-189 (1887).
31. Jean Le Rond d'Alembert, Fondamental, in *Encyclopèdie*, Paris (1754?).
32. Henri Poincaré , Sur la théorie des quanta, *Journal de Physique théorique et appliquée*, 5 serie t. 2:5-34 (1912); repr. *Oeuvres*, vol. IX:626-653.
33. Jean Le Rond d'Alembert, Recherches sur la courbe que forme une corde tenduë mise en vibration (1747-1750), *Histoire de L'Academie Royale des Sciences et Belles Lettres*, Berlin:214-249 (1749) and:355-360 (1752).
34. Tito M. Tonietti, Continuum's Continuous Conflicts, in *Proceedings Conference on Foundations of Mathematics and Physics - Perugia*, U. Bartocci, *et al.* (eds.), Benjamin Wesley Publ., Blumberg:271-288 (1990).
35. Nathalie Charraud, *Infini et Incoscient - Essai sur Georg Cantor*, Anthopos Economica, Paris (1994).
36. R. Cooke, *The History of Mathematics*, Wiley, New York (1997).

INDEX